Animal Husbandry: Perspectives on Production and Management

Animal Husbandry: Perspectives on Production and Management

Edited by Ashlie Archer

SYRAWOOD
PUBLISHING HOUSE

New York

Published by Syrawood Publishing House,
750 Third Avenue, 9th Floor,
New York, NY 10017, USA
www.syrawoodpublishinghouse.com

Animal Husbandry: Perspectives on Production and Management
Edited by Ashlie Archer

International Standard Book Number: 978-1-68286-792-1 (Hardback)

Cataloging-in-Publication Data

Animal husbandry : perspectives on production and management / edited by Ashlie Archer.
      p. cm.
Includes bibliographical references and index.
ISBN 978-1-68286-792-1
1. Animal culture. 2. Domestic animals. 3. Livestock. 4. Livestock--Management.
I. Archer, Ashlie.
SF41 .A55 2019
636--dc23

TABLE OF CONTENTS

Permissions

List of Contributors

Index

PREFACE

The branch of agriculture that deals with the raising of animals for the provision of food, fiber and other products is referred to as animal husbandry. Many commercial sectors are dependent on the optimum production and management of animals. These include the dairy and meat industry, aquaculture production, honey manufacture, silk farming, etc. Animals can be reared both intensively and extensively. In extensive farming, animals are allowed to roam at will. In intensive systems, they are kept in high-density feedlots, climate-controlled buildings, barns, cages, etc. and their forage is provided. There is an increased reliance on intensive animal farming and production systems suited to the land available in modern animal husbandry. Besides these, there is also an increased use of chemical and mechanized methods. Embryo transfer and artificial insemination are common practices today to improve productivity or quality of the animals. This book discusses the fundamentals as well as modern approaches of animal husbandry. It attempts to understand the multiple branches that fall under this discipline and how such concepts have practical applications. For all readers who are interested in modern animal husbandry, the case studies included herein will serve as an excellent guide to develop a comprehensive understanding.

This book has been the outcome of endless efforts put in by authors and researchers on various issues and topics within the field. The book is a comprehensive collection of significant researches that are addressed in a variety of chapters. It will surely enhance the knowledge of the field among readers across the globe.

It gives us an immense pleasure to thank our researchers and authors for their efforts to submit their piece of writing before the deadlines. Finally in the end, I would like to thank my family and colleagues who have been a great source of inspiration and support.

Editor

Methodological impact of starch determination on starch content and ileal digestibility of starch in grain legumes for growing pigs

Dagmar Jezierny[1], Rainer Mosenthin[1], Nadja Sauer[2], Klaus Schwadorf[3] and Pia Rosenfelder-Kuon[1*]

Abstract

Background: Grain legumes represent a valuable energy source in pig diets due to their high starch content. The present study was conducted to determine the content and apparent ileal digestibility (AID) of starch in different grain legume cultivars for pigs by means of both a polarimetric and enzymatic method for starch determination.

Methods: Three experiments were conducted with six barrows each which were fitted with ileal T-cannulas. In total, 18 diets including six different cultivars of faba beans (*Vicia faba L.*) and peas (*Pisum sativum L.*), five different cultivars of lupins (*Lupinus luteus L., Lupinus angustifolius L.*), and one diet with a soybean meal (SBM) were fed.

Results: The starch content of faba beans and peas was greater ($P < 0.05$) when determined polarimetrically than enzymatically (438 vs. 345 g/kg dry matter (DM) in faba beans and 509 vs. 390 g/kg DM in peas, respectively). Considerable lower starch contents were obtained in lupins and SBM, with 82 and 48 g/kg DM (analyzed polarimetrically) and <1.1 and 3 g/kg DM (analyzed enzymatically), respectively. Mean values for contents of neutral detergent fiber (NDF) and acid detergent fiber (ADF) in grain legumes ranged from 111 and 79 g/kg DM in peas to 248 and 207 g/kg DM in lupins, respectively. Contents of condensed tannins in the colored flowered faba bean cultivars ranged from 2.1 to 7.4 g/kg DM. The AID of starch was greater ($P < 0.05$) in pea than in faba bean cultivars, and using the polarimetric starch determination method resulted in greater ($P < 0.05$) digestibility values than using enzymatic starch analysis (84 vs. 80% in faba beans and 86 vs. 83% in peas). Moreover, AID of starch differed ($P < 0.05$) within pea cultivars and starch digestibility in faba beans decreased linearly ($P < 0.05$) as the content of condensed tannins increased. However, there was no relationship between contents of NDF and ADF and AID of starch in pea and faba bean cultivars.

Conclusion: Both contents and AID of starch in grain legumes can vary as influenced by the analytical method used for starch determination. Generally, starch digestibility is greater when measured by polarimetric rather than enzymatic methods.

Keywords: Grain legumes, Growing pigs, Ileal starch digestibility, Starch determination method

* Correspondence: pia.rosenfelder@uni-hohenheim.de
[1]University of Hohenheim, Institute of Animal Science, Emil-Wolff-Strasse 10, 70599 Stuttgart, Germany
Full list of author information is available at the end of the article

Background

The search for alternative protein sources in livestock nutrition has resulted in growing interest in the use of grain legumes as alternatives for animal by-products or oilseed products such as soybean meal (SBM). In addition, grain legumes have proven to be a valuable energy source in animal nutrition due to their high content of starch in faba beans and peas and lipids in lupins [1]. According to NRC [2], average starch contents in faba beans and peas amount to 445 and 493 g/kg dry matter (DM), respectively, which are lower when compared to cereal grains (671, 559, and 708 g/kg DM for wheat, barley, and corn, respectively), but considerably higher than in lupins, SBM, and rapeseed meal (7, 21, and 66 g/kg DM, respectively). The use of grain legumes in livestock nutrition, however, is often limited due to the presence of secondary plant metabolites (tannins, protease inhibitors, alkaloids, lectins, pyrimidine glycosides, saponins), also referred to as anti-nutritional factors (ANF). These metabolites can induce feed refusals (tannins, alkaloids), reduced nutrient digestibility (tannins, protease inhibitors, lectins), or even toxic effects (alkaloids) [3].

In general, legume starch contains 30 to 40% amylose and 60 to 70% amylopectin, whereas cereal starch consists of 20 to 25% amylose and 75 to 80% amylopectin [4]. Hydrolysis of starch by pancreatic α-amylase has been reported to be inversely related to the amylose content, with high amylose starches being particularly resistant [5].

As starches in grain legumes are closely associated with proteins, the determination of starch in grain legumes may be confounded by the analytical method used [6], mainly due to the hydrophobic properties of these proteins and the enclosure of the protein-starch network by cell wall compounds [5]. Commonly used procedures for starch determination in food and feed ingredients are based on gravimetric, iodometric, and polarimetric methods [7, 8]. In the European Union, the polarimetric method is the official method used for starch analysis in feeds [9, 10]. According to Champ et al. [11], however, this method lacks precision due to numerous artifacts (e.g. amino acids, mono- and oligosaccharides). Thus, Mitchell [12] suggests the enzymatic method to be advantageous over the polarimetric method for analyzing starch in complex plant matrices such as mixed diets due to enzyme specificity. Until now, no comparative experiments have been conducted to compare values from enzymatic and polarimetric methods for starch determination in different grain legumes and ileal digesta samples collected from pigs fed these grain legumes. Therefore, the objective of the present study was to determine the starch content in different cultivars of faba beans, peas, and lupins and ileal digesta samples by means of a polarimetric and enzymatic procedure. Furthermore, the effect of these different methods as influenced by the presence of ANF in the feed ingredients on apparent ileal digestibility (AID) of starch in growing pigs was assessed.

Methods

Experimental procedure

A total of 18 feed ingredients including six seed-grade cultivars of faba beans (*Vicia faba L.*) and peas (*Pisum sativum L.*), five seed-grade cultivars of lupins (*Lupinus L.* spp.), and one commercially available SBM were used in three consecutive experiments with six growing barrows (German Landrace × Piétrain) each. The experiment was arranged as a row-column design. Each of the three experiments included six pigs (rows) and six periods (columns). Within each of the three experiments, the animals were randomly allocated to the 18 assay feed ingredients in periods one to three and four to six, respectively, resulting in two replications per experiment and a total of six observations per assay feed ingredient throughout all three experiments. The average initial and final body weight (BW) of the pigs at the beginning and end of the experiment was 23 ± 2 kg and 45 ± 4 kg, respectively. The pigs were surgically fitted with T-cannulas at the distal ileum according to the procedures described by Li et al. [13]. After a 7-day recuperation period from surgery, the pigs were fed twice daily their semi-synthetic diets in two equal meals (0700 and 1900 h) at a daily level of 30 g/kg (as-fed) of their individual BW, determined on d 1 of every experimental period. Ileal digesta samples were collected for a total of 24 h from 1900 to 0700 h on d 5 and from 0700 to 1900 h on d 6. The individual samples of digesta of each pig were pooled separately after every sampling period, freeze-dried, and ground to 0.5 mm prior to analysis.

The research protocol was approved by the German Ethical Commission for Animal Welfare. Care of the animals used in this experiment was in accordance with the EEC directive 86/609 [14]. The diets based on casein and corn starch contained one of the feed ingredients each (grain legumes or SBM), and were supplemented with (on as-fed basis) 100 g/kg dextrose, 50 g/kg cellulose, 20 g/kg plant oil, and L-cystine, L-threonine, L-lysine-HCl, vitamins, and minerals to fulfil NRC [15] nutrient requirements for pigs from 20 to 50 kg BW. Titanium dioxide was used as a digestibility marker. Further details on diet composition and nutrient contents have been reported by Jezierny et al. [16].

Analytical procedure

Contents of DM, neutral detergent fiber (NDF), acid detergent fiber (ADF), condensed tannins, and titanium dioxide were determined as described by Jezierny et al. [16].

The starch contents in the feed ingredients, the diets, and in ileal digesta samples were analyzed both polarimetrically and enzymatically. The polarimetric method represents the official starch determination method in the European Union for feed according to the Commission Directive 1999/79/EC [10], and comprises two steps. First, the sample was treated with diluted hydrochloric acid (1.128%). After clarification and filtration, the optical rotation of the solution was measured polarimetrically. In the second step, the sample was extracted with 40% ethanol. After acidifying the filtrate with hydrochloric acid, clarifying, and filtering, the optical rotation was measured according to the same procedure as used in the first step. The difference between these two measurements, multiplied by a factor is equivalent to the starch content of the sample. This factor, however, is not constant as it may vary among various types of starch and feedstuffs. For the quantitative determination of starch by enzymatic analysis, samples had to be pretreated to convert starch into a soluble form. Therefore, homogenized samples were mixed with hydrochloric acid (8 mol/L) and dimethylsulfoxide and incubated at 60 °C for 30 min. Thereafter, water was added and the solution was adjusted to pH 4–5. After filtration, soluble starch was quantified by enzymatic analysis. For this purpose, starch first was hydrolyzed to D-glucose in the presence of the enzyme amyloglucosidase, followed by phosphorylation of D-glucose to D-glucose-6-phosphate by means of ATP (adenosine-5'-triphosphate) in the presence of hexokinase with simultaneous formation of ADP (adenosine-5'-diphosphate). Then, in the presence of the enzyme glucose-6-phosphate dehydrogenase, the D-glucose-6-phosphate was oxidized by NADP (nicotinamide-adenine dinucleotide phosphate) to D-gluconate-6-phosphate with the formation of reduced NADPH. The amount of NADPH formed is stoichiometric to the amount of D-glucose formed by hydrolysis of starch and was quantified spectrophotometrically at 340 nm (type Lambda 25, PerkinElmer Inc., Waltham, MA) [17].

Calculations

The AID of starch in the diets was calculated using the following equation:

$$ID_{Ai} = 100\% - [(I_{Ai} \times S_{Di}) / (S_{Ai} \times I_{Di})] \times 100\%$$

where ID_{Ai} = apparent ileal starch digestibility in the ith diet (%), I_{Ai} = marker concentration in the ith diet (g/kg DM), S_{Di} = starch content in digesta of the ith diet (g/kg DM); S_{Ai} = starch content in the ith diet (g/kg DM), and I_{Di} = marker concentration in digesta of the ith diet (g/kg DM).

The AID of starch in the feed ingredients was calculated according to the following equation:

$$ID_I = (ID_{Ai} - ID_C \times C_C) / C_I$$

where ID_I = apparent ileal starch digestibility in the feed ingredient (%), ID_{Ai} = apparent ileal starch digestibility in the diet (%), ID_C = apparent ileal digestibility of corn starch (%), C_C = contribution level of starch from corn starch to the diet (%) and C_I = contribution level of starch from the feed ingredient to the diet (%).

Statistical analysis

Homogeneity of variances and normal distribution of the data were confirmed and experimental data were analyzed using the MIXED procedure of SAS [18]. The linear model included the fixed effects of grain legume species, experiment, grain legume cultivar, analysis, species × cultivar (cultivar is nested within species), experiment × animal, experiment × replication, species × analysis and the random effects experiment × replicate × period and experiment × replicate × animal. Multiple comparisons among cultivars within species were performed using a t-test with degrees of freedom determined by the Kenward-Roger method [19]. This test was performed only when a preliminary F-test [20] revealed significant differences within a species (SLICE = species option in MIXED). Significant differences between treatments were indicated by different superscript letters using the algorithm for letter-based representation of all pair-wise comparisons according to Piepho [21]. The significance level was set at $\alpha = 0.05$. Furthermore, the effects of NDF and ADF contents in all grain legumes and of tannin contents in faba beans on AID of starch were modelled by a linear regression.

Results
Chemical composition of grain legumes

Analyzed contents of starch, both determined polarimetrically and enzymatically, and contents of NDF and ADF in grain legume cultivars as well as condensed tannins in faba bean cultivars are presented in Table 1.

Within pea and faba bean cultivars, starch contents ranged between 495 and 535 g/kg DM for cultivars Rocket and Hardy, respectively, and between 421 and 456 g/kg DM for cultivars Fuego and Aurelia, respectively, when determined polarimetrically. Using the enzymatic approach, starch contents within pea cultivars ranged from 372 to 423 g/kg DM for cultivars Santana and Hardy, and within faba bean cultivars from 311 to 393 g/kg DM for cultivars Espresso and Limbo. Within lupin cultivars, starch content ranged from 56 in Bornal to 96 g/kg DM in Boregine when determined polarimetrically. However, starch content for all lupin cultivars was below the detection limit of 1.1 g/kg DM when analyzed enzymatically. The starch content in SBM amounted to 48.0 and 3.0 g/kg DM, based on

Table 1 Analyzed contents of carbohydrates and condensed tannins of feed ingredients (g/kg DM)

Species	Cultivar	DM	Starch		NDF	ADF	Condensed tannins
			polarimetric	enzymatic			
Pisum sativum	Santana[1]	881	503	372	105	81	ND
	Jutta[1]	869	505	389	104	79	ND
	Phönix[1]	877	518	392	106	68	ND
	Harnas[1]	871	496	385	109	86	ND
	Rocket[1]	874	495	378	126	83	ND
	Hardy[1]	872	535	423	115	75	ND
Vicia faba	Aurelia[1]	878	456	358	126	101	ND
	Divine[2]	883	452	371	128	112	2.1
	Gloria[1]	886	425	314	127	111	ND
	Limbo[2]	890	442	393	138	116	7.0
	Fuego[2]	876	421	321	165	137	7.4
	Espresso[2]	872	432	311	156	134	4.2
Lupinus spp.	Probor[3]	902	95	<1.1[5]	224	185	ND
	Bornal[4]	892	56	<1.1[5]	252	208	ND
	Boregine[3]	909	96	<1.1[5]	247	195	ND
	Boruta[3]	901	90	<1.1[5]	261	230	ND
	Idefix[3]	905	74	<1.1[5]	258	219	ND
SBM		905	48	3	114	74	ND

Abbreviations: ADF acid detergent fiber, DM dry matter, ND not determined, NDF neutral detergent fiber, SBM soybean meal
[1]White flowered cultivar
[2]Colored flowered cultivar
[3]L. angustifolius
[4]L. luteus
[5]values were below the detection limit of 1.1 g/kg DM

polarimetric and enzymatic measurements, respectively. Contents of NDF and ADF in faba bean cultivars ranged from 126 to 165 and 101 to 137 g/kg DM, respectively, and in the pea cultivars from 104 to 126 and 68 to 86 g/kg DM, respectively. Condensed tannins contents in faba beans amounted to 2.1, 4.2, 7.0, and 7.4 g/kg DM in Divine, Espresso, Limbo, and Fuego, respectively. In peas and lupins, condensed tannins were not detected.

Average starch contents in pea, faba bean, and lupin cultivars are presented in Table 2. Among grain legume species, pea cultivars had greatest average starch contents,

Table 2 Effect of starch determination method on starch content in grain legumes#

Species	Starch content, g/kg DM		P-value
	polarimetric	enzymatic	
Pisum sativum	509 ± 8.0[aC]	390 ± 8.0[bC]	<0.001
Vicia faba	438 ± 8.0[aB]	345 ± 8.0[bB]	<0.001
Lupinus spp.	82 ± 8.8[A]	<1.1[*A]	<0.001

#LSmeans ± standard error of the means
Abbreviations: DM dry matter
a,bWithin a row, means without a common superscript differ (P < 0.05)
A,B,CWithin a column, means without a common superscript differ (P < 0.05)
*values were below the detection limit of 1.1 g/kg DM

followed by faba beans, and lowest contents in lupins, irrespective of the method used for determination (P < 0.05). Average starch contents in pea, faba bean, and lupin cultivars amounted to 509, 438, and 82 g/kg DM when analyzed polarimetrically and 390, 345, and <1.1 g/kg DM, respectively, when analyzed enzymatically. The mean polarimetric analyzed starch values in grain legumes were greater when compared to the mean starch contents that were analyzed enzymatically (P < 0.001) (Table 2).

Ileal starch digestibility in grain legumes

The AID of starch of faba beans and peas is shown in Table 3 using either polarimetric or enzymatic analysis for starch determination. The AID of starch for lupins and SBM could not be determined by means of the difference method due to their low starch contents, resulting in a low contribution level to the total starch contents in the diet.

Within pea cultivars, AID of starch was greater (P < 0.05) for cultivar Harnas (89%) in comparison to cultivars Phönix (83%), Rocket (83%), and Hardy (85%), when the polarimetric method was used for starch analysis. Based on the enzymatic procedure, AID of starch within peas was greater (P < 0.05) for cultivar Jutta (88%) when compared to

Table 3 Apparent ileal starch digestibility in faba beans and peas

Species	Cultivar	Ileal starch digestibility (in %)		SEM	P-value
		polarimetric	enzymatic		
Pisum sativum	Santana[1]	86[abc]	85[ab]		
	Jutta[1]	89[ab]	88[a]		
	Phönix[1]	83[c]	80[bc]		
	Harnas[1]	89[a]	84[ab]		
	Rocket[1]	83[c]	77[c]		
	Hardy[1]	85[bc]	82[bc]		
	LSmeans	86[d]	83[eB]	0.8	<0.001
Vicia faba	Aurelia[1]	86	81		
	Divine[2]	86[*]	82		
	Gloria[1]	85	77		
	Limbo[2]	81[*]	75		
	Fuego[2]	84[*]	80		
	Espresso[2]	85[*]	82		
	LSmeans	84[d]	80[eA]	0.8	<0.001

Abbreviations: SEM standard error of the means

[1]White flowered cultivar

[2]Colored flowered cultivar

[a,b,c]Within a column, means without a common superscript differ within species (P < 0.05)

[d,e]Within a row, means without a common superscript differ between starch determination methods (P < 0.05)

[A,B]Within a column, means without a common superscript differ between species (P < 0.05)

[*]linear effect of dietary content of condensed tannins (P < 0.05)

cultivars Hardy (82%), Phönix (80%), and Rocket (77%). Apparent ileal starch digestibility in faba beans did not differ (P > 0.05) between cultivars, and ranged from 81 (Limbo) to 86% (Aurelia, Divine) using the polarimetric starch determination method and from 75 (Limbo) to 82% (Divine, Espresso) with the enzymatic starch determination procedure.

Mean values for AID of starch did not differ between faba bean and pea cultivars (84 vs. 86%) when the polarimetric method was used (Table 3). However, average AID of starch was lower (P < 0.05) in faba beans when compared to pea cultivars (80 vs. 83%), if starch contents were analyzed enzymatically. Similarly, mean values for AID of starch of faba beans and peas were lower (P < 0.05) when starch content was determined enzymatically rather than polarimetrically.

Discussion

Mean values for polarimetrically analyzed starch contents of peas were within the range of previously reported data [22, 23]. Based on this approach, starch contents in faba beans were greater and starch contents in blue lupins lower when compared to values reported by Berk et al. [23] and Moschini et al. [22]. However, the polarimetric method resulted in a starch content of 56 g/kg DM in the yellow lupin cultivar Bornal which

was greater than average values published by Berk et al. [23] for yellow lupins (35 g/kg DM). Starch content of SBM was in good agreement with tabulated values published by DLG [24], which have been determined polarimetrically as this procedure is used as official method for feed analysis in Germany [7]. Using the enzymatic starch determination procedure, average starch contents in the grain legumes were consistently lower when compared to literature data, with up to 111, 160, and 12 g/kg DM lower contents for peas [25–27], faba beans [25, 28], and lupins [29], respectively. Comparative studies on measurements of starch contents in grain legumes by using polarimetric and enzymatic procedures have not been published, but the results reported herein were confirmed by Obuchowski et al. [30] in different triticale grain samples. The authors determined up to 109 g/kg DM lower starch contents when using the enzymatic method rather than the polarimetric procedure. According to Beutler [8], results obtained by the polarimetric method may be biased by the presence of substances, such as mono-, oligo-, and polysaccharides or fiber in feed and food ingredients. Thus, greater starch values obtained by the use of the polarimetric method can be attributed to the fact that both starch and other carbohydrates such as pentosans and β-glucans are hydrolyzed [30]. Similarly, Priepke et al. [29] reported that measurement of starch contents in lupins according to the polarimetric approach were confounded by the presence of mono-, oligo-, and polysaccharides. The enzymatic starch analysis method, however, is based on the principle that starch is hydrolyzed to glucose. Complete hydrolysis of starch to glucose and specific measurement of the released glucose are required to obtain accurate starch values. However, incomplete hydrolysis of starch to glucose or incomplete assay of released glucose due to inadequate gelatinization or conditions that reduce enzymatic activity, will result in lower starch values when the enzymatic method is used [6].

Differences in starch contents within grain legume cultivars may be due to differences in genetics, but also to varying growing and harvesting conditions [31, 32]. Jørgensen et al. [33] and Prolla et al. [34], however, observed no differences in starch content of peas and common beans, respectively, between different years of harvest, Therefore, it cannot be ruled out that in the present study both variations in harvest and growing conditions and genetics may be responsible for the differences in starch content within grain legume cultivars.

In the present study, AID of starch of faba beans was up to 13 percentage units lower compared to AID values obtained by Jansman et al. [35] in piglets. Results may be biased by the use of different methods for starch analysis as these authors used the polarimetric and enzymatic starch determination method for feed and digesta samples, respectively. For raw peas, AID of starch determined with the enzymatic method was up to 13 percentage units lower in comparison with results by Stein and Bohlke [36] who reported AID of starch of 90% in raw peas using the same analytical procedure. However, in a study with growing pigs using an enzymatic procedure for starch determination, Sun et al. [37] obtained AID of starch coefficients between 78 and 80% in raw peas, measured in different experimental periods. These values were lower than those for most pea cultivars in the present study.

Hydrolysis of starch in the gastrointestinal tract is influenced by a number of host factors, in addition to differences in physical characteristics of the ingested food and starch. Host factors, for example, include intensity of chewing the feed, the availability of pancreatic α-amylases, and digesta retention time in the small intestine [38]. Starch depolymerization is affected by several digestive enzymes that cleave the α-(1–4-) and α-(1–6-) glycosidic bonds. In monogastric species, the main starch hydrolyzing enzymes are α-amylases, amyloglucosidases, and isoamylases [39, 40]. There are differences in starch digestibility between different categories of feed ingredients such as digestibility of starch in peas in comparison to e.g. cereal starch is lower at the ileal level ($P < 0.05$), but there are no differences over the entire digestive tract [41]. Consequently, more starch from peas than from cereals will enter the large intestine. According to Wiseman [41], up to 20% of starch digestion from peas will take place in the large intestine compared with only 4 to 7% for cereal starch.

Differences in AID of starch have been attributed to the existence of an enzyme-resistant starch fraction, differences in starch digestion rate, or both [42]. It is generally acknowledged that starch digestibility is affected by the crystal structure of amylopectin present in starch granules [42]. Accordingly, A-type pattern found in cereal starches have proven to be highly digestible, whereas B-type pattern exhibited by starches from tubers, such as potatoes, renders them more resistant to digestion by pancreatic amylase; and C-type pattern found in grain legume starches being intermediate between the A- and B-type patterns revealing some resistance to hydrolysis by α-amylases [37, 43, 44]. Gernat et al. [45] hypothesized that the C-type pattern is a mixture of the A- and B-type patterns with varying ratios. According to these authors, pea starch is composed of 61.4% type A and 38.6% type B, in comparison to starch of beans with 83.0% type A and 17.0% type B. However, these structural differences in starch granules among grain legumes do not explain the greater AID of starch in peas when compared to faba beans in the present study. Besides variations in crystallinity of amylopectin, the amylose content is another important factor

affecting starch digestibility [46, 47]. Unlike amylopectin that is highly branched, amylose polymers have less surface area and more intramolecular hydrogen bonds [42]. Therefore, amylose is depolymerized at a slower rate and extent than amylopectin due to decreased accessibility for α-amylase [48]. According to Sun et al. [37], digestibility of starch is generally inversely proportional to its amylose content, because the amylase starts hydrolyzing in the amorphous regions of the amylopectin [49]. However, these findings could not be confirmed by the results of the present study, where faba beans exhibited lower digestibility values than peas despite greater amylose contents in peas (44.4%) than in faba beans (40.0%) [50]. According to the results of a recently conducted literature review [44], there is little information on the structure of grain legume starches at different levels of structural organization (granular, supramolecular, and molecular). Consequently, it is difficult in many instances to explain the variation in characteristics among grain legume starches reported in the literature. Moreover, there is limited information on grain legume starches with respect to susceptibility towards acid and enzyme hydrolysis, polymorphic composition, rate and extent of retrogradation, and on the levels of rapidly digestible starch, slowly digestible starch, and resistant starch. As many of the studies on grain legume starches have been conducted on individual cultivars, it is difficult to conclude, whether characteristics of structure and their relationships reported for an individual cultivar are truly representative for the whole species [44]. Finally, it has to be acknowledged that particle size may also contribute to differences in starch digestibility among feed ingredients of the same origin. For example, according to Owsley et al. [51], reduction in particle size of sorghum resulted in improved apparent ileal and total digestibility of DM, starch, protein, and GE of growing pigs.

Grain legume starches are generally part of the protein matrix [5]. Most of these proteins are quite hydrophobic, and the protein-starch network is surrounded by cell walls. The starch tends therefore to be kept in the interior of the particles protected from water. Starches from tubers and legumes are particularly well protected from the polar environment of luminal fluids [5] resulting in limited access of host enzymes including α-amylase.

Differences in AID of starch between pea cultivars in the present study may also be attributed to the presence of cell wall components surrounding the starch granules, thereby inhibiting access of amylases to the granule. According to results of in vitro digestion of starch in red kidney beans (*Phaseolus vulgaris*), amylolysis was enhanced by wet homogenization and pepsin pretreatment [52]. These findings suggest that disruption of cell walls is important for efficient starch digestion, however, no relationship between AID of starch and NDF or ADF contents in the diets could be established (data not shown), likely due to low variation among cultivars.

In faba beans, condensed tannins reduce AID of starch in pigs (e.g. [53]). Condensed tannins strongly inhibit the activity of α-amylase, maltase, sucrase, and lactase [54]. As these enzymes are the most important carbohydrases, condensed tannins may interfere with in vivo carbohydrate digestion and absorption, thus increasing the proportion of starch reaching the large intestine. This is in accordance with the results of the present study, where the AID of starch (determined polarimetrically) of the tannin containing faba bean cultivars Divine, Espresso, Limbo, and Fuego decreased ($P < 0.05$) with increasing tannin content of the diets (amounting to 0.9, 1.8, 2.7, and 3.0 g/kg DM for diets containing the cultivars Divine, Limbo, Fuego, and Espresso, respectively). Differences between AID of starch in faba beans and peas or within pea cultivars may be attributed to variations in contents of resistant starch. Moreover, variations in amylose to amylopectin ratio of grain legumes may, at least in part, explain the observed differences in AID of starch.

Conclusions

Starch contents and AID of starch in grain legumes obtained by means of the polarimetric method are greater than starch values determined enzymatically. This methodical difference is apparently due to the presence of other feed components such as carbohydrates that interfere with the polarimetric starch analysis and (or) has to be attributed to incomplete starch hydrolysis when the enzymatic procedure is applied. Differences in AID of starch between faba beans and peas are confined to the use of the enzymatic method for starch determination. Within pea cultivars, differences in AID of starch were observed for both starch analysis methods. There was a negative linear effect of condensed tannins in the diet on AID of starch in faba beans when the polarimetric method was used for starch determination. The polarimetric approach is the official method used in the European Union for starch determination in feed. However, there is increasing evidence that the enzymatic method is superior over the polarimetric method due to its enzyme specificity for starch.

Abbreviations
ADF: Acid detergent fiber; AID: Apparent ileal digestibility; ANF: Anti-nutritional factors; BW: Body weight; DM: Dry matter; NDF: Neutral detergent fiber; SBM: Soybean meal

Acknowledgements
The authors are grateful to H. Brehm and M. Steffl (University of Hohenheim, Stuttgart, Germany) for their excellent work with animal surgery.

Funding
This project was financially supported by the H. W. Schaumann Stiftung (Hamburg, Germany). The authors would like to thank Agri Obtentions (Guyancourt, France), Lochow-Petkus GmbH (Bergen, Germany), Norddeutsche Pflanzenzucht Lembke KG (Holtsee, Germany), Saatzucht Gleisdorf (Gleisdorf, Austria), Saatzucht Steinach GmbH (Bocksee, Germany), and Südwestsaat GbR (Rastatt, Germany) for providing the grain legumes.

Authors' contributions
DJ, NS and RM conceived and performed the study, performed the statistical data analyses and drafted the manuscript. PR participated in drafting the manuscript. KS conducted the starch analysis. All authors read and approved the final manuscript.

Competing interests
The authors declare that they have no competing interests.

Author details
[1]University of Hohenheim, Institute of Animal Science, Emil-Wolff-Strasse 10, 70599 Stuttgart, Germany. [2]Present address: Agricultural Analytic and Research Institute Speyer, Obere Langgasse 40, 67346 Speyer, Germany. [3]University of Hohenheim, Core Facility Hohenheim, Emil-Wolff-Strasse 12, 70599 Stuttgart, Germany.

References
1. Salgado P, Freire JPB, Mourato M, Cabral F, Toullec R, Lallès JP. Comparative effects of different legume protein sources in weaned piglets: nutrient digestibility, intestinal morphology and digestive enzymes. Livest Prod Sci. 2002;74:191–202. doi:10.1016/S0301-6226(01)00297-4.
2. NRC. Nutrient requirements of swine. 11th ed. Washington DC: National Academies Press; 2012.
3. Jezierny D, Mosenthin R, Bauer E. The use of grain legumes as a protein source in pig nutrition: a review. Anim Feed Sci Technol. 2010;157:111–28. doi:10.1016/j.anifeedsci.2010.03.001.
4. Kozlowska H. Nutrition. In: Hedley CL, editor. Carbohydrates in grain legume seeds: improving nutritional quality and agronomic characteristics. Wallingford, Oxon, UK, New York, NY, USA: CABI Publishing; 2001. p. 61–88.
5. Bach Knudsen KE. Triennial growth symposium: effects of polymeric carbohydrates on growth and development in pigs. J Anim Sci. 2011;89: 1965–80. doi:10.2527/jas.2010-3602.
6. Hall MB, Jennings JP, Lewis BA, Robertson JB. Evaluation of starch analysis methods for feed samples. J Sci Food Agric. 2001;81:17–21. doi:10.1002/1097-0010(20010101)81:1<17::AID-JSFA758>3.0.CO;2-B.
7. VDLUFA. Verband Deutscher Landwirtschaftlicher Untersuchungs- und Forschungsanstalten.: Methodenbuch Band III. Die chemische Untersuchung von Futtermitteln. Mit Ergänzungslieferungen 1983, 1988, 1993, 1997, 2004, 2006, 2007, 2012. Darmstadt: VDLUFA-Verlag; 1976.
8. Beutler H. Enzymatische Bestimmung von Stärke in Lebensmitteln mit Hilfe der Hexokinase-Methode. Starch/staerke. 1978;30:309–12. doi: 10.1002/star.19780300906.
9. Commission Regulation 152/2009. Commission Regulation of 27 January 2009 laying down the methods of sampling and analysis for the official control of feed. Off J Eur Union. 2009;L54:1-130.
10. Commission Directive 1999/79/EC. Amending the third commission directive 72/1999/EEC of 27 April 1972 establishing community methods of analysis for the official control of feedingstuffs. Off J Eur Union. 1999;209:23–7.
11. Champ M, Martin L, Noah L, Gratas M. Analytical methods for resistant starch. In: Cho SS, Prosky L, Dreher ML, editors. Complex carbohydrates in foods. New York: Marcel Dekker; 1999. p. 169–87.
12. Mitchell GA. Methods of starch analysis. Starch/staerke. 1990;42:131–4.
13. Li S, Sauer WC, Fan MZ. The effect of dietary crude protein level on ileal and fecal amino acid digestibility in early-weaned pigs. J Anim Physiol Anim Nutr. 1993;70:117–28. doi:10.1111/j.1439-0396.1993.tb00314.x.
14. Council Directive 86/609. European Communities Council Directive of 24 November 1986 on the approximation of laws, regulations and administrative provisions of the member states regarding the protection of animals used for experimental and other scientific purposes; Off J Eur Communities. 1986;L358:1-28.

15. NRC. Nutrient requirements of swine. 10th ed. Washington DC: National Academies Press; 1998.

16. Jezierny D, Mosenthin R, Sauer N, Roth S, Piepho HP, Rademacher M, et al. Chemical composition and standardised ileal digestibilities of crude protein and amino acids in grain legumes for growing pigs. Livest Sci. 2011;138:229–43. doi:10.1016/j.livsci.2010.12.024.

17. Boehringer-Mannheim/R-Biopharm/Roche, Darmstadt, Germany. Commercially available test kit "Test-Combination Starch - UV-method", for the determination of native starch and of partially hydrolized starch in foodstuffs and other materials; Cat. No. 10207748035.

18. SAS. SAS User's Guide: Statistics. Inst. Inc., Cary, NC; 2008.

19. Kenward MG, Roger JH. Small sample inference for fixed effects from restricted maximum likelihood. Biometrics. 1997;53:983–97.

20. Winer BJ, Brown DR, Michels KM. Statistical principles in experimental design. 3rd ed. New York: McGraw Hill; 1991.

21. Piepho HP. An algorithm for a letter-based representation of All-pairwise comparisons. J Comput Graph Stat. 2004;13:456–66. doi:10.1198/1061860043515.

22. Moschini M, Masoero F, Prandini A, Fusconi G, Morlacchini M, Piva G. Raw pea (pisum sativum), raw faba bean (vicia faba var. Minor) and raw lupin (lupinus albus var. Multitalia) as alternative protein sources in broiler diets. Ital J Anim Sci. 2005;4:59–69.

23. Berk A, Bramm A, Böhm H, Aulrich K, Rühl G. The nutritive value of lupins in sole cropping systems and mixed intercropping with spring cereals for grain production. New Zealand: Canterbury; 2008.

24. DLG, editor. Kleiner Helfer für die Berechnung von Futterrationen: Wiederkäuer und Schweine. Frankfurt am Main. Germany: DLG-Verlag; 1999.

25. Liponi GB, Casini L, Martini ML, Gatta D. Faba bean (Vicia faba minor) and pea seeds (Pisum sativum) as protein sources in lactating ewes' diets. Ital J Anim Sci. 2007;6. doi:10.4081/ijas.2007.1s.309.

26. Montoya CA, Leterme P. Effect of particle size on the digestible energy content of field pea (Pisum sativum L.) in growing pigs. Anim Feed Sci Techno. 2011;169:113–20. doi: 10.1016/j.anifeedsci.2011.06.004.

27. Montoya CA, Leterme P. Validation of an in vitro technique for determining ileal starch digestion of field peas (Pisum sativum) in pigs. Anim Feed Sci Technol. 2012;177:259–65. doi:10.1016/j.anifeedsci.2012.06.008.

28. Zijlstra RT, Lopetinsky K, Beltranena E. The nutritional value of zero-tannin faba bean for grower-finisher pigs. Can j anim sci. 2008;88:293–302. doi:10.4141/CJAS07146.

29. Priepke A, Matthes A, Schubert C. Futterwert und Einsatzmöglichkeiten von Blauen Lupinen und Nebenprodukten aus der Energiepflanzenproduktion in der Mastschweinefütterung. 2nd ed. Mecklenburg-Vorpommern; Landesforschungsanstalt für Landwirtschaft und Fischerei. 2009.

30. Obuchowski W, Banaszak Z, Makowska A, Łuczak M. Factors affecting usefulness of triticale grain for bioethanol production. J Sci Food Agric. 2010;90:2506–11. doi:10.1002/jsfa.4113.

31. Mossé J, Baudet J. Crude protein content and aminoacid composition of seeds: variability and correlations. Plant Food Hum Nutr. 1983;32:225–45. doi:10.1007/BF01091188.

32. Wang N, Daun JK. Effect of variety and crude protein content on nutrients and certain antinutrients in field peas (Pisum sativum). J Sci Food Agric. 2004;84:1021–9. doi:10.1002/jsfa.1742.

33. Jørgensen H, Bach Knudsen KE, Lauridsen C. Influence of different cultivation methods on carbohydrate and lipid compositions and digestibility of energy of fruits and vegetables. J Sci Food Agric. 2012;92:2876–82. doi:10.1002/jsfa.5755.

34. Prolla IRD, Barbosa RG, Veeck APL, Augusti PR, da Silva LP, Ribeiro ND, et al. Cultivar, harvest year, and storage conditions affecting nutritional quality of common beans (haseolus vulgaris L.). Ciênc Tecnol Aliment. 2010;30:96–102.

35. Jansman AJM, Huisman J, van der Poel AFB. Ileal and faecal digestibility in piglets of field beans (Vicia faba L.) varying in tannin content. Anim Feed Sci Technol. 1993;42:83–96. doi:10.1016/0377-8401(93)90025-F.

36. Stein HH, Bohlke RA. The effects of thermal treatment of field peas (Pisum sativum L.) on nutrient and energy digestibility by growing pigs. J Anim Sci. 2007;85:1424–31. doi:10.2527/jas.2006-712.

37. Sun T, Nygaard Lærke H, Jørgensen H, Bach Knudsen KE. The effect of extrusion cooking of different starch sources on the in vitro and in vivo digestibility in growing pigs. Anim Feed Sci Technol. 2006;131:67–86. doi:10.1016/j.anifeedsci.2006.02.009.

38. Englyst HN, Hudson GJ. The classification and measurement of dietary carbohydrates. Food Chem. 1996;57:15–21. doi:10.1016/0308-8146(96)00056-8.

39. Gray MG. Starch digestion and absorption in nonruminants. J Nutr. 1992;122:72–7.

40. Copeland L, Blazek J, Salman H, Tang MC. Form and functionality of starch. Food Hydrocoll. 2009;23:1527–34. doi:10.1016/j.foodhyd.2008.09.016.

41. Wiseman J. Variations in starch digestibility in non-ruminants. Anim feed sci technol. 2006;130:66–77. doi:10.1016/j.anifeedsci.2006.01.018.

42. Singh J, Dartois A, Kaur L. Starch digestibility in food matrix: a review. Trends food sci technol. 2010;21:168–80. doi:10.1016/j.tifs.2009.12.001.

43. Gallant DJ, Bouchet B, Buléon A, Pérez S. Physical characteristics of starch granules and susceptibility to enzymatic degradation. Eur J Clin Nutr. 1992;46:S3–S16.

44. Hoover R, Hughes T, Chung HJ, Liu Q. Composition, molecular structure, properties, and modification of pulse starches: a review. Food Res Int. 2010;43:399–413. doi:10.1016/j.foodres.2009.09.001.

45. Gernat C, Radosta S, Damaschun G, Schierbaum F. Supramolecular structure of legume starches revealed by X-Ray scattering. Starch/staerke. 1990;42:175–8. doi:10.1002/star.19900420504.

46. Regmi PR, Metzler-Zebeli BU, Gänzle MG, van Kempen TATG, Zijlstra RT. Starch with high amylose content and low in vitro digestibility increases intestinal nutrient flow and microbial fermentation and selectively promotes bifidobacteria in pigs. J Nutr. 2011;141:1273–80. doi:10.3945/jn.111.140509.

47. Regmi PR, van Kempen TATG, Matte JJ, Zijlstra RT. Starch with high amylose and low in vitro digestibility increases short-chain fatty acid absorption, reduces peak insulin secretion, and modulates incretin secretion in pigs. J Nutr. 2011;141:398–405. doi:10.3945/jn.110.132449.

48. Batlle N, Carbonell JV, Sendra JM. Determination of depolymerization kinetics of amylose, amylopectin, and soluble starch by aspergillus oryzae α-amylase using a fluorimetric 2-p-toluidinylnaphthalene-6-sulfonate/flow-injection analysis system. Biotechnol Bioeng. 2000;70:544–52. doi:10.1002/1097-0290(20001205)70:5<544::AID-BIT9>3.0.CO;2-5.

49. Rooney LW, Pflugfelder RL. Factors affecting starch digestibility with special emphasis on sorghum and corn. J Anim Sci. 1986;63:1607–23.

50. Cai J, Cai C, Man J, Zhou W, Wei C. Structural and functional properties of C-type starches. Carbohydr Polym. 2014;101:289–300. doi:10.1016/j.carbpol.2013.09.058.

51. Owsley WF, Knabe DA, Tanksley Jr TD. Effect of sorghum particle size on digestibility of nutrients at the terminal ileum and over the total digestive tract of growing-finishing pigs. J Anim Sci. 1981;52:557.

52. Tovar J, Björck IM, Asp N. Analytical and nutritional implications of limited enzymic availability of starch in cooked red kidney beans. J Agric Food Chem. 1990;38:488–93. doi:10.1021/jf00092a034.

53. van der Poel AFB, Gravendeel S, van Kleef DJ, Jansman AJM, Kemp B. Tannin-containing faba beans (Vicia faba L.): effects of methods of processing on ileal digestibility of protein and starch for growing pigs. Anim Feed Sci Technol. 1992;36:205–14. doi:10.1016/0377-8401(92)90057-D.

54. Carmona A, Borgudd L, Borges G, Levy-Benshimol A. Effect of black bean tannins on in vitro carbohydrate digestion and absorption. J Nutr Biochem. 1996;7:445–50. doi:10.1016/0955-2863(96)00077-0.

2

Relationship between HSP90a, NPC2 and L-PGDS proteins to boar semen freezability

Julián Valencia, Germán Gómez, Walter López, Henry Mesa and Francisco Javier Henao*

Abstract

Background: The purpose of this study was to determine the association of three proteins involved in sperm function on the freezability of porcine semen: the heat shock protein 90 alpha (HSP90a), the Niemann-Pick disease type C2 protein (NPC2), and lipocalin-type prostaglandin D synthase (L-PGDS).
Six adult boars (each boar was ejaculated three times, 18 in total) were classified by freezability based on the percentage of functionally competent sperm. The male semen with highest freezability (MHF) and the male semen with lowest freezability (MLF) were centrifuged immediately after collection to separate seminal plasma and spermatozoa to make four possible combinations of these two components and to incubate them for 3 h, adjusting the temperature to 17 °C, to freeze them afterwards. The quantification of proteins was performed in two stages: at zero and at 3 h after incubation of the four combinations.

Results: The spermatozoa × incubation time (IT) interaction only had effect ($P < 0.01$) on HSP90a levels; this protein increased in seminal plasma, after 3 h of incubation, in larger quantity ($P < 0.05$) in combinations with MLF spermatozoa. In relation with the NPC2 protein, two isoforms of 16 and 19 kDa were identified. The 19 kDa isoform was affected ($P < 0.01$) only by the seminal plasma × IT interaction, with superior values ($P < 0.01$) both at zero and three hours of incubation, in the combinations with MHF seminal plasma; and 16 kDa isoform was affected ($P < 0.01$) only by the IT with reduction after 3 h of incubation. The levels of L-PGDS was affected ($P < 0.01$) only by the spermatozoa × IT interaction, which reduced ($P < 0.01$) in combinations with MLF spermatozoa after 3 h of incubation.

Conclusions: It is possible to consider that the three proteins evaluated were associated with freezability of boar semen due, especially, to the fact that mixtures with MLF spermatozoa showed greater increase levels of the HSP90a protein and reduction of L-PGDS in plasma. In addition, the seminal plasma of MHF had higher concentration of the NPC2 of 19 kDa protein, which was reduced by incubating with MHF spermatozoa.

Keywords: Boar freezability, HSP90a, L-PGDS, NPC2, Semen preservation, Seminal plasma

Background

The quality of frozen-thawed boar semen shows strong variability associated with differences in freezability between individuals [1], apparently determined by both seminal plasma and sperm characteristics [2]. Some studies have reported significant associations between several seminal proteins and freezability of boar semen, some located in the seminal plasma such as fibronectin type 1 [3] and β subunit of N-acetyl-β-hexosaminidase [4], and others in the sperm cell, such as the heat shock protein 90 alpha [5], the acrosin binding protein and the triosephosphate isomerase protein [6].

From the work done by Pursel et al. [7] it was suggested that incubating semen for a period of 3 to 5 h at temperatures above 15 °C increases the resistance of spermatozoa to cold shock. With this same approach, Casas and Althouse [8] found greater stability and integrity of the sperm membrane after the thermal shock at 5 °C, and Yeste et al. [9] demonstrated greater sperm resistance to freezing-thawing processes when semen was previously maintained at 17 °C. This period has been called "holding time". These findings and the aforementioned proteins allow formulating conjectures about biological phenomena that determine the differences

* Correspondence: fhenao@ucaldas.edu.co
Universidad de Caldas, Faculty of Agricultural Sciences, A.A. 275, Manizales, Caldas, Colombia

in freezability between individuals, which could occur between semen collection and holding time at 17 °C thanks to the interaction between seminal plasma and sperm.

Among the most important effects of freezing on semen quality that stand out is the reduceded sperm survival, and membrane changes and loss of its selective permeability to ions such as calcium, because of a phenomenon called cryocapacitation, which has events that are similar, but no identical, to true sperm capacitation [10]. True sperm capacitation is an event characterized by reordering of the sperm membrane, altered motility pattern, and changes in metabolic activity which allow the acrosome reaction and finally the penetration of the oocyte [11]. Other changes in true capacitation process such as the release of seminal plasma proteins from the surface of the spermatozoa, the loss of membrane cholesterol and signaling induced by reactive oxygen species (ROS), are induced in a disordered manner during freezing [12]. In this respect, there is evidence that ROS can induce sperm capacitation, in a balanced and controlled process, by the activation of adenylate cyclase and cyclic adenosine monophosphate (cAMP); however, freezing produces an excessive production of ROS, which generate lipid peroxidation of membrane fatty acids, DNA damage and macromolecules damage such as proteins and lipids [13].

As a result of cryopreservation, plasma membrane destabilizes and becomes more permeable to Ca^{2+} influx [14]. As a consequence the levels of cAMP [11] and the kinase-32 protein activity increase, resulting in phosphorylation on tyrosine residues of certain proteins among which it is important to mention the binding to proacrosin [15]. In the movement of cholesterol from the plasma membrane, specific molecules of high affinity intervene with it such as the Niemann-Pick type C2 protein which is closely related to the lipid structure of the plasma membrane [16, 17]. It has been confirmed that this protein is the most secreted into the epididymis [17], and it is also secreted into seminal vesicles, prostate and vas deferens [18]. NPC2 protein is considered a decapacitating factor [19] which binds more efficiently those spermatozoa with higher content of cholesterol [18].

Heat shock proteins (HSP) are also involved in the sperm resilience to withstand cryopreservation, possibly as molecular chaperones of starting proteins of the phosphorylation cascades represented by threonine/serine kinases or tyrosine kinase [20]. The above approach is based on the fact that the primary function of HSP is the conservation of these enzymes by correcting defects occurring during folding or damage by denaturation due to thermal or oxidative stress, thus preventing their aggregation in an irreversible way [21]. Among the HSPs, the most studied in boar semen is the HSP90a, which is found intracellularly in lower quantities in low freezability than in higher freezability spermatozoa [5, 22], possibly due to their exit from the spermatozoa into the extracellular space because of the loss in the integrity of the plasma membrane during the cooling [23]. The association between HSP90a and freezability can be attributed to its participation in capacitation and apoptosis processes [5, 22]. Considering that freezing induces apoptosis of sperm reducing the time of viability of these cells [24].

Prostaglandin D synthase lipocalin type (L-PGDS) is an associate protein in boars and bulls with the acrosome reaction [25], with the transformation of prostaglandin H_2 in prostaglandin D2 [26] and with the transport of lipophilic substances such as retinoid, with high affinity with retinoic acid and retinol [27]. These two latter molecules are involved in the early capacitation [14], because they change the permeability of the plasma membrane [28]. Additionally, L-PGDS is present in the acrosomal membrane of freshly ejaculated spermatozoa and then disappears within the sperm acrosome reaction [25], and the in vitro incubation of sperm with this protein increases their binding capacity to the zona pellucida [29].

The purpose of this study was to determine the dynamics of HSP90a, the NPC2 protein and L-PGDS in seminal plasma during the period between collection and cooling to 17 °C and their effect on freezability of boar semen.

Methods

Animals
Six boars between 18 and 43-month-old, five Pietrain and one Landrace × Large White of an insemination center in the central-western Colombia, with normal semen analysis, optimal sanitary and fertility conditions were used for the study. They were kept in individual 2.6 m × 2.5 m pens, fed 2 kg/animal/d of a commercial feed for boars, water at will, and ejaculated every 8 d between 0700 and 0900 h.

Semen collection and freezing
The six boars were ejaculated three consecutive times (18 ejaculates in total) using the gloved hand technique. The rich fraction of each male was separated visually, diluted 1:1.5 with Androhep® plus (Minitüb, Germany) and transported to the Laboratory of Biology Reproduction at Universidad de Caldas for freezing and seminal evaluation. The transport was performed at a temperature greater than equal to 17 °C.

The freezing protocol used was Minitüb®, developed as follows:

The sperm concentration was determinate by photometry, and an aliquot at 30×10^6 spermatozoa/mL was prepared for semen analysis. The remaining semen was centrifuged at 17 °C at 800 × g for 20 min, then the pellet was adjusted at a concentration of 2×10^9 spermatozoa/mL with cooling diluent (Cryo Androstar plus® + 20% egg yolk) at 17 °C, and spermatozoa were cooled down to 5 °C at 1 °C every 7.5 min. Next, spermatozoa were diluted at 1×10^9 spermatozoa/mL with freezing extender (Cryo Androstar plus® + 20% egg yolk + 6% glycerol) at 5 °C. Spermatozoa were finally packed in 0.5 mL french plastic straws, in a controlled environment at 5 °C. The straws were frozen by exposure to liquid nitrogen vapors (5 cm above the level of nitrogen) for 20 min, and stored into liquid nitrogen. The semen was thawed at 38 °C for 20 s in a water bath, and diluted in Androhep® plus at 26 °C at a rate of 7.5 mL of diluent per straw.

Evaluation of sperm variables

Sperm variables evaluated were: sperm motility by computer assisted semen analysis (CASA) and software Sperm Class Analyzer (Ver. 5.4, 2013; Microptic S.L., Barcelona, Spain), membrane structural integrity (MSI) by staining with SYBR-14 and propidium iodide (LIVE/DEAD® Sperm Viability Kit L-7011, Molecular Probes, Europe), incubation for 5 min at 37 °C in a water bath, and visualization of 100 spermatozoa at 1,000× magnification in a epifluorescence microscope (Nikon Eclipse 80i, Japan) with BV-2A filter [30], membrane functional integrity (MFI) through the short hypoosmotic swelling test (sHOST) and examination of 100 spermatozoa at 400× magnification in a phase-contrast microscope (Nikon Eclipse 50i, Japan) [31], acrosome integrity in sperm fixed with 2% glutaraldehyde, without staining, and examination of 100 spermatozoa at 400× magnification in a phase-contrast microscope (Nikon Eclipse 50i, Japan) [32] and sperm morphology by counting 100 spermatozoa in semen fixed with formalin saline solution. without staining, at 400× magnification in a phase-contrast microscope (Nikon Eclipse 50i, Japan) [32]. Furthermore, to add credibility to the evaluation [33] an sHOST combined with a viability test [34] was performed as follows: two vials were prepared, vial 1 with 1.5 mL semen at 17 °C and vial 2 with 1 mL of hypoosmotic solution at 75 mOsm/kg, 5 μL of SYBR-14 and 5 μL of PI which were progressively incubated in a water bath at 37 °C; then, 500 μL of semen from vial 1 were transferred to vial 2, which was incubated for 5 min at 37 °C and fixed with 10 μL of 2% glutaraldehyde in BTS to evaluate acrosome domain [32]. Visualization was performed by the counting of 100 spermatozoa in a differential interference contrast and epifluorescence microscope (Nikon Eclipse 80i, Japan) with BV-2A filter at 1,000× magnification. The result of the test was: percentage of functionally competent sperm (PFCS) and percentage of cryocapacitated sperm (CS). PFCS or

sperm having the major prerequisites for the penetration through the oocyte [35], except for motility; that is: the total sperm alive, morphologically normal, with membrane functional integrity or sHOST positive, and with acrosome integrity, after membranes stress which emulates part of the effects of freezing. The CS were those sperm sHOST positive alive and with true acrosome reaction [34]. Both populations were adjusted for previous agglutination and coiled tail values. The sperm not functionally competent were those having one or more of the following characteristics: morphological abnormalities, loss of membrane functional integrity or sperm sHOST negative, loss of acrosome integrity, agglutination and cell death.

Determination of freezability

Three ejaculates from each of the six males once subjected to freezing and thawing processes were evaluated based on PFCS to determine freezability of males. The highest freezability semen is that with highest PFCS after freezing-thawing and the lowest freezability semen is that with lowest PFCS after freezing-thawing.

Combinations of seminal plasma and sperm with different freezability

The rich fraction of the male with the highest freezability (MHF) and the male with the lowest freezability (MLF), were collected and centrifuged at 800 × g for 10 min at room temperature, immediately after collection to separate the seminal plasma and spermatozoa (SPZ). The seminal plasma obtained was filtered through a 0.2 μm membrane pore size to remove non- separated spermatozoa. With the SPZ and the seminal plasma obtained, four combinations were made, mixing one part spermatozoa with six plasma: Low freezability spermatozoa + High freezability seminal plasma (LFS + HFSP), High freezability spermatozoa + Low freezability seminal plasma (HFS + LFSP), Low freezability spermatozoa + Low freezability seminal plasma (LFS + LFSP) and High freezability spermatozoa + High freezability seminal plasma (HFS + HFSP). These mixtures were added with Androhep® (1:1.5) for a concentration of 213×10^6 spermatozoa/mL each combination, and incubated for 3 h adjusting the temperature to 17 °C to be frozen later. The analysis of sperm variables was performed before and after freezing, in frozen semen stored for 15 d.

Identification and quantification of proteins in seminal plasma

The seminal plasma intended for quantification of proteins was obtained in two stages: on the farm immediately after the collection (0 h) by centrifugation at 800 × g for 10 min a room temperature, and in the laboratory after 3 h of incubation. Both seminal plasma samples were stored at −20 °C.

The total protein concentration estimated by the Bradford method (BIO-RAD Laboratories, Inc., Hercules, CA) at 595 nm wavelength in a UV-visible spectrophotometer (Perkin Elmer Lambda XLS, UK). Protein separation was carried out by sodium dodecyl sulfate polyacrylamide gel electrophoresis (SDS-PAGE) discontinuous (7% for stacking and 16% for resolution) as follows: 1) Incubation at 95 °C for 5 min, of a 1: 3 dilution of seminal plasma and Laemmli buffer denaturing solution; 2) calculation and sowing of 10 µg total protein per well; and 3) gel run with buffer TRIS-Glycine 50 V for 20 min and then at 100 V for 4 h. The transfer was carried out at constant current of 25 mA for 12 h, to polyvinylidene difluoride (PVDF) membranes previously treated with absolute methanol.

Incubation with the antibodies was conducted as follows: 1) washing of the membrane in Tween 20 at 0.1% in phosphate buffered saline (PBST) for 10 min and with phosphate buffered saline (PBS) for 10 min (twice); 2) blocking with 3% bovine serum albumin (BSA) in PBST for 12 h at 4 °C; 3) washing and addition of primary antibodies (produced in rabbits) diluted in 1% BSA solution in PBST as follows: anti-HSP90a at 1:10,000 dilution, anti-L-PGDS at 1:10,000 dilution and anti-NPC2 at 1:100 dilution, to incubate for 12 h at 4 °C; 4) washing and addition of the secondary antibody conjugated with horseradish peroxidase enzyme (produced in goat, anti-rabbit) at 1:5,000 dilution to hatch for 2 h at room temperature 20 to 22 °C. The development was conducted as recommended by the Opti-4CN Kit (BIO-RAD Laboratories, Inc., USA), and the identification and quantification of the bands obtained was conducted with the ImageLab-4 BIO-RAD software based on density and volume, using Gel Doc^{TM} XR+ (BIO-RAD Laboratories, Inc., Hercules, CA). BSA was used as internal concentration control and the bands were normalized through a measurement of volume and density after staining with Ponceau red prior to incubation with the antibodies, with the specifications for normalization of the ImageLab-4 BIO-RAD software.

Statistical analysis
All statistical analyses were performed using SAS (SAS Inst. Cary, NC). All sperm quality and protein variables were tested for normality using the Shapiro-Wilk test. Homogeneity of variance for each effect included in the analyses was tested with the Levene (group means) and the Brown and Forsythe (group medians) tests. Variables with normal distribution were analyzed using the GLM procedure, while variables with non-normal distributions with the GENMOD procedure. Correlation analysis were performed with the CORR procedure.

The association between levels of heat shock protein 90 alpha, the Niemann-Pick disease protein type C2 and lipocalin type prostaglandin D synthase, with freezability based on PFCS was analyzed as a factorial arrangement 2×2×2 (seminal plasma and SPZ high and low freezability, and zero and three hours of incubation) with three replicates in a completely randomized design. Variance analysis and multiple comparisons were adjusted using the Tukey's test. Spearman correlation analysis was carried out among protein concentrations and variables of post-freezing seminal quality. Other seminal variables were analyzed using a factorial 2×2 design (high and low freezability seminal plasma and SPZ), with three replicates in a completely randomized design. The effects of seminal plasma and SPZ, and their interaction were assessed in a log-linear model, consisting of a Poisson regression with a logarithmic function. The results were expressed as least square mean and standard error.

Results and discussion
Evaluation of males by freezability
The difference in the freezability of the six boars evaluated based on the PFCS was found in this work (Table 1). Freezability level is a condition that is maintained throughout the productive life of boars [1] allowing their characterization with the evaluation of one or a few ejaculates [36]. The findings of this study are consistent with the above, where the evaluation of three ejaculates per male with an 8 d of interval, allowed identifying males with high and low freezability as seen in Table 1, where is shown that boar 1 (male with the highest freezability), in the three evaluations, presented PFCS higher than boar 6 (male with the lowest freezability).

Relationship of the dynamics of the three plasma proteins with freezability
The interaction of three factors seminal plasma, SPZ and IT, and seminal plasma × SPZ, did not affect ($P > 0.05$) the concentration of the proteins studied. The interaction SPZ × IT was highly significant ($P < 0.01$) on the

Table 1 PFCS in frozen-thawed semen of six boars in an insemination center of central-western Colombia

Boars	Evaluation 1	Evaluation 2	Evaluation 3	Average
1	5.00	6.72	5.60	5.77
2	6.79	4.44	6.10	5.77
3	4.40	5.34	2.84	4.19
4	3.52	4.20	4.10	3.94
5	3.48	3.12	2.72	3.10
6	1.74	2.70	2.73	2.39

HSP90a levels, a significant effect ($P < 0.05$) on the levels of the L-PGDS, and had no effect ($P > 0.05$) on the concentration of the two isoforms of NPC2. The interaction seminal plasma × IT only affected ($P < 0.01$) the NPC2 levels of 19 kDa. The seminal plasma factor did not affect the HSP90a and L-PGDS levels, and the SPZ factor did not affect the NPC2 of 19 kDa ($P > 0.05$). The NPC2 16 kDa was affected ($P < 0.01$) only by IT, with a reduction after 3 h of incubation.

Heat shock protein alpha (HSP90a)

SPZ × IT interaction had a highly significant effect ($P < 0.01$) on the HSP90a levels, while the seminal plasma factor was not significant ($P > 0.05$). Tukey's multiple comparisons showed that the concentration of HSP90a on seminal plasma was higher after 3 h of incubation in combinations with SPZ of MLF (Figs. 1 and 2). This protein has multiple cellular functions such as: protection to toxic agents, heat stress [21] and oxidative stress during freezing [22]; modulation of apoptosis [37]; and participation in motility [23] and in sperm maturation [22]. Casas et al. [5] associated high intracellular concentration of this molecule prior to freezing with high freezability males, and Huang et al. [23] mention the possibility of leaking of this protein from spermatozoa because of reduction of the membrane integrity which could explain the results found in this work.

The concentration of the protein HSP90a at 0 h of incubation had a positive correlation not significant with structural membrane integrity (0.51; $P = 0.08$) (Table 2). The concentration of the protein HSP90a after 3 h of incubation had negative and significant correlation with progressive motility (-0.64; $P < 0.05$), negative and not significant correlation with total motility (-0.54; $P = 0.06$) (Table 2). In this regard, Huang et al. [23] found

that the decrease of HSP90a in the sperm is associated with low post-freezing motility and Casas et al. [5] showed that high intracellular concentration of this protein pre-freezing has a positive correlation with post-freezing motility.

Niemann disease protein type C2 protein (NPC2)

After performing the Western blot of NPC2 protein two isoforms were visualized: 16 and 19 kDa (Fig. 3b) which coincide with that reported by Okamura et al. [17] who found these two isoforms when using polyclonal antibodies to the aforementioned protein.

The interaction IT × seminal plasma was highly significant ($P < 0.01$) on the NPC2 levels of 19 kDa, while the SPZ factor was not significant ($P > 0.05$), and NPC2 16 kDa was affected ($P < 0.01$) only by the IT with a reduction after 3 h of incubation (Fig. 4). Multiple comparisons by Tukey allowed to establish differences in concentration of the NPC2 of 19 kDa isoform by effect of the interaction seminal plasma × IT, which presented a higher concentration in the combinations with seminal plasma of MHF both at zero and 3 h of incubation ($P < 0.01$) with a lower concentration at the end of incubation (Fig. 5). In this regard, it is considered that NPC2 is one of the proteins with higher affinity to cholesterol [16] and high levels of this protein in seminal plasma are related to high cholesterol ratio in the sperm membrane [18]. In this way, it is possible that reduction of NPC2 in seminal plasma after 3 h of incubation is due to union of this protein to sperm membrane.

In this work it was shown that high freezability males have a higher level of NPC2 of 19 kDa in the plasma, which can be associated with a higher level of cholesterol in the sperm membrane and with greater efficiency in the protein binding to spermatic cholesterol. Therefore, this type of male might have a better mechanism of prevention of cryocapacitation. This is based in the fact that the release of cholesterol from the membrane is a factor that triggers the cryocapacitation process and the procedures performed to the semen during freezing can promote this release [38]. Furthermore, NPC2 of 19 kDa, in mixtures with seminal plasma of MLF showed no reduction in the concentration at 3 h of incubation (Fig. 5) which can be related to the loss of its function by oxidation due to the effect of reactive oxygen species in this type of plasma [39].

Lipocalin type Prostaglandin D synthase (L-PGDS)

SPZ × IT interaction was significant ($P < 0.05$) on the levels of this protein, while the seminal plasma factor was not significant ($P < 0.05$). Multiple comparisons by Tukey showed that the concentration of L-PGDS was lower ($P < 0.01$) in SPZ of MLF after 3 h of incubation

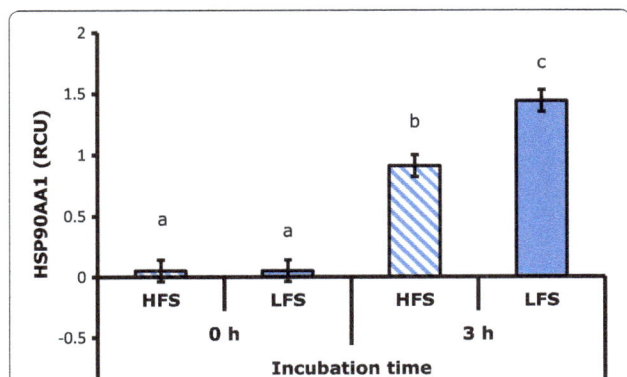

Fig. 1 Effect of interaction spermatozoa per incubation time on the concentration of HSP90a in seminal plasma. RCU: relative concentration units. Different letters indicate highly significant differences ($P < 0.01$)

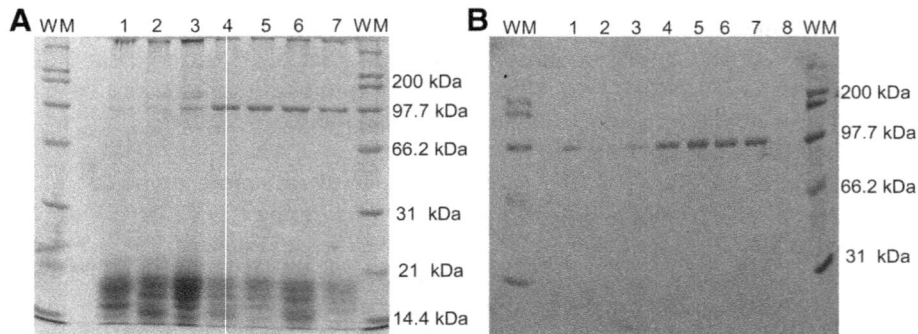

Fig. 2 A, Polyacrylamide gel. B, PVDF membrane with anti-HST90a antibody. WM: weight marker, 1: positive control, 2: seminal plasma of MHF, 3: seminal plasma of MLF, 4: HFS + HFSP, 5: LFS + HFSP, 6: HFS + LFSP, 7: LFS + LFSP, 8: negative control with BSA

(Figs. 6 and 7). It is considered that this protein is a carrier of retinoid such as retinoic acid and retinol [27] which affect the permeability of the plasma membrane while interacting with phospholipids, resulting in greater input of ions from the outside [28]. This phenomenon may be related to cryocapacitation due to access of ions such as Ca^{2+} which turns out to be acrosome and hypermotility reaction [14]. Additionally, in vitro incubation of sperm with L-PGDS protein increases its binding to the zona pellucida [29] an action that is possible only after changes in membrane that occur during the capacitation process [38].

It is possible to consider that the three evaluated proteins were associated with freezability of boar semen due, especially, to the fact that mixtures with spermatozoa of the low freezability males recorded higher increase in levels of HSP90a protein, and reduction of L-PGDS in plasma once the incubation period was completed. While combinations with spermatozoa from the high freezability males showed lower HSP90a reduction and no change in the concentration of L-PGDS in the same period, which can ensure greater protection to thermal and oxidative stress by HSP90a [5] and smaller amounts of retinoic acid and retinol that promote changes in the membrane permeability [28]. In addition, the seminal plasma of males of high freezability had greater concentration of the NPC2 protein of 19 kDa which was reduced by incubating with high freezability spermatozoa, which in turn could be associated with a better prevention mechanism of

Table 2 Spearman correlation coefficients between three seminal proteins with any post-freezing sperm variables (correlation coefficients/probability value)

Time of incubation	Proteins	Membrane structural integrity	Membrane functional integrity	Acrosome integrity	Progressive motility	Total motility	Percentage of functionally competent sperm
0 h	HSP90a	0.51	0.21	−0.15	0.28	0.22	0.08
		0.08	0.50	0.62	0.36	0.48	0.79
	NPC2 19 kDa	−0.44	0.02	0.27	−0.11	−0.01	0.20
		0.15	0.93	0.38	0.72	0.96	0.52
	NPC2 16 kDa	−0.50	0.09	0.04	−0.24	0.02	0.21
		0.09	0.75	0.89	0.45	0.93	0.50
	L-PGDS	−0.10	0.27	−0.13	−0.22	−0.20	−0.10
		0.73	0.37	0.68	0.48	0.51	0.95
3 h	HSP90a	−0.33	0.34	−0.47	−0.64	−0.54	−0.21
		0.29	0.26	0.12	0.02	0.06	0.50
	NPC2 19 kDa	−0.16	0.05	0.01	−0.10	−0.12	−0.10
		0.59	0.86	0.96	0.74	0.69	0.75
	NPC2 16 kDa	−0.14	0.39	−0.18	−0.26	−0.07	0.15
		0.66	0.19	0.55	0.40	0.81	0.64
	L−PGDS	−0.05	−0.42	0.43	0.04	0.29	0.31
		0.87	0.16	0.15	0.89	0.34	0.32

HSP90a Heat shock protein 90 alpha, *NPC2* Niemann-Pick disease type C2 protein, *L-PGDS* Lipocalin-type prostaglandin D synthase

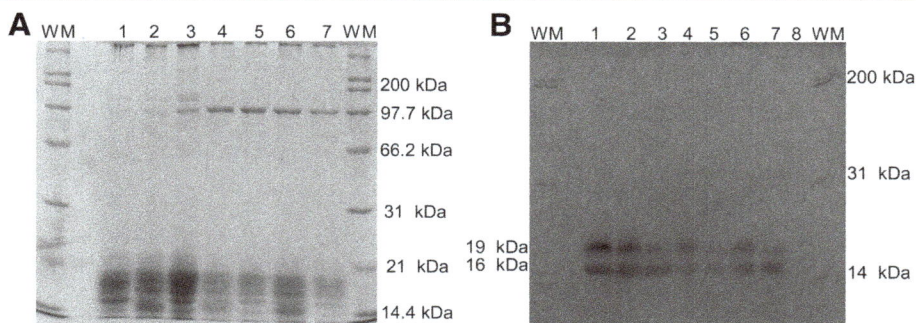

Fig. 3 A, Polyacrylamide gel. B, PVDF membrane with anti-NPC2 antibody. WM: weight marker, 1: Positive control, 2: seminal plasma of MHF, 3: seminal plasma of MLF, 4: HFS + HFSP, 5: LFS + HFSP, 6: HFS + LFSP, 7: LFS + LFSP, 8: negative control with BSA

cryocapacitation of these sperm by effect of plasma [38]. However, it is important to mention that the correlation analysis, the concentration of L-PGDS, NPC2 kDa of 19 kDa, and NPC2 of 16 kDa proteins at zero and three hours of incubation had no significant correlation ($P > 0.05$) with any post-freezing sperm variable (Table 2). There are evidences that the incubation of the semen at 17 °C for 24 h minimizes the alterations in the membrane fluidity and architecture, and generate resistance of the spermatozoa to cold shock, possibly related to the effect of seminal plasma proteins [8]. Although, in the present study, the incubation in this period was only of 3 h, were found changes in the concentration of the three proteins studied, related to the origin of seminal plasma and spermatozoa (both coming of male with highest or lowest freezability). Although these results provide relevant information on the mechanism by which the holding time reduces the effects of cold shock, it is important to conduct studies with longer incubation periods, including evaluation of the lipid architecture and membrane fluidity, and the effect of seminal plasma proteins in preventing the cholesterol efflux by measuring it in the membrane.

Effect of sperm and seminal plasma on semen quality

In the descriptive analysis of frozen-thawed seminal quality variables from the four combinations, it was found that acrosome integrity, progressive motility, total motility and PFCS mean values, were highest in the combinations that had high freezability spermatozoa; and the number of cryocapacitated spermatozoa and membrane functional integrity mean values were lowest in these same combinations (Table 3).

The SPZ × seminal plasma interaction did not affect ($P > 0.05$) any sperm variable; there was seminal plasma effect only on acrosome integrity ($P < 0.05$) and highly significant effect of SPZ ($P < 0.01$) on total motility, acrosome integrity and CS, and significant ($P < 0.05$) on MFI and progressive motility. There was not significant effect ($P > 0.05$) of seminal plasma and SPZ factors on membrane structural integrity and Percentage of functionally competent sperm. When testing for multiple comparisons for acrosome integrity it was observed that a higher value ($P < 0.05$) of 20.80 ± 0.09 with seminal plasma of MHF was

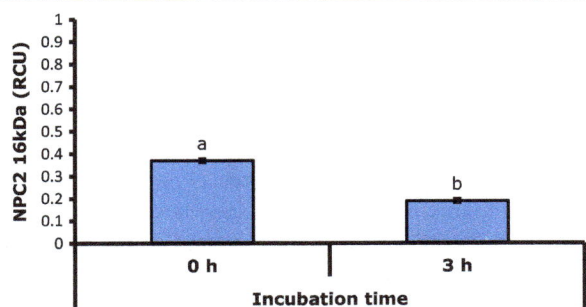

Fig. 4 Effect of incubation time on the concentration of NPC2 of 16 kDa. RCU: relative concentration units. Different letters indicate highly significant differences ($P < 0.01$)

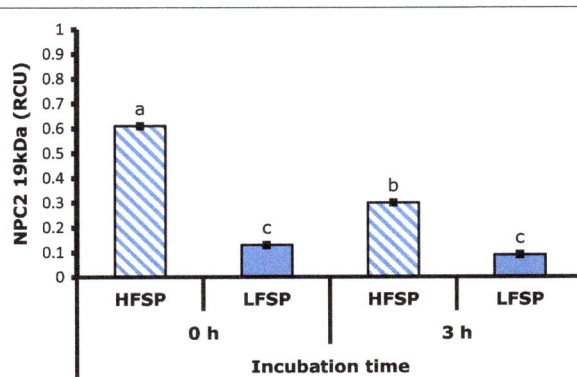

Fig. 5 Effect of seminal plasma interaction per incubation time on the concentration NPC2of 19 kDa. RCU: relative concentration units. Different letters indicate highly significant differences ($P < 0.01$)

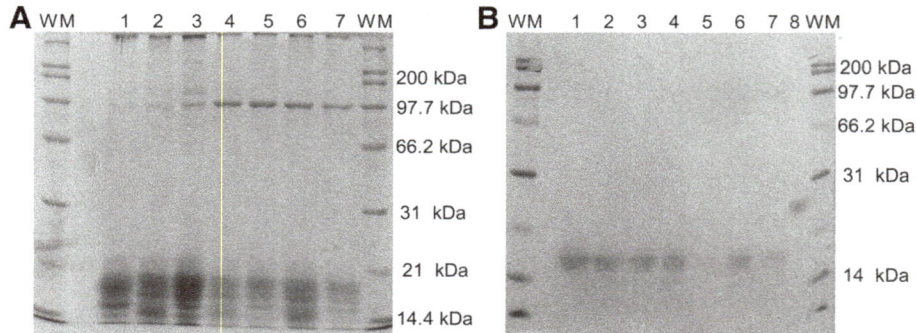

Fig. 6 A, Polyacrylamide gel. B, PVDF membrane with anti-L-PGDS antibody. WM: weight marker, 1: positive control, 2: seminal plasma of MHF, 3: seminal plasma of MLF, 4: HFS + HFSP, 5: LFS + HFSP, 6: HFS + LFSP, 7: LFS + LFSP, 8: negative control with BSA

obtained compared to seminal plasma found with MLF of 14.51 ± 0.11.

Likewise, in assessing the effect of SPZ on this same variable, combinations with SPZ of MLF showed lower acrosome integrity ($P < 0.01$) than mixtures with SPZ of MHF (10.4 ± 0.12 vs 28.87 ± 0.07). According to the above, there are reports of proteins related to acrosome integrity as the N-acetyl-β-hexosaminidase [4] and proteins of sperm surface, called Aspartic acid-Glutamine-Histidine, DQH [40] with acrosome protective function. In addition, this finding may be related to the action of L-PGDS in the acrosome membrane. In the case of total and progressive motility, in combinations with SPZ of MLF, lower values ($P < 0.01$) total motility (22.37 ± 0.08) were found than those found in combinations with SPZ of MHF (31.76 ± 0.07); likewise the progresive motility in SPZ of MLF combinations (10.72 ± 0.12) was lower ($P < 0.05$) than that found with SPZ of MHF (15.05 ± 0.10). There exist reports that associate high intracellular levels of HSP90a with motility [5, 23], and in the present study the SPZ of MLF combinations showed higher amount of this protein in the seminal

plasma after 3 h incubation, possibly due to leakage from inside the cell. Other sperm proteins recently associated with freezability and motility are voltage-dependent anion channel 2 which regulate the permeability of the membrane [41], the acrosin binding protein and the triosephosphate isomerase protein connected to the capacitation process [6].

In the analysis of multiple comparisons, the CS values were higher ($P < 0.01$) in the SPZ of MLF (18.35 ± 0.09) compared to the SPZ of MHF (10.39 ± 0.12). This may be related to cholesterol levels in the sperm membrane, leading to a greater or lesser resistance to freezing, or to the difference in the amount of poly-unsaturated fatty acids [42], associated to the level of lipid peroxidation by ROS, molecules that can cause cryocapacitation [43].

Conclusions

The concentration of the HSP90a protein increased in seminal plasma after 3 h of incubation, and this increase was greater in combinations with spermatozoa of low freezability, a fact which is possibly associated with reduction of membrane integrity or degradation of the protein by heat shock.

Two isoforms (16 and 19 kDa) of the NPC2 protein were found in this work. The concentration of 19 kDa was higher in the combinations with seminal plasma of MHF, both at 0 and 3 h of incubation with a lower concentration at the end of incubation, which may be associated with a better prevention mechanism of the cryocapacitation.

The L-PGDS protein levels decreased after 3 h of incubation in the mixtures containing spermatozoa of the male of low freezability, which may be related to the affinity of this protein to retinol and with effect of these on the permeability of the membrane that would lead the start of the cryocapacitation and subsequent acrosome reaction.

The mixtures with low freezability spermatozoa showed greater increase levels of the HSP90a protein and

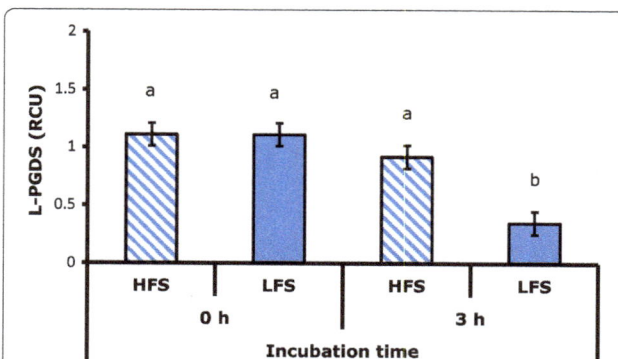

Fig. 7 Effect of spermatozoa interaction times incubation time on L-PGDS concentration. RCU: relative concentration units. Different letters indicate highly significant differences ($P < 0.01$)

Table 3 Frozen- thawed seminal quality variables from the four combinations of sperm and seminal plasma of males with high and low freezability

Combinations	Replicates	Membrane structural integrity	Membrane functional integrity	Acrosome integrity	Progressive motility	Total motility	Percentage of functionally competent sperm	Cryocapacitated sperm
HFS + HFSP	1	51.00	38.07	28.00	18.65	36.30	3.24	14.58
	2	47.00	33.75	41.00	20.15	39.07	3.75	6.00
	3	48.00	40.32	26.00	12.88	29.16	3.80	7.60
	Average	48.67	37.38	31.67	17.23	34.84	3.60	9.39
LFS + HFSP	1	54.00	47.84	16.00	11.54	23.08	2.76	23.00
	2	41.00	56.40	12.00	6.14	15.48	2.82	15.98
	3	37.00	33.44	13.00	10.82	24.48	1.76	16.72
	Average	44.00	45.89	13.67	9.50	21.01	2.45	18.57
HFS + LFSP	1	50.00	33.60	30.00	17.47	36.90	5.60	10.40
	2	50.00	47.88	25.00	9.75	25.62	3.36	14.28
	3	46.00	31.16	24.00	12.26	24.34	1.64	9.84
	Average	48.67	37.55	26.33	13.16	28.95	3.53	11.51
LFS + LFSP	1	51.00	51.15	4.00	14.96	27.03	1.86	19.53
	2	45.00	40.42	10.00	12.99	28.57	2.82	15.98
	3	51.00	40.94	10.00	8.37	15.88	2.58	18.92
	Average	49.00	44.17	8.00	12.11	23.83	2.42	18.14

HFS + LFSP High freezability spermatozoa + Low freezability seminal plasma, *LFS + HFSP* Low freezability spermatozoa + High freezability seminal plasma, *FS + HFSP* High freezability spermatozoa + High freezability seminal plasma, *LFS + LFSP* Low freezability spermatozoa + Low freezability seminal plasma

reduction of L-PGDS in seminal plasma after 3 h of incubation; also, the seminal plasma of MHF had higher concentration of the NPC2 of 19 kDa, which was reduced by incubating with MHF spermatozoa. According to the above, it is possible to consider that the three proteins evaluated were associated with freezability of boar semen.

Abbreviations
BSA: Bovine serum albumin; cAMP: Cyclic adenosine monophosphate; CASA: Computer assisted seminal analysis; CS: Cryocapacitated sperm; HFS + HFSP: High freezability spermatozoa + High freezability seminal plasma; HFS + LFSP: High freezability spermatozoa + Low freezability seminal plasma; HSP: Heat shock proteins; HSP90a: Heat shock protein 90 alpha; IT: Incubation time; LFS + HFSP: Low freezability spermatozoa + High freezability seminal plasma; LFS + LFSP: Low freezability spermatozoa + Low freezability seminal plasma; L-PGDS: Lipocalin-type prostaglandin D synthase; MFI: Membrane functional integrity; MHF: Male with high freezability; MLF: Male with low freezability; MSI: Membrane structural integrity; NPC2: Niemann-Pick disease type C2 protein; PBS: Phosphate buffered saline; PBST: Tween 20 at 0.1% in PBS; PFCS: Percentage of functionally competent sperm; PVDF: Polyvinylidene difluoride; RCU: Relative concentration units; ROS: Reactive oxygen species; SDS-PAGE: Sodium Dodecyl Sulfate Polyacrylamide Gel Electrophoresis; sHOST: Short hypoosmotic swelling test; SPZ: Spermatozoa

Acknowledgements
Dr. Gonzalo Barreneche and the Pig farming Cooperative from the Coffee Triangle CERCAFE.

Funding
Project funded by the Vice-Rector's Office at Universidad de Caldas.

Authors' contributions
All authors read and approved the final manuscript.

Competing interests
The authors declare that they have no competing interests.

References
1. Roca J, Hernandez M, Carvajal G, Vazquez JM, Martinez EA. Factors influencing boar sperm cryosurvival. J Anim Sci. 2006;84:2692–9.
2. Henao FJ, Valencia J, Mesa H, Gómez G. Effect of seminal plasma and sperm of boars valued by freezability on seminal cryopreservation. RevFacNacAgron. 2016;69:7903–10.
3. Vilagran I, Yeste M, Sancho S, Castillo J, Oliva R, Bonet S. Comparative analysis of boar seminal plasma proteome from different freezability ejaculates and identification of Fibronectin 1 as sperm freezability marker. Andrology. 2015;3:345–56.
4. Wysocki P, Orzolek A, Strzezek J, Koziorowska-Gilun M, Zasiadczyk L, Kordan W. The activity of N-acetyl-beta-hexosaminidase in boar seminal plasma is linked with semen quality and its suitability for cryopreservation. Theriogenology. 2015;83:1194–202.
5. Casas I, Sancho S, Ballester J, Briz M, Pinart E, Bussalleu E, et al. The HSP90AA1 sperm content and the prediction of the boar ejaculate freezability. Theriogenology. 2010;74:940–50.
6. Vilagran I, Castillo J, Bonet S, Sancho S, Yeste M, Estanyol JM, et al. Acrosin-binding protein (ACRBP) and triosephosphate isomerase (TPI) are good markers to predict boar sperm freezing capacity. Theriogenology. 2013;80:443–50.
7. Pursel VG, Johnson LA, Schulman LL. Effect of dilution, seminal plasma and incubation period on cold shock susceptibility of boar spermatozoa. J Anim Sci. 1973;37:528–31.
8. Casas I, Althouse GC. The protective effect of a 17° C holding time on boar sperm plasma membrane fluidity after exposure to 5° C. Cryobiology. 2013;66:69–75.
9. Yeste M, Estrada E, Rivera Del Alamo MM, Bonet S, Rigau T, Rodriguez-Gil JE. The increase in phosphorylation levels of serine residues of protein HSP70 during holding time at 17° C is concomitant with a higher cryotolerance of boar spermatozoa. PLoS One. 2014;9:e90887.
10. Green CE, Watson PF. Comparison of the capacitation-like state of cooled boar spermatozoa with true capacitation. Reproduction. 2001;122:889–98.

11. Harayama H. Viability and protein phosphorylation patterns of boar spermatozoa agglutinated by treatment with a cell-permeable cyclic adenosine 3',5'-monophosphate analog. J Androl. 2003;24:831–42.

12. Leahy T, Gadella BM. Sperm surface changes and physiological consequences induced by sperm handling and storage. Reproduction. 2011;142:759–78.

13. Aitken RJ, Baker MA. Oxidative stress and male reproductive biology. Reprod Fertil Dev. 2004;16:581–88.

14. Breitbart H. Intracellular calcium regulation in sperm capacitation and acrosomal reaction. Mol Cell Endocrinol. 2002;187:139–44.

15. Tardif S, Dube C, Bailey JL. Porcine sperm capacitation and tyrosine kinase activity are dependent on bicarbonate and calcium but protein tyrosine phosphorylation is only associated with calcium. Biol Reprod. 2003;68:207–13.

16. Manaskova-Postlerova P, Davidova N, Jonakova V. Biochemical and binding characteristics of boar epididymal fluid proteins. J Chromatogr B Analyt Technol Biomed Life Sci. 2011;879:100–6.

17. Okamura N, Kiuchi S, Tamba M, Kashima T, Hiramoto S, Baba T, et al. A porcine homolog of the major secretory protein of human epididymis, HE1, specifically binds cholesterol. Biochim Biophys Acta. 1999;1438: 377–87.

18. Legare C, Thabet M, Gatti JL, Sullivan R. HE1/NPC2 status in human reproductive tract and ejaculated spermatozoa: consequence of vasectomy. Mol Hum Reprod. 2006;12:461–68.

19. Lasserre A, Barrozo R, Tezon JG, Miranda PV, Vazquez-Levin MH. Human epididymal proteins and sperm function during fertilization: un update. Biol Res. 2001;34:165–78.

20. Ecroyd H, Jones RC, Aitken RJ. Tyrosine phosphorylation of HSP-90 during mammalian sperm capacitation. Biol Reprod. 2003;69:1801–7.

21. Wang P, Shu Z, He L, Cui X, Wang Y, Gao D. The pertinence of expression of heat shock proteins (HSPs) to the efficacy of cryopreservation in HELAs. Cryo Letters. 2005;26:7–16.

22. Casas I, Sancho S, Briz M, Pinart E, Bussalleu E, Yeste M, et al. Freezability prediction of boar ejaculates assessed by functional sperm parameters and sperm proteins. Theriogenology. 2009;72:930–48.

23. Huang SY, Kuo YH, Lee WC, Tsou HL, Lee YP, Chang HL, et al. Substantial decrease of heat-shock protein 90 precedes the decline of sperm motility during cooling of boar spermatozoa. Theriogenology. 1999;51:1007–16.

24. Pena FJ, Saravia F, Johannisson A, Wallgren M, Rodriguez-Martinez H. Detection of early changes in sperm membrane integrity pre-freezing can estimate post-thaw quality of boar spermatozoa. Anim Reprod Sci. 2007;97:74–83.

25. Gerena RL, Irikura D, Eguchi N, Urade Y, Killian GJ. Immunocytochemical localization of lipocalin-type prostaglandin D synthase in the bull testis and epididymis and on ejaculated sperm. Biol Reprod. 2000;62:547–56.

26. Fouchecourt S, Metayer S, Locatelli A, Dacheux F, Dacheux JL. Stallion epididymal fluid proteome: qualitative and quantitative characterization; secretion and dynamic changes of major proteins. Biol Reprod. 2000;62:1790–803.

27. Tanaka T, Urade Y, Kimura H, Eguchi N, Nishikawa A, Hayaishi O. Lipocalin-type prostaglandin D synthase (beta-trace) is a newly recognized type of retinoid transporter. J Biol Chem. 1997;272:15789–95.

28. Stillwell W, Wassall SR. Interactions of retinoids with phospholipid membranes: optical spectroscopy. Methods Enzymol. 1990;189:373–82.

29. Goncalves RF, Barnabe VH, Killian GJ. Pre-treatment of cattle sperm and/or oocyte with antibody to lipocalin type prostaglandin D synthase inhibits in vitro fertilization and increases sperm-oocyte binding. Anim Reprod Sci. 2008;106:188–93.

30. Garner DL, Johnson LA. Viability assessment of mammalian sperm using SYBR-14 and propidium iodide. Biol Reprod. 1995;53:276–84.

31. Perez-Llano B, Lorenzo JL, Yenes P, Trejo A, Garcia-Casado P. A short hypoosmotic swelling test for the prediction of boar sperm fertility. Theriogenology. 2001;56:387–98.

32. Pursel VG, Johnson LA. Glutaraldehyde fixation of boar spermatozoa for acrosome evaluation. Theriogenology. 1974;1:63–8.

33. Johnson LA, Weitze KF, Fiser P, Maxwell WM. Storage of boar semen. Anim Reprod Sci. 2000;62:143–72.

34. Perez-Llano B, Sala R, Reguera G, Garcia-Casado P. Changes in subpopulations of boar sperm defined according to viability and plasma and acrosome membrane status observed during storage at 15° C. Theriogenology. 2009;71:311–17.

35. Bhattacharyya AK, Kanjilal S. Assessment of sperm functional competence and sperm-egg interaction. Mol Cell Biochem. 2003;253:255–61.

36. Hernandez M. Criopreservación espermática en la especie porcina: variabilidad individual. Doctoral thesis. Universidad de Murcia, Departamento de medicina y cirugía animal. 2007. http://hdl.handle.net/10201/157.

37. Powers MV, Clarke PA, Workman P. Dual targeting of HSC70 and HSP72 inhibits HSP90 function and induces tumor-specific apoptosis. Cancer Cell. 2008;14:250–62.

38. Leahy T, Gadella BM. Capacitation and capacitation-like sperm surface changes induced by handling boar semen. Reprod Domest Anim. 2011;46 Suppl 2:7–13.

39. Intasqui P, Antoniassi MP, Camargo M, Nichi M, Carvalho VM, Cardozo KH, et al. Differences in the seminal plasma proteome are associated with oxidative stress levels in men with normal semen parameters. Fertil Steril. 2015;104:292–301.

40. Manaskova P, Peknicova J, Elzeinova F, Ticha M, Jonakova V. Origin, localization and binding abilities of boar DQH sperm surface protein tested by specific monoclonal antibodies. J Reprod Immunol. 2007;74:103–13.

41. Vilagran I, Yeste M, Sancho S, Casas I, del Alamo MM R, Bonet S. Relationship of sperm small heat-shock protein 10 and voltage-dependent anion channel 2 with semen freezability in boars. Theriogenology. 2014;82:418–26.

42. Waterhouse KE, Hofmo PO, Tverdal A, Miller Jr RR. Within and between breed differences in freezing tolerance and plasma membrane fatty acid composition of boar sperm. Reproduction. 2006;131:887–94.

43. Tortolero, Arata-Bellabarba G, Osuna JA, Gómez R, Regadera J. Estrés oxidativo y función espermática: Revisión. Rev Venez Endocrinol Metab. 2005;3:12–9.

3

New approach to assess sperm DNA fragmentation dynamics: Fine-tuning mathematical models

Isabel Ortiz[1][*] ⓘ, Jesús Dorado[1], Jane Morrell[2], Jaime Gosálvez[3], Francisco Crespo[4], Juan M. Jiménez[5] and Manuel Hidalgo[1]

Abstract

Background: Sperm DNA fragmentation (sDF) has been proved to be an important parameter in order to predict in vitro the potential fertility of a semen sample. Colloid centrifugation could be a suitable technique to select those donkey sperm more resistant to DNA fragmentation after thawing. Previous studies have shown that to elucidate the latent damage of the DNA molecule, sDF should be assessed dynamically, where the rate of fragmentation between treatments indicates how resistant the DNA is to iatrogenic damage. The rate of fragmentation is calculated using the slope of a linear regression equation. However, it has not been studied if sDF dynamics fit this model. The objectives of this study were to evaluate the effect of different after-thawing centrifugation protocols on sperm DNA fragmentation and elucidate the most accurate mathematical model (linear regression, exponential or polynomial) for DNA fragmentation over time in frozen-thawed donkey semen.

Results: After submitting post-thaw semen samples to no centrifugation (UDC), sperm washing (SW) or single layer centrifugation (SLC) protocols, sDF values after 6 h of incubation were significantly lower in SLC samples than in SW or UDC. Coefficient of determination (R^2) values were significantly higher for a second order polynomial model than for linear or exponential. The highest values for acceleration of fragmentation (aSDF) were obtained for SW, followed by SLC and UDC.

Conclusion: SLC after thawing seems to preserve longer DNA longevity in comparison to UDC and SW. Moreover, the fine-tuning of models has shown that sDF dynamics in frozen-thawed donkey semen fit a second order polynomial model, which implies that fragmentation rate is not constant and fragmentation acceleration must be taken into account to elucidate hidden damage in the DNA molecule.

Keywords: Colloid centrifugation, Dynamics, Fine-tuning, Mathematical models, Sperm DNA fragmentation

Background

The importance of the assessment of sperm chromatin to predict the potential fertility has been shown in humans and animals [1–5]. The crucial role that sperm DNA fragmentation (sDF) plays in sperm analysis is due to its relationship with infertility problems in spite of apparently normal values for routine sperm parameters such as motility, morphology or integrity of sperm membranes [6]. The assessment of this parameter is even more critical when sperm quality is limited or compromised, as happens in some subfertile males, or in cool-shipped or frozen-thawed semen samples [7]. Therefore, it becomes of the utmost importance to select spermatozoa with intact DNA in order to achieve a higher success in pregnancy rates [8].

Previous studies have evaluated the effect of different centrifugation techniques to select frozen-thawed donkey sperm [9, 10] concluding that, although sperm quality was improved when colloid centrifugation was performed, this procedure did not select intact DNA spermatozoa when performing a static analysis of sDF (baseline value). Nevertheless, it has been shown in several studies [11, 12] that a dynamic assessment of sDF is more accurate to simulate ex vivo sperm maintenance and to evaluate latent chromatin damage than only

* Correspondence: v52orjai@uco.es
[1]Veterinary Reproduction Group, Department of Animal Medicine and Surgery, Faculty of Veterinary Medicine, University of Cordoba, 14071 Cordoba, Spain
Full list of author information is available at the end of the article

considering baseline values. In these studies semen samples were submitted to thermal stress and sDF values were recorded at different times. Subsequently, a linear regression equation was calculated and the rates of fragmentation (the slope of the linear regression equation, sDF%/time) of the treatments were compared.

This approach solved the issue of the cryptic DNA damage; however, another question arose: do DNA dynamics fit to a linear regression model? Although linear regression is the simplest model, it entails that DNA damage is a simple process with constant speed. Before accepting this statement as an actual fact, a fine-tuning of mathematical models for DNA fragmentation over time should be carried out. Thus, the objectives of this study were to evaluate the effect of different after-thawing centrifugation protocols on sperm DNA fragmentation and elucidate the most accurate mathematical model for describing DNA fragmentation dynamics in frozen-thawed donkey semen.

Methods
Animals and semen collection
Six healthy Andalusian donkeys (aged from 6 to 15) were used for this study. Semen collection was performed using a Missouri artificial vagina with an in-line gel filter (Minitüb, Tiefenbach, Germany) in the presence of a jenny in natural or induced estrus. Three ejaculates per animal were collected, obtaining a total number of 18 ejaculates. All animal procedures were performed in accordance with the Spanish laws for animal welfare and experimentation.

Sperm freezing and thawing
Sperm was frozen and thawed following the methodology described by Ortiz et al. [10]. Briefly, seminal plasma was removed by centrifugation ($400 \times g$ for 7 min) and the sperm pellet was resuspended with a commercial freezing medium containing egg-yolk and glycerol (Gent; Minitüb, Tiefenbach, Germany). Then, semen was slowly cooled for 2 h, loaded into 0.5 mL straws, placed 2.5 cm above the surface of the liquid nitrogen (LN_2) for 5 min and plunged in LN_2. Thawing was performed in a water bath at 37 °C for 30 s.

Sperm processing after thawing
After thawing, each semen sample was divided into three aliquots and submitted to different centrifugation protocols.

Uncentrifuged diluted control (UDC)
After thawing, sperm was extended in a physiologically balanced solution that would support sperm viability (INRA96; IMV Technologies, L'Aigle, France) to a final

concentration of 25×10^6 sperm/mL. Sperm parameters were analyzed as described below.

Sperm washing (SW)
Sperm was thawed, diluted at the 1:1 ratio and centrifuged ($400 \times g$ for 7 min). The sperm pellet was resuspended in INRA96 to 25×10^6 sperm/mL for sperm analysis.

Single layer centrifugation (SLC)
Sperm selection was carried out using the colloid Androcoll-E-Small (Swedish University of Agricultural Sciences, Uppsala, Sweden) as described by Ortiz et al. [10]. In short, 2 mL of thawed sperm were carefully placed on 4 mL of colloid. The suspension was centrifuged ($300 \times g$ for 20 min) and the pellet was resuspended in INRA96 to a final concentration of 25×10^6 sperm/mL. Then, sperm parameters were assessed as described in the following.

Sperm analysis after thawing
Sperm motility and membrane integrity
Total (TM, %) and progressive (PM, %) sperm motility were evaluated by computer-assisted sperm analysis (CASA) using the Sperm Class Analyzer (SCA 2011 v.5.0.1; Micimetric S.L., Barcelona, Spain) with the settings described by Ortiz et al. [13]. Membrane integrity was assessed using Vital Test kit (Halotech DNA, Madrid, Spain) following the manufacturer's instructions [10]. Red-stained sperm were considered as membrane-damaged and green sperm were considered as membrane-intact sperm (MIS, %).

Sperm DNA fragmentation (sDF) analysis
The degree of DNA damage in each sample was quantified using the sperm DNA fragmentation index. This parameter was assessed using the Halomax kit (Halotech DNA SL, Madrid, Spain) as previously described by Ortiz et al. [10]. This test is based on the dispersion of the chromatin (halo) after an exposure to a lysing solution. In order to evidence the halos of chromatin, samples were stained with a commercial kit for green fluorescence (Halotech DNA SL). Those sperm with large halos (at least double diameter than the core) were considered to have fragmented DNA. At least 300 spermatozoa per sample were counted and the percentage of fragmented DNA cells was recorded (sDF, %).

Experimental design
Experiment 1: Effect of UDC, SW and SLC on DNA fragmentation dynamics
A dynamic assessment of DNA fragmentation was carried out by incubating an aliquot from UDC, SW and SLC samples for 24 h at 37 °C. The sDF was evaluated

at T0 (baseline), T3, T6 and T24 h and compared between and within treatments.

Experiment 2: Comparison between the coefficient of determinations (R^2) of linear, exponential and polynomial regression in sDF dynamics

The accuracy of three different regression models (linear, exponential, and polynomial) was evaluated by comparing the coefficient of determination (R^2). Then, sDF dynamics were compared among treatments using the most accurate regression model.

Statistical analysis

In Experiment 1 statistical analysis was performed using the Statistical Analysis Systems software (SAS v.9.0; SAS Institute Inc., Cary, NC, USA). A general linear model (PROC MIXED) with animals, ejaculates, treatments and time as fixed effects was performed. Differences between treatments (UDC vs. SW vs. SLC) and times (T0 vs. T3 vs. T6 vs. T24) were assessed.

In Experiment 2, sDF (%) values (y coefficient) at 0, 3, 6 and 24 h (x coefficient) were adjusted to linear, exponential and second order polynomial models. The R^2 was calculated for each replicate and model using Microsoft Excel for Mac v.14 (Microsoft Corporation, Redmond, WA, USA) and R^2 was compared separately for each treatment (UDC, SW and SLC), among models (linear, exponential and polynomial) with PROC MIXED (SAS) using animals and ejaculates as fixed effects.

Since second order polynomial functions are parabolic lines, the derivative function $\frac{d\,sDF}{dt}\left(\frac{\%}{h}\right)$ was calculated for each treatment (UDC, SW and SLC). Afterwards, a graphic was represented using the rate of change of sDF (%/h, DNA fragmentation velocity) of the polynomial function $\frac{d\,sDF}{dt}\left(\frac{\%}{h}\right)$ as y coefficient and time (0, 3, 6 and 24 h) as x coefficient. The slopes of these straight lines (DNA fragmentation acceleration, %/h^2) were compared between treatments (UDC vs. SW vs. SLC) by ANCOVA using GraphPad Prism v.6 for Mac OS v.6 (GraphPad Software, San Diego, CA, USA).

All values were expressed as the mean ± standard error of the mean (SEM). Significant differences were considered when $P < 0.05$. Duncan *post hoc* test was carried out to assess differences between treatments.

Results

Sperm parameters after thawing

Mean values of sperm parameters obtained immediately after thawing were as follows: TM = 58.31 ± 4.57, PM = 47.66 ± 4.07, MIS =57.53 ± 2.71, and sDF = 12.98 ± 1.52.

Sperm DNA fragmentation (sDF) dynamics

A comparison between UDC, SW, and SLC after thawing up to 24 h of incubation at 37 °C is shown in Fig. 1. Significantly lower values of sDF ($P < 0.001$) were found for SLC after 24 h of incubation.

Table 1 illustrates the sDF dynamics over time for each treatment (UDC, SW and SLC). Significantly higher values ($P < 0.001$) of sDF were obtained after 6 h of incubation for UDC and SW. However, in SLC-selected aliquots there were not significant differences for 24 h of incubation at 37 °C.

Regression models fit to the data

The R^2 for SDF dynamics was significantly higher ($P = 0.001$) in polynomial regression models in comparison to linear and exponential models for UDC (0.9699 ± 0.0087 vs. 0.8694 ± 0.0335 vs. 0.8014 ± 0.343), SW (0.9667 ± 0.0120 vs. 0.9324 ± 0.0190 vs. 0.8828 ± 0.0251), and SLC (0.9706 ± 0.0097 vs. 0.8326 ± 0.0605 vs. 0.0826 ± 0.0581) (Fig. 2).

sDF dynamics in polynomial regression

Figure 3 shows the graphical representation of the polynomial regression models for UDC, SW and SLC.

Since second order polynomial functions are graphically represented with parabolic lines, they cannot be compared with each other as a whole. The derivative of the polynomial function $\frac{d\,sDF}{dt}\left(\frac{\%}{h}\right)$ is the rate of change of the function (fragmentation rate, %/h). Figure 4 represents the velocity of fragmentation with respect to time for UDC, SW and SLC. The slopes of these lines are the acceleration of fragmentation (aSDF, fragmentation rate/time, %/h^2). Significant differences between the slopes of UDC ($-0.0683 ± 0.0265$), SW ($0.0106 ± 0.0130$) and SLC ($-0.0073 ± 0.0141$) were obtained ($P = 0.0141$).

Discussion

The results of this study indicate that DNA in frozen-thawed donkey sperm selected by SLC is more resistant to a stressor (incubation at 37 °C up to 24 h) than control or SW. In order to compare sDF values obtained after each centrifugation procedure (UDC, SW and SLC), semen samples were submitted to incubation at 37 °C for 24 h. Afterwards, a static and a dynamic assessment of the DNA fragmentation dynamics were carried out. The static analysis of the sDF dynamics, which consisted of a comparison of treatments (UDC vs. SW vs. SLC) immediately finishing the centrifugation protocols (T0) and after 3 h (T3), 6 h (T6) and 24 h (T24) of incubation at 37 °C. No differences in sDF values were seen between treatments up to 24 h. The stability of the DNA molecule in each centrifugation protocol (T0 vs. T3 vs. T6 vs. T24) was evaluated. In this

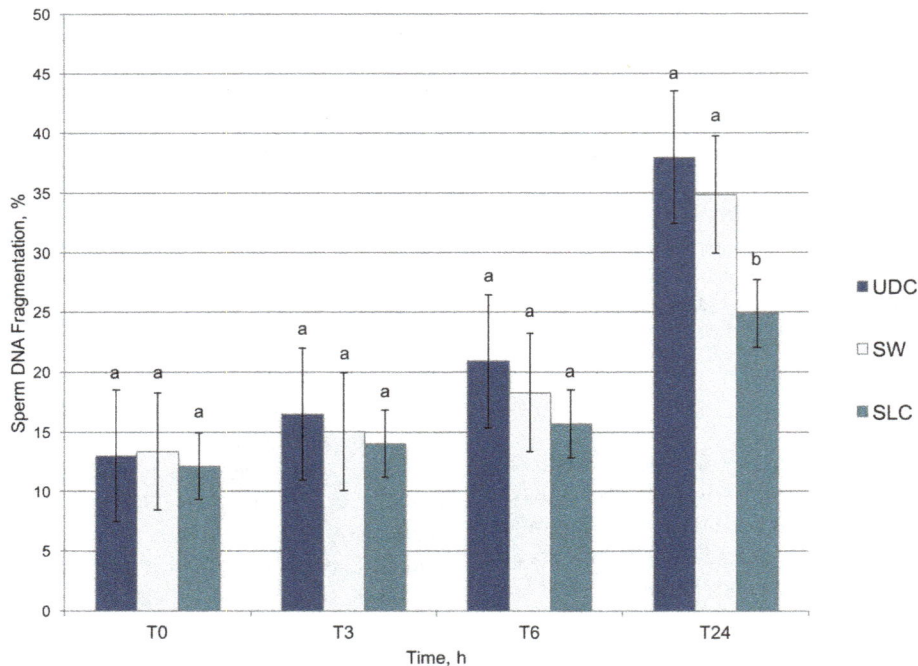

Fig. 1 Effect of centrifugation (UDC, SW, and SLC) on the percentage of sperm DNA fragmentation (sDF, %) of frozen-thawed donkey sperm for 24 h of incubation at 37 °C. Values are expressed as means (*bars*) ± SEM (*error bars*). Different superscripts letters indicate significant differences ($P < 0.05$). UDC = Uncentrifuged diluted control; SW = Sperm washing; SLC = Single Layer Centrifugation; T0, T3, T6, T24 = Incubation for 0, 3, 6, 24 h at 37 °C

regard, sDF values did not increase until 6 h of incubation for UDC and SW. However, sDF remained stable for up to 24 h of incubation in SLC samples. Previous studies performed in stallion semen and donkey semen have not found differences in DNA fragmentation baseline values (T0) after colloid centrifugation [10, 14]. However, according to other studies, sDF must be studied dynamically by submitting the semen sample to a stressor in order to find possible cryptic damage in the DNA [11, 12, 15]. When a dynamic assessment was performed, other studies showed that SLC was able to select those stallion sperm more resistant to DNA fragmentation [16, 17]. Thus, the static and dynamic assessments of sDF dynamics seem to agree that the DNA fragmentation process in SLC samples is slower in comparison to UDC and SW.

Table 1 Effect of time of incubation of sperm at 37 °C on sperm DNA fragmentation (sDF, %) within each centrifugation procedure (UDC, SW and SLC)

	T0	T3	T6	T24	P-value
UDC	12.98 ± 1.52[c]	16.47 ± 1.40[c]	20.90 ± 1.34[b]	35.86 ± 3.21[a]	<0.001
SW	13.35 ± 1.17[c]	15.01 ± 1.27[bc]	18.28 ± 2.21[b]	34.84 ± 2.60[a]	<0.001
SLC	11.48 ± 1.67[b]	14.02 ± 2.16[b]	15.70 ± 2.26[b]	24.91 ± 2.19[a]	<0.001

T0, T3, T6, T24 incubation for 0, 3, 6, 24 h at 37 °C, UDC uncentrifuged diluted control, SW sperm washing, SLC single layer centrifugation
Values are expressed as mean ± standard error of the mean (SEM)
Different letters indicate significant differences

Dynamic processes (such as sperm DNA fragmentation) are commonly used in biology as they provide insight into how a force (e. g. incubation at 37 °C over time) acts to change a cell, an organism, a population, or an assemblage of species [18]. Since we want to know how sDF changes over time *(t)*, that is, *sDF(t)*, a dynamic model is appropriate. There are two main types of dynamic models, "discrete time" and "continuous time", depending on whether time is represented in discrete steps or along continuous axis. In practice, it is not possible to evaluate sDF continually: instead, we must divide the sample into aliquots at intervals of time. Thus, it is crucial to choose a suitable time scale for our study. Previous reports in donkeys [19] and horses [12] have shown that significant changes in sDF between treatments and individuals occurred from 6 to 24 h of incubation at 37 °C when using a chromatin dispersion test, or 4 h of incubation for the sperm chromatin structure assay (SCSA) by flow cytometry [20]. However, since one of the objectives of this study was to fine-tune a model, in order to adjust the model to reality, we needed to have as many points as possible. Therefore, we set the points at 0, 3, 6 and 24 h of incubation.

Once the variable (sDF) and the units of time (0, 3, 6 and 24 h) are stablished, their relationship was evaluated. The most common single-variable functions or mathematical models which regularly arise in all areas of biology are linear, exponential and polynomial. It is common to think of a relationship between two variables as

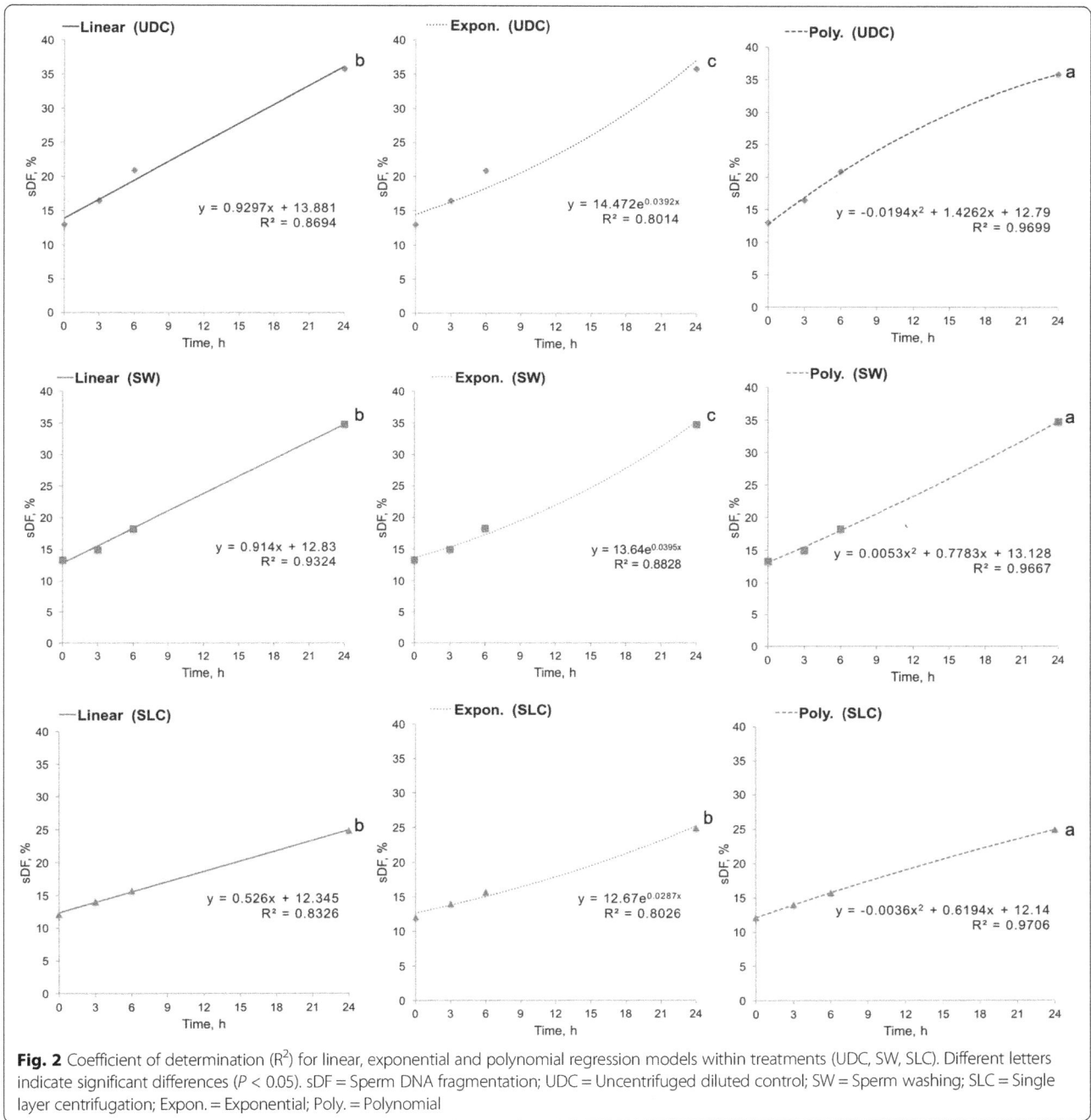

Fig. 2 Coefficient of determination (R^2) for linear, exponential and polynomial regression models within treatments (UDC, SW, SLC). Different letters indicate significant differences ($P < 0.05$). sDF = Sperm DNA fragmentation; UDC = Uncentrifuged diluted control; SW = Sperm washing; SLC = Single layer centrifugation; Expon. = Exponential; Poly. = Polynomial

a straight line; in fact, it has been taken for granted that sDF follows a linear regression equation

$$y = f(x) = ax + b \; ;$$

where y = sDF (%), x = time(h); a = fragmentation rate (%/h); b = intercept (%). This model has been applied to calculate the sDF rate (rSDF) using the slope of the linear regression line. Nonetheless, a linear regression model, as its name indicates, is a straight line. This would mean that rSDF is constant over time which causes some limitations. In this sense, a phenomenon called the "Plateau effect" has been previously described when using linear regression to assess sDF dynamics in stallion sperm [12]. This so-called Plateau effect is a change in the slope of the regression line, due to a change in the velocity of fragmentation. This singularity leads to confounding results if the equation is not divided into two different lines, representing 0–6 h and 6–24 h of incubation. Nonetheless, that provisional adjustment shows that the sDF dynamics does not fit a linear regression equation.

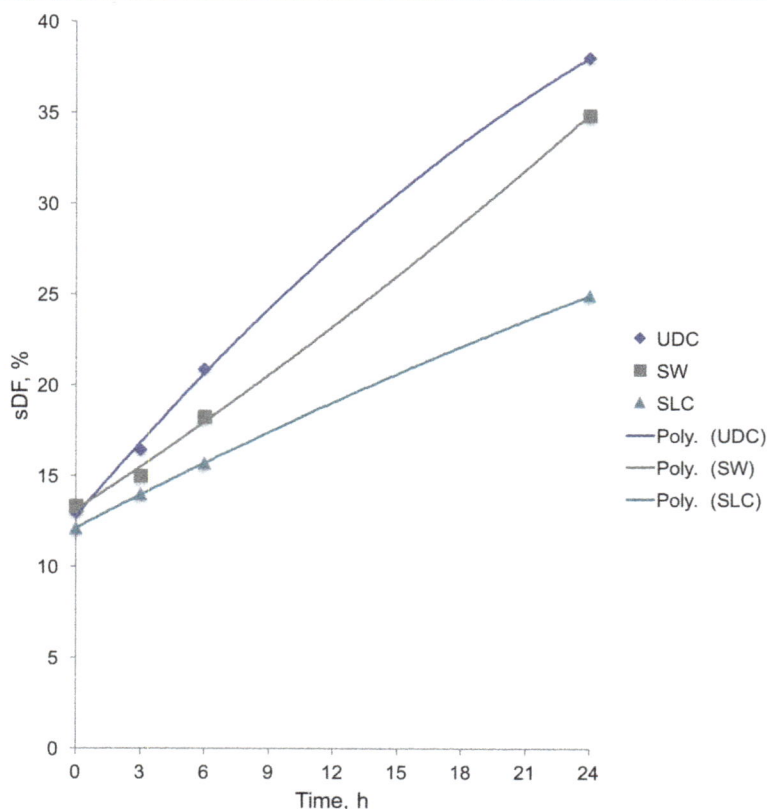

Fig. 3 Polynomial regression lines for UDC, SW and SLC. sDF = Sperm DNA fragmentation; UDC = Uncentrifuged diluted control; SW = Sperm washing; SLC = Single layer centrifugation; Poly. = Polynomial

Non-linear equations are those which can be represented as curves. In this sense, exponential and polynomial functions are very popular for modelling biological functions. Exponential function is probably one of the most important function in dynamic models of biology, it is represented as

$$y = f(x) = b \cdot e^{ax};$$

its main application is describing fast growth [21]. The exponential model has previously been applied to explain the behavior of sDF dynamics in human ejaculates [22]; nevertheless, this explanation was merely descriptive since there was no statistical comparison between other mathematical models. Last but not least, second order polynomial function (or quadratic),

$$y = f(x) = ax^2 + bx + c;$$

has a wide variety of important uses in biology. It describes the rate of growth when resources are limited [23]. Although previous studies have shown sDF curves which might fit a polynomial model [24, 25], to date, no study has tried to explained sDF dynamics using this model. Actually, to the best of the authors' knowledge, a

fine-tuning of models has never been performed for sDF dynamics. Furthermore, this quadratic model explains by itself the behavior of sDF from 0 to 24 h of incubation, without the need to split the 0–6 h and 6–24 h as it occurs with the traditional assessment by linear regression [12].

On the whole, mathematical models are mainly used to make predictions about the behavior of a variable at any time. Logically, the closer to reality the model is the more accurate the prediction will be. In statistics, R^2 provides a measure of how well observed outcomes are replicated by the model, based on the proportion of total variation of outcomes explained by the model [26]. Surprisingly, second order polynomial achieved the significantly highest R^2 in all the treatments studied (UDC, SW and SLC), becoming a more accurate model than linear regression or exponential to predict sDF over time. The more than acceptable R^2 mean values obtained for UDC, SW and SLC (0.9699, 0.9667 and 0.9706, respectively) indicate that the conclusions obtained from this model are very close to reality.

The fact that our model is a parabola implies that sDF rate is not constant, which also means that there is sDF acceleration, an observation that has never been

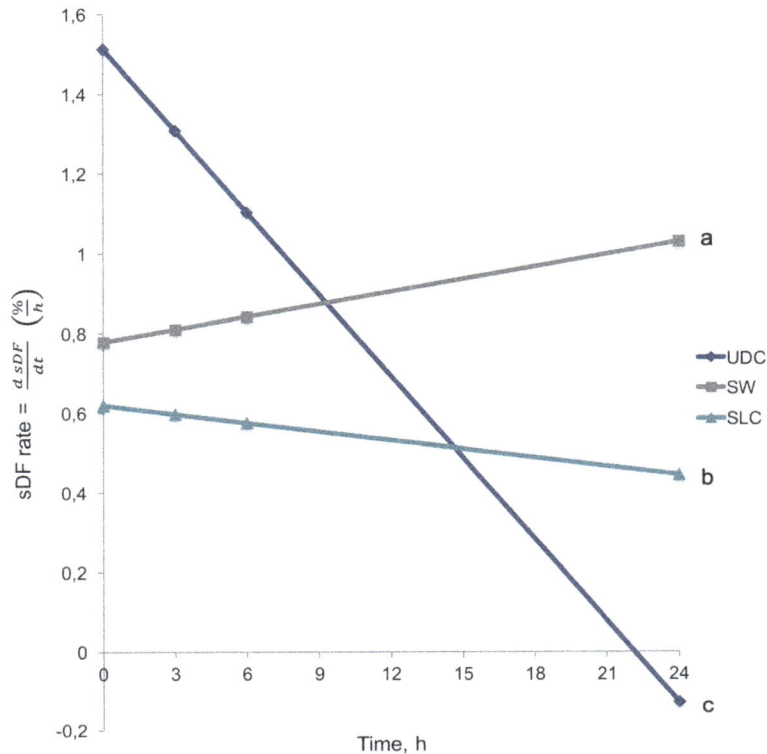

Fig. 4 Sperm DNA fragmentation rate (%/h) in relation to time of incubation at 37 ℃ for UDC, SW and SLC. Different letters indicate significant differences between slopes ($P < 0.05$). sDF = Sperm DNA fragmentation; UDC = Uncentrifuged diluted control; SW = Sperm washing; SLC = Single layer centrifugation

described before. In quadratic functions, we can track the rate of change of the function using a difference equation. A difference equation specifies how much a variable changes from one time unit to the next. In quadratic functions, the rate of change over time is expressed as follows:

$$y' = f'(x) = \frac{dy}{dx} = 2ax + b;$$

in our function, the rate of change of the function is the rate of change of rSDF. Therefore, sDF rate (rSDF) in quadratic functions is expressed as

$$rSDF = \frac{dsDF}{dt} \left(\frac{\%}{h}\right),$$

and the fragmentation acceleration (aSDF) is the rate of change over time, or the slope of the derivative of the quadratic function:

$$aSDF = \frac{dsDF/dt}{t} \left(\frac{\%}{h^2}\right).$$

Surprisingly, when representing the rates of change for UDC, SW and SLC, the three lines obtained were very different. Although UDC showed faster rSDF for about 10 h, it was also the treatment with significantly lower

acceleration ($aSDF_{UDC} = -0.0683 \pm 0.0265$); this marked deceleration explains the Plateau effect described in stallions. In SW samples, rSDF values were lower than UDC until 10 h of incubation, but a higher acceleration ($aSDF_{SW} = 0.0106 \pm 0.0130$) increased rSDF from that point on. The SLC also showed a negative acceleration ($aSDF_{SLC} = -0.0073 \pm 0.0141$), but not as marked as in UDC samples. On the one hand, centrifuged samples after thawing (SW and especially SLC) showed lower values than control (UDC) for 10 and 15 h, respectively. On the other hand, centrifugation increased the acceleration of fragmentation, in particular in SW samples. It could be possible that post-thawing centrifugation, mainly SW, damaged the fixing mechanism of the DNA molecule [27, 28]. However, sDF values are lower during the time studied for SW and SLC. In this sense, if DNA fragmentation processes also occur in vivo in a non-linear manner, the initially rapid fragmentation rate may reduce the effective and fertile sperm concentration before they have a chance to colonize the oviducts. Therefore, delaying the early fragmentation events could be beneficial in this respect and may help to improve fertility. Further studies involving fertility and the relationship between DNA fragmentation timing and the timing of sperm transport and fertilization are needed to characterize the mechanism of action of this damage.

Nonetheless, we need to keep in mind that rSDF values have been obtained from a model, i.e. they are expected rather than observed values. In order to fit the model even more, further studies with more frequent assessments (between 6 and 24 h and after 24 h) would be needed to obtain more accurate predictions. It is of the utmost importance to fit data to an accurate model in order to know exactly how this molecule behaves. In this sense, DNA cannot have been correlated to any other sperm parameter [19, 29] using linear regression to work with sDF dynamics. Hopefully, this study provides new tips so that correlation between DNA fragmentation dynamics and sperm quality is focused from a new perspective.

Conclusions

SLC after thawing seems to preserve DNA longevity for longer in comparison to UDC and SW. Moreover, the fine-tuning of models has shown that sDF dynamics in frozen-thawed donkey semen fits a second order polynomial model, which implies that fragmentation rate is not constant and fragmentation acceleration must be taken into account to elucidate hidden damage in the DNA molecule.

Abbreviations

aSDF: Sperm DNA fragmentation acceleration; CASA: Computer-assisted sperm analysis; Expon.: Exponential; h: Hours of incubation at 37 °C; LN$_2$: Liquid nitrogen; min: Minutes; MIS: Membrane intact sperm; PM: Progressive sperm motility; Poly.: Second order polynomial; R^2: Coefficient of determination; rSDF: Sperm DNA fragmentation rate; s: Seconds; SAS: Statistical Analysis Systems Software; sDF: Sperm DNA fragmentation; SEM: Standard error of the mean; SLC: Single layer centrifugation; SW: Sperm washing; T0-T3-T6-T24: 0-3-6-24 h of incubation at 37 °C; TM: Total sperm motility; UDC: Uncentrifuged diluted control

Acknowledgements

The authors are extremely grateful to Fundación Casa del Burro (Rute, Córdoba, Spain) for providing the animals.

Funding

This study was partially supported by grants RZ2009-00006-00-00 (Instituto Nacional de Investigación y Tecnología Agraria y Alimentaria, Ministerio de Ciencia e Innovación, Spain) and AGL-2013-42726-R (Secretaría de Estado de Investigación, Desarrollo e Innovación, Ministerio de Economía y Competitividad, Spain). I. Ortiz was supported by a Ph.D. fellowship from the ceiA3 (Andalucía, Spain) with funding provided by Banco Santander through its Global Division, Santander Universidades. J M Morrell is funded by the Swedish Foundation for Equine Research, Stockholm, Sweden (H14-47-008).

Authors' contributions

IO contributed to all sections. JD and MH contributed to the study design, data analysis and interpretation, preparation of the manuscript and final approval of the manuscript. JMM, JG and FC collaborated in the experimental design. JMM provided the colloid and revised the English. JMJ revised modelling and calculation. All authors read and approved the final manuscript.

Competing interests

The authors declare that they have no competing interests. J M Morrell is the inventor and one of the patent holders of Androcoll-E and SLC.

Author details

[1]Veterinary Reproduction Group, Department of Animal Medicine and Surgery, Faculty of Veterinary Medicine, University of Cordoba, 14071 Cordoba, Spain. [2]Department of Clinical Sciences, Division of Reproduction, Swedish University of Agricultural Sciences, Box 7054 SE-75007 Uppsala, Sweden. [3]Department of Biology, Universidad Autónoma de Madrid, 28049 Madrid, Spain. [4]Department of Reproduction, Centro Militar de Cría Caballar (FESCCR-Ministry of Defense), 05005 Ávila, Spain. [5]Department of Chemistry and Physics of Materials, University of Salzburg, Hellbrunner Straße 34/III, A-5020 Salzburg, Austria.

References

1. Evenson DP. The Sperm Chromatin Structure Assay (SCSA®) and other sperm DNA fragmentation tests for evaluation of sperm nuclear DNA integrity as related to fertility. Anim Reprod Sci. 2016;169:56–75.
2. Esteves SC. Novel concepts in male factor infertility: clinical and laboratory perspectives. J Assist Reprod Genet. 2016;33:1319–35.
3. Ozkosem B, Feinstein SI, Fisher AB, O'Flaherty C. Advancing age increases sperm chromatin damage and impairs fertility in peroxiredoxin 6 null mice. Redox Biol. 2015;5:15–23.
4. Ni K, Spiess AN, Schuppe HC, Steger K. The impact of sperm protamine deficiency and sperm DNA damage on human male fertility: a systematic review and meta-analysis. Andrology. 2016;4:789–99.
5. Johnston SD, López-Fernández C, Arroyo F, Gosálbez A, Cortés Gutiérrez EI, Fernández J-L, et al. Reduced sperm DNA longevity is associated with an increased incidence of still born; evidence from a multi-ovulating sequential artificial insemination animal model. Reprod Genet. 2016; 33(9):1231–8.
6. Oleszczuk K, Augustinsson L, Bayat N, Giwercman A, Bungum M. Prevalence of high DNA fragmentation index in male partners of unexplained infertile couples. Andrology. 2013;1:357–60.
7. Brinsko SP, Rigby SL, Lindsey AC, Blanchard TL, Love CC, Varner DD. Pregnancy rates in mares following hysteroscopic or transrectally-guided insemination with low sperm numbers at the utero-tubal papilla. Theriogenology. 2003;59:1001–9.
8. Morrell JM, Johannisson A, Dalin AM, Hammar L, Sandebert T, Rodriguez-Martinez H. Sperm morphology and chromatin integrity in Swedish warmblood stallions and their relationship to pregnancy rates. Acta Vet Scand. 2008;50:2.
9. Ortiz I, Dorado J, Acha D, Gálvez MJ, Urbano M, Hidalgo M. Colloid single-layer centrifugation improves post-thaw donkey (Equus asinus) sperm quality and is related to ejaculate freezability. Reprod Fertil Dev. 2015;27:332–40.
10. Ortiz I, Dorado J, Morrell JM, Crespo F, Gosálvez J, Gálvez MJ, et al. Effect of single-layer centrifugation or washing on frozen–thawed donkey semen quality: Do they have the same effect regardless of the quality of the sample? Theriogenology. 2015;84:294–300.
11. Gosalvez J, Lopez-Fernandez C, Fernandez JL, Gouraud A, Holt WV. Relationships between the dynamics of iatrogenic DNA damage and genomic design in mammalian spermatozoa from eleven species. Mol Reprod Dev. 2011;78:951–61.
12. Lopez-Fernandez C, Crespo F, Arroyo F, Fernandez JL, Arana P, Johnston SD, et al. Dynamics of sperm DNA fragmentation in domestic animals II. The stallion. Theriogenology. 2007;68:1240–50.
13. Ortiz I, Dorado J, Ramírez L, Morrell JM, Acha D, Urbano M, et al. Effect of single layer centrifugation using Androcoll-E-Large on the sperm quality parameters of cooled-stored donkey semen doses. Animal. 2014;8:308–15.
14. Costa AL, Martins-Bessa A, de Andrade AR, Guimaraes T, Rebordao MR, Gaivao I, et al. Single Layer Centrifugation with Androcoll-ETM improved progressive motility and percentage of live spermatozoa with intact acrosome of chilled stallion semen but did not have an effect on DNA integrity. Open J Anim Sci. 2012;2:159–65.
15. Gosalvez J, Fernandez JL, Gosalbez A, Arrollo F, Agarwal A, Lopez-Fernandez C. Dynamics of sperm DNA fragmentation in mammalian species as assessed by the SCD methodology. Fertil Steril. 2007;88:S365.
16. Crespo F, Gosalvez J, Gutiérrez-Cepeda L, Serres C, Johnston SD. Colloidal centrifugation of stallion semen results in a reduced rate of sperm DNA fragmentation. Reprod Domest Anim. 2013;48:e23–5.
17. Gutiérrez-Cepeda L, Fernández Á, Crespo F, Ramírez MÁ, Gosálvez J, Serres C. The effect of two pre-cryopreservation single layer colloidal centrifugation protocols in combination with different freezing extenders on

the fragmentation dynamics of thawed equine sperm DNA. Acta Vet Scand. 2012;54:72.

18. Otto SP, Day T. A biologist's guide to mathematical modeling in ecology and evolution. Princenton: Princenton University Press; 2007.

19. Cortes-Gutierrez EI, Crespo F, Gosalvez A, Davila-Rodriguez MI, Lopez-Fernandez C, Gosalvez J. DNA fragmentation in frozen sperm of Equus asinus: Zamorano-Leones, a breed at risk of extinction. Theriogenology. 2008;69:1022–32.

20. Sabatini C, Mari G, Mislei B, Love C, Panzani D, Camillo F, et al. Effect of post-thaw addition of seminal plasma on motility, viability and chromatin integrity of cryopreserved donkey jack (Equus asinus) spermatozoa. Reprod Domest Anim. 2014;49:989–94.

21. Ferretti AC, Joyce GF. Kinetic properties of an RNA enzyme that undergoes self-sustained exponential amplification. Biochemistry. 2013;52:1227–35.

22. Gosalvez J, Nunez R, Fernandez JL, Lopez-Fernandez C, Caballero P. Dynamics of sperm DNA damage in fresh versus frozen-thawed and gradient processed ejaculates in human donors. Andrologia. 2011;43:373–7.

23. Martín-Landrove R, Guillén N, Martín-Landrove M. A kinetic model for tumor survival curves: its relation to the linear-quadratic model. 3rd ed. Munich: World Congress on Medical Physics and Biomedical Engineering: Radiation Protection and Dosimetry, Biological Effects of Radiation; 2009. p. 508–11.

24. Gosálvez J, López-Fernández C, Hermoso A, Fernández JL, Kjelland ME. Sperm DNA fragmentation in zebrafish (Danio rerio) and its impact on fertility and embryo viability — Implications for fisheries and aquaculture. Aquaculture. 2014;433:173–82.

25. Gosalvez J, Ramirez MA, Lopez-Fernandez C, Crespo F, Evans KM, Kjelland ME, et al. Sex-sorted bovine spermatozoa and DNA damage: II. Dynamic features. Theriogenology. 2011;75:206–11.

26. Colin Cameron A, Windmeijer FAG. An R-squared measure of goodness of fit for some common nonlinear regression models. J Econ. 1997;77:329–42.

27. Gunes S, Al-Sadaan M, Agarwal A. Spermatogenesis, DNA damage and DNA repair mechanisms in male infertility. Reprod BioMed Online. 2015;31:309–19.

28. Marcon L, Boissonneault G. Transient DNA strand breaks during mouse and human spermiogenesis:New insights in stage specificity and link to chromatin remodeling. Biol Reprod. 2004;70:910–8.

29. Lewis SEM. Should sperm DNA fragmentation testing be included in the male infertility work-up? Reprod BioMed Online. 2015;31:134–7.

4

Artificial rearing influences the morphology, permeability and redox state of the gastrointestinal tract of low and normal birth weight piglets

Hans Vergauwen[1†], Jeroen Degroote[2†], Sara Prims[1], Wei Wang[2,3], Erik Fransen[4], Stefaan De Smet[3], Christophe Casteleyn[1], Steven Van Cruchten[1], Joris Michiels[2] and Chris Van Ginneken[1*]

Abstract

Background: In this study the physiological implications of artificial rearing were investigated. Low (LBW) and normal birth weight (NBW) piglets were compared as they might react differently to stressors caused by artificial rearing. In total, 42 pairs of LBW and NBW piglets from 16 litters suckled the sow until d19 of age or were artificially reared starting at d3 until d19 of age. Blood and tissue samples that were collected after euthanasia at 0, 3, 5, 8 and 19 d of age. Histology, ELISA, and Ussing chamber analysis were used to study proximal and distal small intestine histo-morphology, proliferation, apoptosis, tight junction protein expression, and permeability. Furthermore, small intestine, liver and systemic redox parameters (GSH, GSSG, GSH-Px and MDA) were investigated using HPLC.

Results: LBW and NBW artificially reared piglets weighed respectively 40 and 33% more than LBW and NBW sow-reared piglets at d19 ($P < 0.01$). Transferring piglets to a nursery at d3 resulted in villus atrophy, increased intestinal FD-4 and HRP permeability and elevated GSSG/GSH ratio in the distal small intestine at d5 ($P < 0.05$). GSH concentrations in the proximal small intestine remained stable, while they decreased in the liver ($P < 0.05$). From d5 until d19, villus width and crypt depth increased, whereas PCNA, caspase-3, occludin and claudin-3 protein expressions were reduced. GSH, GSSG and permeability recovered in artificially reared piglets ($P < 0.05$).

Conclusion: The results suggest that artificial rearing altered the morphology, permeability and redox state without compromising piglet performance. The observed effects were not depending on birth weight.

Keywords: Milk replacer, Oxidative stress, Small intestine, Suckling period, Tight junction proteins

Background

The neonatal period in the pig's life is accompanied with high morbidity and mortality [1, 2]. In addition, increasing litter sizes in modern swine production have led to higher rates of piglets born with a low birth weight (LBW) [3]. Both newborn LBW human [4–6] and LBW piglets [7–10] seem to have a lower capacity to mount an antioxidant response. Newborns transitioning from

maternal mediated respiration to autonomous pulmonary respiration outside the uterus are suddenly exposed to O_2-derived free radicals [11]. This increased the production of reactive species in various organs [7]. A redox imbalance affects cellular signaling, protein synthesis, and enhances proteolysis that can ultimately lead to a dysfunctional intestinal barrier and suboptimal regenerative potential as shown in vitro [12–15]. Consequently, the observed redox imbalance and the downstream effects could explain the abnormal absorption and metabolism of nutrients, reduced growth and impaired development of the small intestine, liver, and muscle observed in LBW piglets [10, 16–21]. This redox imbalance appears to persist beyond weaning [9]. Wang et al.

* Correspondence: chris.vanginneken@uantwerpen.be
†Equal contributors
[1]Laboratory of Applied Veterinary Morphology, Department of Veterinary Sciences, Faculty of Biomedical, Pharmaceutical and Veterinary Sciences, University of Antwerp, Campus Drie Eiken, Universiteitsplein 1, D.U.015, 2610 Wilrijk, Belgium
Full list of author information is available at the end of the article

observed that mRNA expression of occludin, heme oxygenase 1, catalase, thioredoxin reductase genes and occludin protein expression continued to be lower in LBW pigs during the suckling period [21]. This apparently conflicts with the observation that LBW piglets that survive the critical first days after birth show an intestinal morphology, digestive capacity, cytokine production, intestinal motility and permeability that is comparable with those seen in normal birth weight (NBW) littermates [22–24].

The increasing incidence of supernumerary and LBW piglets has raised an urgent need for innovative rearing strategies [1]. Next to cross-fostering [25], supplementing piglets [26] and split nursing [27], piglets can be transferred to a nursery and artificially reared [28–30]. Similar to conventional weaning, the artificially reared piglet encounters psychological and physical stressors including maternal and littermate separation, abrupt changes in diet composition and environment, lower intake of bioactive substances, as well as unfamiliar drinking nipples, increased exposure to pathogens and antigens, comingling and establishment of social hierarchy with unfamiliar pigs from different litters. The physiological responses to the strategy of full artificial rearing are largely unknown. Conventional weaning is associated with the induction of intestinal oxidative stress [31–33] and LBW piglets have more difficulties to maintain a balanced redox state when exposed to weaning stressors [9]. It is unknown at present if the response to artificial rearing, since it includes similar stressors as conventional weaning, is different in LBW piglets—which have a lower antioxidant capacity [21]—compared to NBW piglets.

Therefore, we aimed to investigate the impact of artificial rearing on piglet performance, proximal and distal small intestinal (SI) morphology, mitosis, apoptosis, and tight junction protein expression, permeability, and SI, liver and systemic redox state development compared to conventional rearing. Given the similarities between conventional weaning stressors and artificial rearing stressors we hypothesized that artificial rearing results in a redox imbalance and negatively affects intestinal morphology and functionality. Secondly, given the differences observed between NBW and LBW piglets during the suckling period, we hypothesized that in view of their affected redox state, the morphology and functionality of the small intestine is suboptimal in LBW piglets and that artificial rearing has a greater negative impact in this birth weight category.

Methods
Pig model and tissue collection
Eighty-four piglets (Topigs hybrid × Piétrain) were selected during two consecutive farrowing rounds. Only

sows with 14 or more live-born piglets were selected and no cross-fostered piglets were included in the experiment. Piglets were tagged and weighed within 12 h after parturition. A LBW piglet was defined as a pig having a birth weight between 0.75 and 0.90 kg and belonging to the lower quartile of litter birth weights. A NBW piglet had a birth weight within 1 SD unit of the mean birth weight of the whole litter. The average birth weight of all the piglets used in this study was 0.81 ± 0.01 kg and 1.26 ± 0.02 kg for LBW and NBW piglets, respectively. Forty-two pairs of LBW and NBW gender-matched littermates were either left to suckle the sow or transferred to a nursery at the age of 3 d and had ad libitum access to a milk replacer that was refreshed twice a day (Table 1). Milk replacer was made by mixing 200 g of milk replacer powder into 1 L of water, preheated at 42 °C. This resulted in a dry matter (DM) content of 193 g/kg milk. This is comparable with the DM content of sow milk, which is around 170 to 190 g/kg milk between 3 d and 42 d of lactation [34]. In Table 1, the calculated nutrient composition of some important nutrients is shown, expressed on DM basis. When account for the dilution factor, it is possible to compare this milk formula with sow milk. For instance, it is clear

Table 1 Composition of the milk replacer used for piglets from 3 d of age until weaning at d 19

Ingredient composition, %	
Coco fat filled whey 50/50	42.00
Skimmed milk powder	17.61
Whey permeate	8.29
Soy protein concentrate Soycomil K	10.00
Cheddar whey powder	8.29
Whey protein concentrate80, DVN	7.00
Spray dried blood plasma P80	4.00
Dicalciumphosphate 18% P	0.32
DL-Methionine	0.31
Citric acid	0.30
L-Tryptophan	0.08
Vitamin and mineral premix[a]	1.8
Calculated nutrient levels	
NEv(1997) MJ/kg	15.48
CP, g/kg	249
CF, g/kg	110
dLYS, g/kg	18.2
dMET + CYS, g/kg	11.5
dTHR, g/kg	11.2
dTRY, g/kg	4.1

[a]The mineral and vitamin premix supplied as the following (per kg diet): Vitamin A, 30, 000 IU; Vitamin D₃, 5000 IU; Vitamin E; 75 mg; Fe^{2+}, 120 mg; Zn^{2+}, 35 mg; Cu^{2+}, 135 mg;Mn^{2+}, 45 mg; Se^{6+}, 350 μg; Ⲓ, 1 mg, BHT, 75 mg/kg

that the crude protein level in the formula (49.8 g/kg milk) is low but comparable with what can be expected in sow milk (51–64 g/kg milk) ([34–36]). The crude fat content was calculated at 22 g/kg milk, which is considerably lower than the crude fat content of sow milk (5.3–6.5 g/kg milk) [34–36]. In contrast, the total amount of lactose in the liquid milk replacer was 80 g/kg milk. Values for sow milk range from 48 to 59 g/kg milk, and thus are lower than what was offered to the formula fed piglets. From this, it might be clear that the milk formula was not completely comparable to the nutrient composition of porcine milk. However, this tailor made milk formula was meant as a compromise of many existing, commercially available milk replacer formulas, and as a copy of sow milk as such. A maximum of 4 NBW-LBW gender matched littermate pairs were co-housed. The starting ambient temperature was 32 °C and linearly decreased to 28 °C towards d19 post-natal. Heating and ventilation was automatically controlled in function of the temperature settings. Six pairs of sow-reared LBW and NBW piglets were sampled on d0, d3, d8 and d19 of age, and 6 pairs of artificially reared LBW and NBW piglets were sampled on d5, d8 and d19 of age. Piglets sampled on d0 were removed from the sow between 12 and 24 h after parturition. At d3, intramuscular iron injections were given (Iron(III) Dextran, 200 mg/piglet, Uniferon). The selected animals did not receive any antibiotic treatment prior or during the experiment.

Pigs were killed by exsanguination by severing the carotid arteries and jugular veins following induction of terminal anesthesia by intramuscular injection of ketamine (15 mg/kg BW) combined with xylazine (2 mg/kg BW). All piglets were weighed prior to euthanasia.

Blood was collected in EDTA and heparinized tubes containing supplemental bathophenanthroline disulfonate sodium salt. Erythrocytes were isolated by centrifuging (3000 × g, 15 min) 0.5 mL of unclotted, heparinized blood. After removing the plasma, erythrocytes were lysed by adding 100 μL of a 70% metaphosphoric acid solution, 600 μL milli Q and intense vortexing. These extracts were then centrifuged (3000 × g, 15 min), and 0.5 mL of the remaining acid extract was transferred to a vial containing 50 μL of a γ-glu-glu internal standard solution. After opening the abdomen, the liver was isolated and samples of the left lateral lobe were dissected for acid and phosphate buffered aqueous extraction, as described for the small intestinal mucosa. Subsequently, the small intestine (SI), defined as the part of the gastrointestinal tract between the pylorus and the ileocecal valve, was dissected and its length was measured. A 10 cm segment of proximal and distal SI (5 and 75% of total SI length, respectively) was taken for Ussing chamber measurements. In addition, 20 cm segments at 5 and 75% of the total SI length were emptied and carefully flushed with saline. The tissue of these 20 cm segments was placed on an ice-cold surface and the mucosa was retrieved by gently scraping the mucosal surface with a glass slide. Aliquots of the mucosa were either used instantaneously for acid and phosphate buffered aqueous extracts or transferred to plastic 2 mL screw-capped tubes, snap-frozen in liquid nitrogen and stored at -80 °C pending redox state analysis. Furthermore, 5 cm segments at 5 or 75% of the total SI length were taken, flushed with saline, snap-frozen in liquid nitrogen and stored at -80 °C pending protein expression analysis. Finally, a 5 cm segment at 5 or 75% of the total SI length was flushed with saline, divided in smaller pieces of max 1.5 cm in length and fixated for 2 h in 4% freshly prepared paraformaldehyde (in 0.01 mol/L phosphate-buffered saline) (volume tissue/volume fixative: 1/5) and routinely processed for paraffin-embedding [37].

Small intestinal histo-morphological measurements
In brief, 4 μm sections of paraffin-embedded samples were mounted on slides and stained with hematoxylin-eosin. Villus height, mid-villus width, and crypt depth were measured at 10× magnification using an Olympus BX61 microscope and image analysis software (analySIS Pro, Olympus Belgium, Aartselaar, Belgium) in 1–3 well-oriented villi and associated crypts in at least 12–15 sections per tissue sample, to yield 30 measurements per small intestinal region.

Small intestinal protein expression profile analysis
The concentration of specific tight junction proteins and markers for apoptosis and mitosis of the proximal and distal intestinal tissue samples was investigated using commercially available enzyme-linked immunosorbent assays (ELISA) of occludin (SEC228Hu), claudin-3 (SEF293Hu), proliferating cell nuclear antigen (PCNA) (SEA591Hu) and caspase-3 (SEA626Hu) (Cloud-Clone Corporation®, Houston, TX, USA). All tissue samples were crushed, dissolved in phosphate-buffered saline solution (PBS, pH 7.4, 0.01 mol/L), sonicated 6 times for 5 s at 4 °C (Sonics Vibracell™, VCX130, Newtown CT, USA), and kept on ice for 30 min. Subsequently the samples were centrifuged for 2 min at 13,400 rpm at 4 °C (Heraeus X3R with TX-750, Thermo Scientific, Rockford, USA), after which the supernatant was isolated, total protein concentration was determined using a Pierce TM BCA Protein Assay Kit (Thermo Scientific, Rockford, USA) and finally the samples were diluted to a total protein concentration of 10 ng/μL. Then, samples were processed on a sandwich ELISA plate and the experiment was performed according to the manufacturer's instructions. Absorbance was

measured using an Infinite M200 Pro spectrophotometer with X-Fluor software at 450 nm at 25 °C (Tecan Group Ltd., Männedorf, Switzerland). Values of protein expression were determined per gram of total protein in a sample, measured using a Pierce™ BCA Protein Assay Kit (ThermoFisher Scientific, Belgium), and expressed as fmol/mg.

Ex vivo measurement of small intestinal permeability

Intestinal mucosal permeability was assessed ex vivo by measuring the translocation of macromolecular markers using the Ussing chamber technique. The segments were first rinsed with saline. The mucosal layer was stripped from the seromuscular layer and pinned onto 1.07 cm^2 sliders that were mounted into modified Ussing chambers (Dipl.-Ing. Mußler Scientific Instruments, Aachen, Germany). All tissues were mounted within 10 min following euthanasia. Tissues were immersed in 6.5 mL Ringer solution (115 mmol/L NaCl, 5 mmol/L KCl, 25 mmol/L $NaHCO_3$, 2.4 mmol/L Na_2HPO_4, 0.4 mmol/L NaH_2PO_4, 1.25 mmol/L $CaCl_2$, 1 mmol/L $MgSO_4$) with 6 mmol/L of mannitol or glucose in the luminal and serosal side, respectively. The system was water-jacketed to 37 °C and oxygenated with 95% O_2 and 5% CO_2. After an equilibration period of 20 min, 4 kDa fluorescein isothiocyanate-dextran (FD-4, Sigma-Aldrich, Bornem, Belgium) and 40 kDa horseradish peroxidase (HRP, type IV, Sigma-Aldrich, Bornem, Belgium) were added to the mucosal side to a final concentration of 0.8 mg/mL of FD-4 and 0.4 mg/mL of HRP. Samples of the buffer solution were taken from the serosal chamber at 20, 40, 60 and 100 min after adding markers. Meanwhile, the same volume of buffer was taken from the mucosal side to keep the volume balance across sides. Fluorescence intensity of FD-4 in the medium was measured at excitation wavelength of 485 nm and emission wavelength of 538 nm using a fluorescence plate reader (Thermo Scientific, Marietta, OH, USA). HRP was measured according to the method described previously [38]. In short, HRP activity was measured by adding a 'start solution' (50 mL of 0.2 mol/L NaH_2PO_4, 1 mL of 0.2 mol/L Na_2HPO_4, 20.4 µL of 30% H_2O_2, 1.7 mL of 1% dianisidine peroxide substrate, made up to 204 mL with water) to the HRP samples and left for 10 min, after which the reaction was terminated by addition of 120 µL of 4% sodium azide and the absorbance was read at 460 nm. The relation between peroxidase concentration and absorbance is linear in the concentration range 10–100 pmol/L. The apparent permeation coefficient (Papp) was calculated as:

$$\text{Papp (cm/s)} : (dc/dt) \times V/c_0/A,$$

Whereby dc/dt is the change of serosal concentration in the 20- to 100-min period (cm/s); V is the volume of the chamber, c_0 is the initial marker concentration in the mucosal reservoirs and A the area of the exposed intestine in the chambers (cm^2).

Mucosal, liver and blood homogenate extracts and biochemical assays

An acid extract was prepared from 1 g of homogenized (Braun homogenizer at 900 rpm) intestinal mucosa or liver that was placed in 10 mL ice-cold perchloric acid (PCA) 10% solution and centrifuged at 15,000 × g for 15 min at 4 °C. The resulting acid extract (0.5 mL) was transferred to tubes containing 50 µL γ-glu-glu internal standard solution. Samples were snap frozen in liquid nitrogen and stored at -80 °C until analysis of GSH and glutathione disulfide (GSSG). The biuret reaction was applied to determine the total protein content. Mucosal GSH and GSSG were measured using a modified high performance liquid chromatography (HPLC) method [39, 40]. The derivation procedure included the reaction of 100 mmol/L iodoacetic acid solution with thiols to form S-carboxymethyl derivatives followed by chromophore derivation of primary amines with dinitrofluorobenzene (DNFB, 1% (v/v) in ethanol). GSH and GSSG were separated through EC250/4.6 Nucleosil 120-7 NH_2 aminopropyl column (Machery-Nagel, Düren, Germany) protected by the same NH_2 guard column (CC8/4). Chromatographic runs were performed at a flow-rate of 1.5 mL/min, starting at 80% solvent A/20% solvent B for 5 min followed by a 10 min linear gradient to 1% solvent A/99% solvent B and a 10 min isocratic period at 1% solvent A/99% solvent B (solvent A: water-methanol solution (1:4, v/v), solvent B: 0.5 mol/L sodium acetate–64% methanol). The column was re-equilibrated to the initial conditions for 15 min while maintaining the column temperature at 40 °C. The UV detector was set at 365 nm for absorption measurements. GSH and GSSG were identified by retention times of authentic standards. Concentrations were determined by using the internal and external standards and expressed as µmol/g protein. In addition, a phosphate buffered aqueous extract was made by mixing approximately 1 g of homogenized mucosa in 10 mL ice cold 1% Triton-X-100 phosphate buffer solution (pH = 7.0), by using an Ultra-Turrax dispensing machine (IKA-Werke GmbH & Co. KG, Staufen, Germany). The supernatant was transferred to 2 mL tubes, snap frozen and stored at -80 °C until analysis. Supernatants were used for the determination of GSH-Px activity and malondialdehyde (MDA; expressed as nmol/g protein) concentration. Assessment of GSH-Px activity (expressed as U/g protein) in EDTA plasma and mucosa was determined spectrophotometrically [41]. The thiobarbituric acid reactive substances (TBARS) method was used to measure MDA concentration in EDTA plasma, liver and mucosa extracts [42].

Statistical analysis

Linear mixed models were fitted to assess the influence of birth weight category (NBW/LBW), feeding (artificially reared/sow-reared) and days postnatal (as a categorical variable) on the quantitative outcome variables. To model the dependence between observations within the same litter, random intercept terms for litter were added to the model. Depending on the research question, separate analyses were carried out in subgroups (e.g. only in sow-reared piglets) or time points were analyzed separately. In the subgroup analyses where no random effect terms was needed, a multiple linear regression model was fitted. Post hoc tests to compare mean values between the different time points (days postnatal) were carried out using Tukey's honestly significant difference. Models were fitted using the Mixed Model procedure of the JMP Pro11 software (SAS Institute, Cary, NC, USA). Significance for the fixed effects was tested using an F-test with Kenward-Roger correction. Data are expressed as means and their standard errors (S.E.), and $P < 0.05$ was considered significant.

Results

Body weight of piglets

Average body weight of LBW and NBW piglets sampled at birth (d0) was 0.77 ± 0.07 and 1.29 ± 0.08 kg, respectively (Fig. 1). The body weight of both birth weight categories did not significantly change during the first 3 days. Afterwards, in both sow- and artificially reared piglets the body weight gradually increased ($P < 0.001$). At each point in time, LBW sow-reared piglets showed lower body

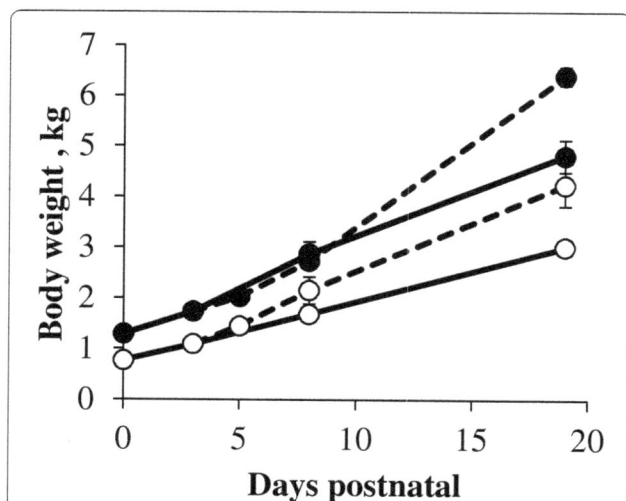

Fig. 1 Body weights (kg) of LBW (open circle) and NBW (*closed circle*) from sow- (*full line*) and artificially reared (*dashed line*) piglets during the suckling period. Values are means ± SE ($n = 6$)

weights compared to their NBW sow-reared littermates ($P < 0.01$). A similar observation was seen in artificially reared piglets at d5 ($P < 0.001$) and d19 ($P < 0.05$). Body weights of LBW and NBW artificially reared piglets were significantly higher, compared to respectively LBW and NBW sow-reared piglets at d19 ($P < 0.05$).

Histo-morphological measurements in the proximal and distal small intestine

The histo-morphological parameters were stable between d0 and d3 (Fig. 2). After d3, villus height significantly decreased ($P < 0.05$; Fig. 2a), whereas villus width ($P < 0.05$; Fig. 2b) and crypt depth significantly increased ($P < 0.05$; Fig. 2c). Furthermore, the transfer of piglets to a nursery caused villus atrophy in the proximal and distal ($P < 0.05$) SI from d3 to d5. Villi were significantly wider (at d19: $P < 0.001$) and crypts deeper (at d8 and d19: $P < 0.01$) in artificially reared piglets compared to sow-reared piglets. In the distal SI of NBW sow-reared piglets villus heights were consistently higher compared to those of their LBW littermates ($P < 0.05$). In contrast, villus heights in the distal SI of NBW artificially reared piglets were on average 193 µm lower at d19 compared to their LBW littermates ($P < 0.05$).

PCNA and caspase-3 protein expression in the proximal and distal small intestine

PCNA protein expression remained at the same level during the suckling period in the proximal and distal SI of sow-reared piglets (Fig. 3a). Transferring the piglets at d3 to a nursery significantly reduced PCNA protein expression in the proximal SI at d5 ($P < 0.05$) as compared to d3, whereas in the distal SI its level remained unaffected. PCNA protein expression in the proximal and distal SI of artificially reared piglets was significantly lower compared to sow-reared piglets at d8 ($P < 0.05$) and at d19 in the proximal SI ($P < 0.001$) (Fig. 3a).

Caspase-3 protein expression significantly decreased in the proximal SI of sow-reared ($P < 0.05$) and artificially reared piglets ($P < 0.001$) piglets from d3 to d19 (Fig. 3b). In contrast, in the distal SI of sow- and artificially reared piglets, no age-related differences were observed. Caspase-3 protein expression was significantly lower in the proximal ($P < 0.05$) and distal ($P < 0.05$) SI of artificially reared piglets compared to sow-reared piglets at d19.

Occludin and claudin-3 protein expression in the proximal and distal small intestine

Occludin expression in the proximal SI of sow-reared piglets decreased from d3 to d19 ($P < 0.001$), whereas in the distal SI its level remained unchanged (Fig. 4a). Occludin expression dropped significantly in the proximal ($P < 0.05$) but not in the distal SI of artificially

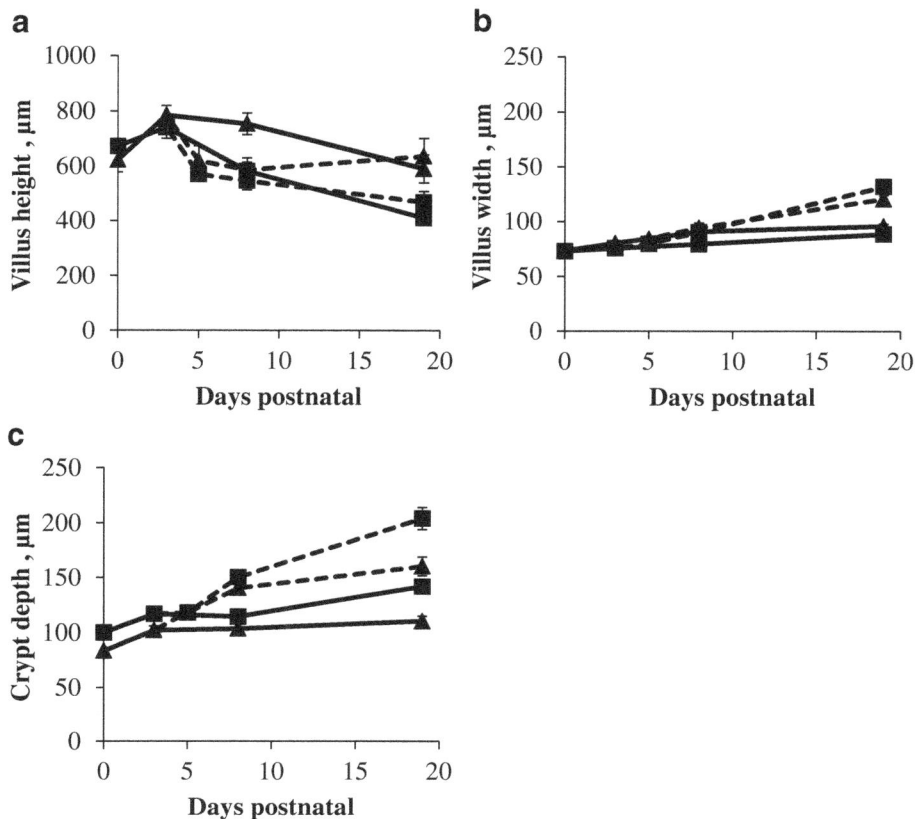

Fig. 2 Villus height (µm) (**a**), villus width (µm) (**b**) and crypt depth (µm) (**c**) in proximal (*square*) and distal (*triangle*) SI of sow- (*full line*) and artificially reared (*dashed line*) piglets expressed as µm. Values are means ± SE (*n* = 12 as LBW and NBW piglets were pooled together)

reared piglets, from d3 to d5, but returned to its initial levels afterwards. The expression of occludin in the SI was comparable for sow- and artificially reared piglets.

Claudin-3 expression in the proximal and in the distal SI of artificially reared piglets dropped significantly from d3 to d5 ($P < 0.01$; Fig. 4b). Claudin-3 expression increased from d5 to d19 and was significantly higher in the SI of artificially reared piglets compared to sow-reared piglets at d19 ($P <$

0.01). Claudin-3 expression in the distal SI of LBW sow-reared piglets was on average 33% higher compared to NBW sow-reared piglets at d3, d8 and d19 ($P < 0.01$).

Ex vivo permeability in the proximal and distal small intestine

FD-4 and HRP permeability in the proximal and distal SI remained stable in sow-reared piglets (Fig. 5a and b). FD-4

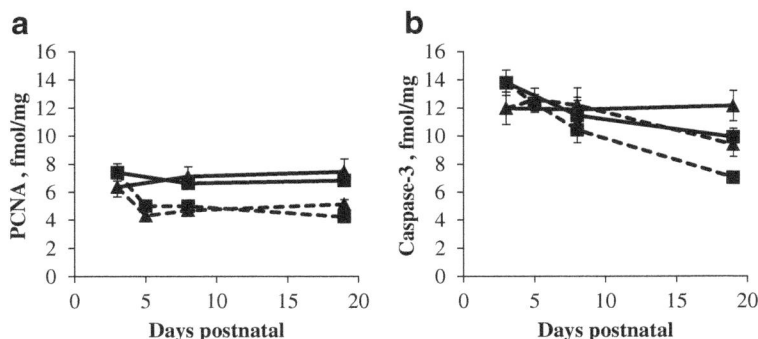

Fig. 3 Relative protein expression of PCNA (fmol/mg) (**a**) and caspase-3 (fmol/mg) (**b**) in the proximal (*square*) and distal (*triangle*) SI of sow- (*full line*) and artificially reared (*dashed line*) piglets. Values are means ± SE (*n* = 12 as LBW and NBW piglets were pooled together)

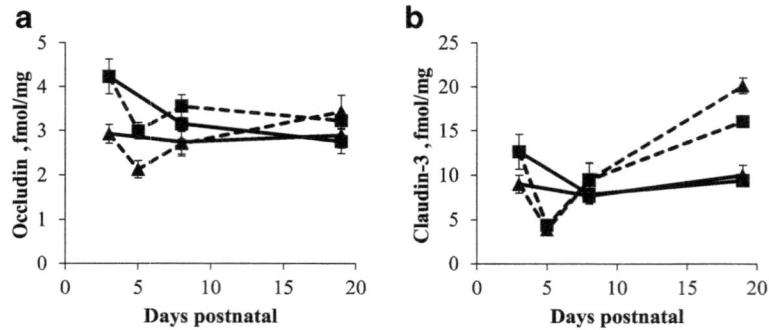

Fig. 4 Relative protein expression of occludin (fmol/mg) (**a**) and claudin-3 (fmol/mg) (**b**) in the proximal (*square*) and distal (*triangle*) SI of sow- (*full line*) and artificially reared (*dashed line*) piglets. Values are means ± SE ($n = 12$ as LBW and NBW piglets were pooled together)

permeability in the proximal SI significantly increased in 5-day-old NBW artificially reared piglets when transferred to a nursery at d3 ($P < 0.05$; Fig. 5a). Given the observation that the Papp of FD-4 was already high at 3 days of age in the proximal SI of LBW piglets, no changes could be observed after the start of artificial rearing. In the distal SI, both FD-4 ($P < 0.05$; Fig. 5a) and HRP ($P < 0.01$; Fig. 5b) permeabilities were significantly increased in both 5-day-old LBW and NBW artificially reared piglets, compared to d3. LBW sow-reared piglets consistently showed significantly higher FD-4 (proximal: on average 0.5×10^{-6} cm/s higher; distal: on average 0.2×10^{-6} cm/s higher) and HRP (proximal: on average 0.09×10^{-6} cm/s higher; distal: on average 0.06×10^{-6} cm/s higher) permeability when compared to their NBW littermates ($P < 0.05$). No such differences were observed in artificially reared piglets.

Mucosal redox state represented by GSH concentration, GSSG concentration, GSSG/GSH ratio, GSH-Px activity and MDA concentration

GSH concentration was constant from birth until d8. In the distal SI mucosa GSH concentrations remained stable SI, whereas in the proximal SI it decreased significantly

after d8 ($P < 0.01$) of sow-reared piglets (Fig. 6a). The concentration of GSSG and the GSSG/GSH ratio did not show any age-related changes in sow-reared piglets ($P > 0.05$; Fig. 6b and c). The activity of GSH-Px in the proximal and distal SI mucosae of sow-reared piglets increased from birth until d3 ($P < 0.05$). After d8, GSH-Px activity decreased significantly ($P < 0.01$), reaching a similar activity at d19 as observed at birth (Fig. 6d).

In the proximal intestine of sow-reared pigs, the concentration of MDA abruptly dropped after birth ($P < 0.001$; Fig. 6e) and remained stable from d3 until d19, whereas in the distal intestine the decrease was more spread out in time ($P < 0.01$). Transferring piglets to a nursery decreased the concentration of GSH, whereas the GSSG concentration and GSSG/GSH ratio were increased in the SI of artificially reared piglets at d5. In the proximal SI of artificially reared piglets GSH concentration peaked, while GSSG concentration and GSSG/GSH ratio showed a minimum at d8 ($P < 0.05$). However by d19, these redox parameters returned to the values noted at birth ($P < 0.05$). GSH-Px activity showed a minimum at d5 ($P < 0.05$), but recovered from d8 to d19 ($P < 0.01$) in the SI of artificially reared piglets.

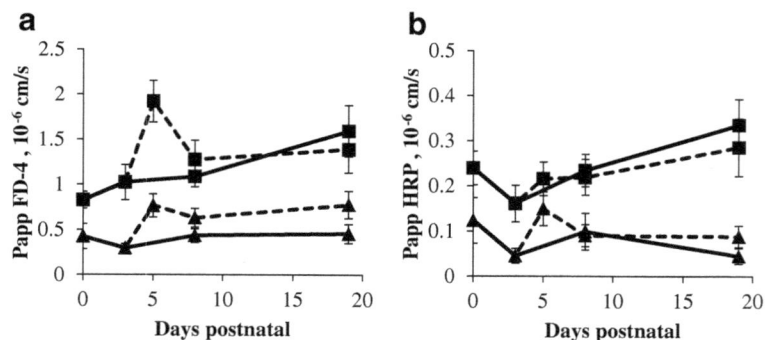

Fig. 5 Intestinal permeability indicated by the Papp of FD-4 (10^{-6} cm/s) (**a**) and HRP (10^{-6} cm/s) (**b**) in the proximal (*square*) and distal SI (*triangle*) of sow- (*full line*) and artificially reared (*dashed line*) piglets. Values are means ± SE ($n = 6$)

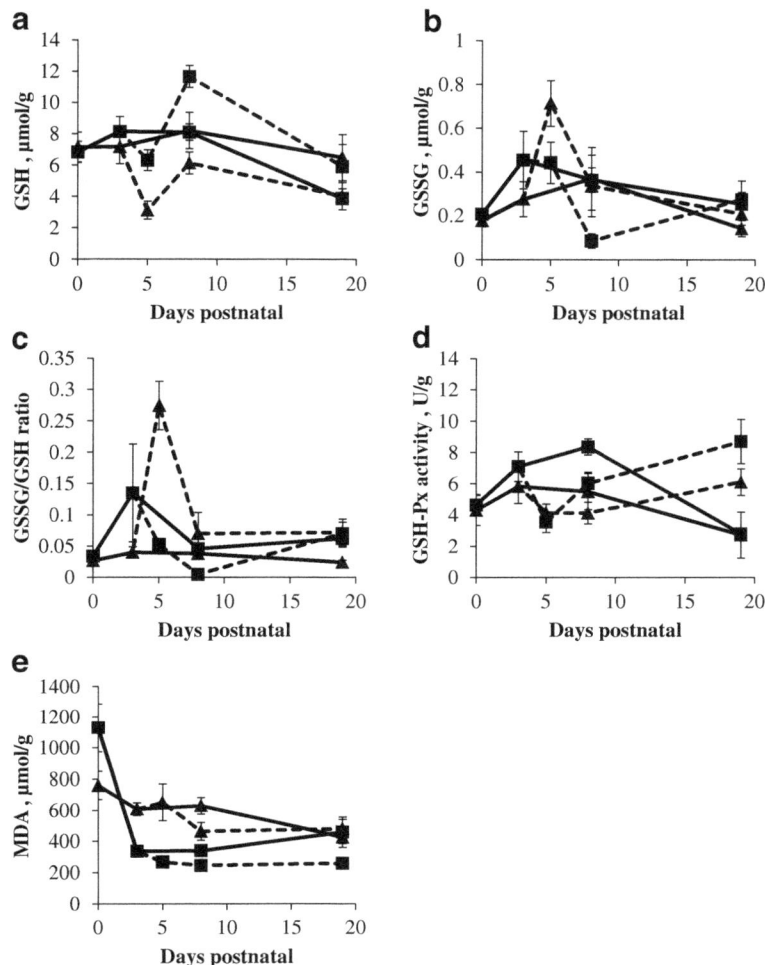

Fig. 6 Mucosal GSH concentration (μmol/g) (**a**), GSSG concentration (μmol/g) (**b**), GSSG/GSH ratio (**c**) and GSH-Px activity (U/g) (**d**), MDA concentration (μmol/g) (**e**) of the proximal (*square*) and distal SI (*triangle*) in sow- (*full line*) and artificially reared (*dashed line*) piglets. Values are means ± SE (*n* = 12 as LBW and NBW piglets were pooled together)

In the proximal SI, GSH concentration was significantly higher ($P < 0.05$), while GSSG concentration ($P < 0.01$) and consequently the GSSG/GSH ratio ($P = 0.001$) was significantly lower in artificially reared piglets compared to sow-reared piglets at d8. GSH-Px activity was significantly lower at d8 ($P < 0.001$) and significantly higher at d19 ($P < 0.001$) in both regions of the SI of artificially reared piglets compared to sow-reared piglets. GSH-Px activity in the SI of LBW sow-reared piglets was significantly higher than their NBW littermates (proximal: on average 1.38 U/g higher; distal: on average 0.64 U/g higher) ($P < 0.001$). A similar observation for both birth weight categories was made in the distal SI of artificially reared piglets (on average 1.24 higher, $P < 0.01$).

At d8, MDA concentration in the proximal SI of sow-reared piglets was significantly higher compared to artificially reared piglets ($P < 0.01$).

Liver redox state represented by GSH concentration, GSSG concentration, GSSG/GSH ratio, GSH-Px activity and MDA concentration

Liver GSH concentration, GSSG/GSH ratio and GSH-Px activity in sow-reared piglets did not significantly change (Fig. 7a, c and d). GSSG concentration decreased significantly from d0 to d3 ($P < 0.05$; Fig. 7b) in sow-reared piglets and returned to concentrations seen at birth afterwards (Fig. 7b). MDA concentration showed a significant increase after d3 in both LBW and NBW piglets ($P < 0.05$; Fig. 7e).

When piglets were introduced to a milk replacer, GSH ($P < 0.05$) and GSSG ($P < 0.01$) concentrations significantly decreased whereas GSSG/GSH ratio ($P < 0.05$) and MDA concentration ($P < 0.001$) significantly increased from d3 to d5. Meanwhile, GSH-Px activity remained unchanged. All redox parameters in the liver were stable from d5 to

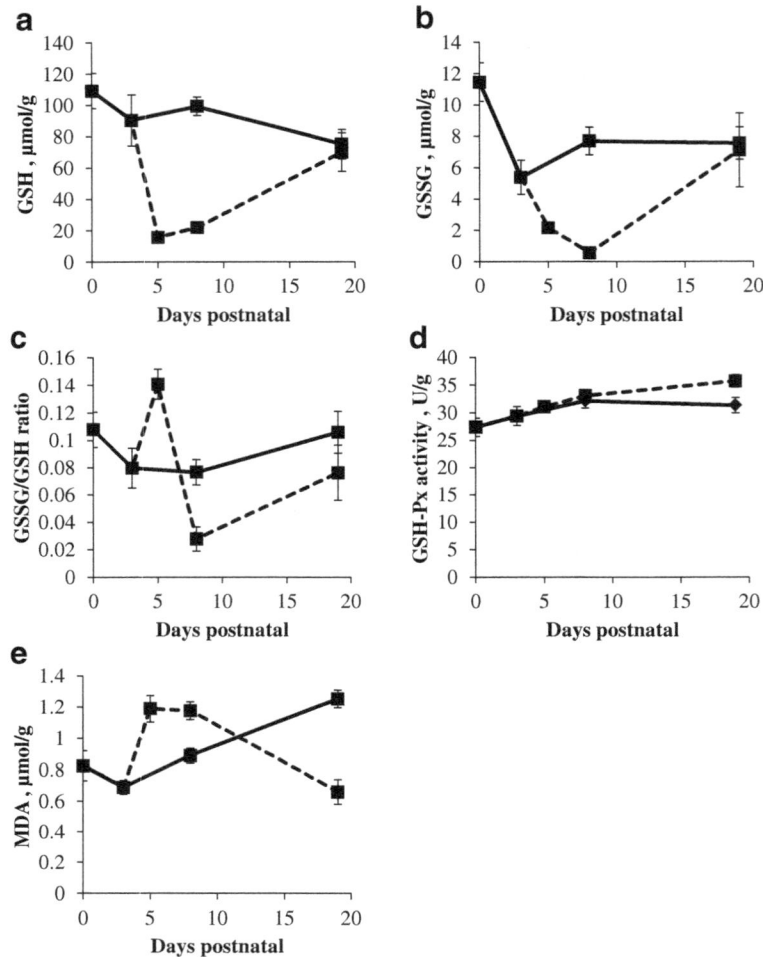

Fig. 7 Liver GSH concentration (µmol/g) (**a**) and GSSG concentration (µmol/g) (**b**), GSSG/GSH ratio (**c**), GSH-Px activity (U/g) (**d**) and MDA (µmol/g) (**e**) in sow- (*full line*) and artificially reared (*dashed line*) piglets. Values are means ± SE ($n = 12$ as LBW and NBW piglets were pooled together)

d8 in artificially reared piglets, except GSSG/GSH ratio that showed a significant decrease ($P < 0.01$). Towards d19, GSH concentration ($P < 0.01$) and GSH-Px activity ($P < 0.05$) significantly increased while MDA concentration ($P < 0.01$) significantly decreased in artificially reared piglets.

At d8, a lower GSH concentration and a higher MDA were observed in artificially reared piglets compared to sow-reared piglets ($P < 0.001$). At d19, higher GSH-Px activities were observed in artificially reared piglets compared to sow-reared piglets ($P < 0.01$).

Systemic redox state represented by GSH concentration, GSSG concentration, and GSSG/GSH ratio, GSH-Px activity and MDA concentration

The GSH-Px activity gradually, significantly increased during the investigated time frame in both sow- and artificially reared piglets ($P < 0.01$; Fig. 8d). GSH concentration

in erythrocytes significantly increased in artificially reared piglets ($P < 0.01$) from d5 to d19, but remained constant in sow-reared piglets from d0 to d19 (Fig. 8a). Plasma MDA concentration decreased 6.77 µmol/g during the first 3 days of life in LBW piglets ($P < 0.01$; Fig. 8e). Plasma GSH-Px activity was significantly lower in artificially reared piglets compared to sow-reared piglets at d8 and d19 ($P < 0.05$). At d19, GSH-Px activity was still lower when LBW piglets were artificially reared ($P < 0.05$). Erythrocyte GSSG concentrations ($P < 0.01$; Fig. 8b) and GSSG/GSH ratio ($P < 0.05$; Fig. 8c) were significantly increased at d5 when artificially reared piglets were transferred to a nursery at d3 (Fig. 8b and c). The concentration of GSH ($P < 0.01$) and GSSG ($P < 0.05$) significantly increased in artificially reared piglets from d5 to d19. At d19, GSH concentration and GSSG concentration were significantly higher in artificially reared when compared to sow-reared piglets ($P < 0.05$). GSH concentration was

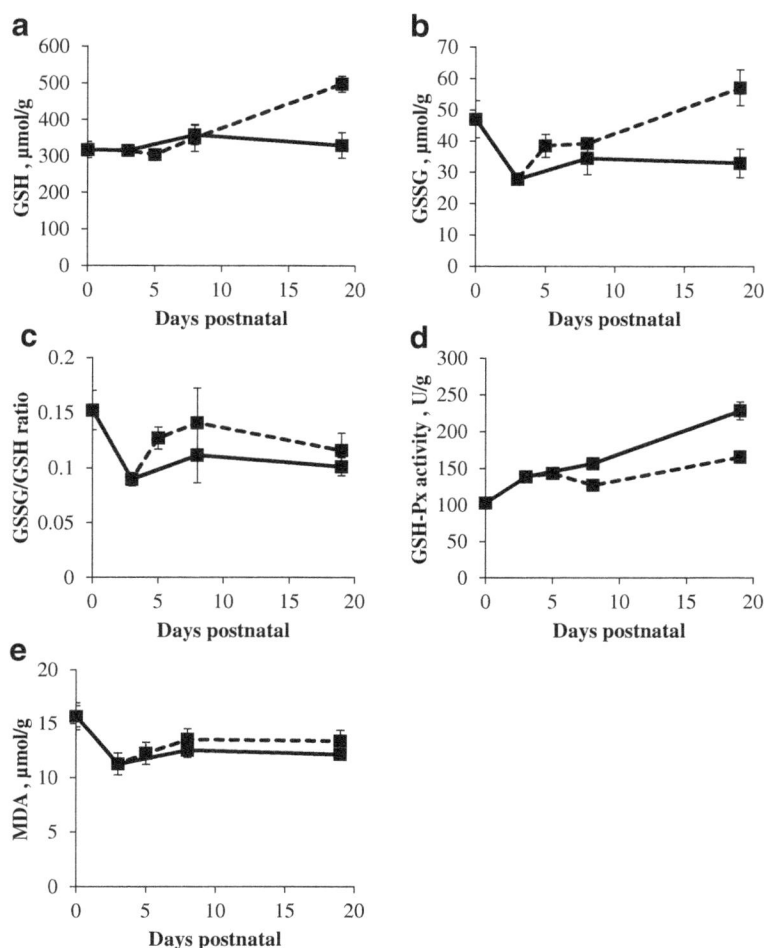

Fig. 8 Systemic GSH concentration (μmol/g) (**a**), GSSG concentration (μmol/g) (**b**), GSSG/GSH ratio (**c**), GSH-Px activity (U/g) (**d**) and MDA concentration (μmol/g) (**e**) in sow- (*full line*) and artificially reared (*dashed line*) piglets. Values are means ± SE ($n = 12$ as LBW and NBW piglets were pooled together)

significantly lower in LBW sow-reared piglets when compared to their NBW littermates. ($P < 0.05$).

Discussion

Given the need for an alternative rearing strategy that lowers the challenges that LBW and supernumerary piglets face during the suckling period, we aimed to investigate the responses to full artificial rearing. Our data demonstrated that artificial rearing beneficially affected piglet performance, notwithstanding impairing effects on small intestinal architecture, permeability and redox state in both LBW and NBW piglets.

Artificial rearing influences piglet performance

This study documents the implications of full artificial rearing of LBW and supernumerary piglets. Under standard rearing conditions these pigs are at high risk of succumbing due to insufficient nutrient intake, increased

disease susceptibility, and physiological deficits (e.g. lower energy reserves) [1]. Our study demonstrated that transitioning piglets to a nursery with ad libitum access to a milk replacer led to significantly higher body weights of LBW and NBW piglets compared to sow-reared piglets at d19. The experiment was terminated on the same day as when the conventionally reared piglets were weaned on the farm with a 3-week batch system. For this specific farm, the average age at weaning is 19.6 d. Furthermore, milk production of the sow strongly decreases towards the end of the suckling phase. Around d 18–19, sow milk starts to be very limiting and becomes hard to compare to the ad libitum access for the artificially reared piglets [43]. Using a weigh-suckle-weigh technique, De Vos et al. [26] showed that piglets with ad libitum access to milk replacer have a higher relative energy intake compared to sow-reared piglets [28]. Our findings confirm this and other previous research where LBW piglets receiving an energy rich diet—comparable with our LBW piglets fed a milk replacer ad libitum—presented a comparable body weight gain as NBW piglets receiving a lower energy intake—comparable with our NBW piglets fed by the sow [44]. In addition, the milk replacer used in our study contained spray-dried plasma which could have contributed further to the higher weight gain in the artificially reared group. Ermer et al. [45] showed that spray-dried porcine plasma increased feed intake. Thus next to ad libitum access to feed, diet composition cannot be neglected.

Artificial rearing influences small intestinal architecture

Small intestinal morphology is one of the major indicators reflecting gut health in pigs [46]. However, caution should be taken when evaluating morphology alone as a measure of gut health. For example, Enterotoxigenic *Escherichia coli*, the major causal agent of neonatal diarrhea, may occur without histological changes in the intestine [47]. Notwithstanding, stereological analysis of small intestinal morphology will provide the most accurate estimation of the intestinal absorptive surface area [48], a proxy of the surface can be calculated using villus height and villus width [47]. In our study, the mucosal surface area of the proximal small intestine at d19 was markedly larger in artificially reared piglets than sow-reared piglets. In this regard, feeding a milk replacer shows promise as the increased mucosal surface suggests a higher ability to absorb nutrients. Moreover, previous research showed an increased activity of maltase and sucrase when piglets are fed a milk replacer [49–51]. Thus, artificial rearing seems to improve the digestive capacity at the level of the small intestine.

The deepening of the crypts could be a response to promote mucus secretion rather than lead to enterocyte maturation and proliferation and thus an increase in PCNA expression. Previous research showed that breast-fed infants showed a delay in the mucin degradation when compared to artificially reared infants [52]. Phillips [53] showed that crypt goblet cells have the ability to restitute the mucus layer and showed a decrease in the percentage of villus epithelial volume occupied by mucin secretory granules.

Transferring piglets to a nursery at d3 exposed them to stressful effects caused by psychological, environmental or nutritional factors similar to those encountered during the conventional weaning process [54]. However, it is difficult to unravel the separate contributions of these factors. Previous studies showed significant villus atrophy at d4 and deeper crypts at d7 in piglets that were separated from the sow and still fed sow's milk compared to unweaned piglets [55, 56]. Similarly, our study showed transient villus atrophy in

piglets transferred to a nursery at d3. This villus height reduction is analogue to the intestinal morphological changes as a result of inadequate food intake immediately after conventional weaning [56–59]. In contrast to conventional weaning, villus length is rapidly restored. This could be related to the inclusion of spray-dried plasma in the milk replacer since this is known to increase villus height [8, 60–63].

Artificial rearing affects small intestinal tight junction protein expression

We hypothesized that artificial rearing influences small intestinal physiology. The intestinal epithelium plays a critical role in the transport of nutrients and macromolecules. At the same time, it has to provide an effective barrier to harmful macromolecules and microorganisms [64]. Epithelial cells constitute a dynamic barrier where large molecules can be transported by transcytosis and this can be measured by HRP [65]. Tight junctions (TJs) are essential components of the physical intercellular barrier and their presence and functionality changes under different physiological and pathological conditions [66, 67]. Well-formed TJs are characterized by low solute permeability which can be determined by measuring FD-4 permeability [65]. The family of junctional adhesion molecules, the claudin and occludin families, are structural transmembrane TJ components that have the potential to mediate cell–cell adhesion [66, 68]. Within TJs, claudins are the main determinants of the selective pore properties [68, 69], while the role of occludin in barrier functioning is more diverse [70, 71]. In mice, the expression of claudin-3 is promoted during the first 3 weeks of life concomitant with the establishment of the intestinal microbiota [72, 73]. Our study showed that claudin-3 expression rose particularly in artificially reared piglets. Possibly the different microbiota fingerprint [74] and the absence of milk born IgA's [34] can be held responsible. Claudin-3 is known to be a "tightening" claudin [75]. This could explain why permeability is seemingly unaffected by the observed increase in claudin-3. However, at the start of artificial rearing, claudin-3 as well as occludin protein levels transiently dropped. This drop is reflected in a concomitantly increased permeability for FD-4 and HRP. Previous studies showed lower abundances of occludin mRNA and protein, claudin-3 protein, and increased lactulose permeability after weaning [50, 76]. Thus artificial rearing induced a similar response as seen after conventional weaning.

Artificial rearing resulted in a redox imbalance

Previously, we investigated the link between oxidative stress, intestinal integrity, and permeability in intestinal epithelial cells in vitro [12] and in vivo during normal suckling [21]. Vergauwen et al. [12] and others showed a redistribution of TJ proteins during times of imposing reactive species and could relate these responses to a compromised permeability [12, 77]. The current study demonstrated that transfer of piglets to a nursery resulted in oxidative stress.

Glutathione (GSH) is an important regulator of the redox status within intestinal epithelial cells [13]. The liver is the major site of GSH biosynthesis and exports GSH via the bile to the proximal SI [78]. Thus GSH originating from the liver supports mucosal GSH by decreasing lipid peroxidation and maintaining the GSSG/GSH redox homeostasis in the proximal intestine [11, 79, 80]. In our study, artificial rearing resulted in the liver in a decreased GSH content, an increased GSSG/GSH ratio and MDA concentration. After this first phase, the redox parameters returned to their initial values. GSH-Px activity in the liver seemed unaffected during the transition period, while GSH-Px activity transiently dropped in the proximal and distal SI when piglets were introduced to a nursery. Previously, conventional weaning caused a drop in GSH-Px activity but GSH-Px activity increased afterwards as part of a feedback mechanism [31]. Our study shows that artificial rearing increases oxidative stress in the SI and as a result the GSH-Px activity increased and was higher than the activity seen in sow-reared piglets. Furthermore, LBW piglets showed a higher GSH-Px activity compared to their NBW littermates. It is clear that the antioxidant capacity of the liver helps protecting the proximal intestine by secreting GSH into the lumen of proximal SI [78, 79]. This could explain the increased mucosal GSH concentration of the proximal SI and the concomitant massive decrease of the liver GSH concentration. On the other hand, a remarkable drop of the GSH concentration in the distal SI was observed from d3 until d5. Consequently, this resulted in a high concentration of MDA and GSSG, resulting in a higher GSSG/GSH ratio in the distal SI. Furthermore, the GSSG/GSH ratio increased systemically upon transfer to a nursery. Degroote et al. [31] already showed that conventional weaning increased the GSSG/GSH ratio.

Our research presents a window of opportunity for antioxidant supplementation to protect piglets from redox imbalance due to artificial rearing. NBW artificially reared piglets were more susceptible to redox imbalance and loss of intestinal integrity upon transferal to a nursery (from d3 to d5). Taken together, these findings favor the transfer of both LBW and NBW piglets to a nursery as a solution for LBW and supernumerary piglets. Further research is necessary to elucidate how these artificially reared LBW and NBW piglets will respond to the introduction to solid food.

Conclusions

In conclusion, we demonstrated that artificial rearing influences morphological and functional parameters in the small intestine, liver and blood in a way similar to what is seen after conventional weaning. Nevertheless, growth performance of artificially reared piglets was positively influenced. In addition, artificially reared piglets rapidly recovered from redox imbalances and restored intestinal permeability within a couple of days. Further research is needed to explore the possibility to supplement LBW and NBW piglets with antioxidants prior to initiating artificial rearing. Thus, artificial rearing is a valuable alternative to raise LBW or supernumerary piglets.

Abbreviations
ELISA: Enzyme linked immunosorbent assay; FD-4: Fluorescein isothiocyanate dextran 4 kDa; GSH: Glutathione; GSH-Px: Glutathione peroxidase; GSSG: Oxidized glutathione; HPLC: High performance liquid chromatography; HRP: Horseradish peroxidase; LBW: Low birth weight; MDA: Malondialdehyde; NBW: Normal birth weight; SI: Small intestine; TJs: Tight junctions

Acknowledgments
We thank K. Huybrechts, G. Vrolix, K. Jennes, M. De Reys, S. Coolsaet, A. Ovyn and T. Van der Eecken for their technical assistance.

Funding
This work was supported by a grant from the government agency for Innovation by Science and Technology (IWT-LO 100856).

Authors' contributions
Author contributions: CVG, JM and SDM concept and design of research; HV, JDG, WW, CVG and JM performed experiments; HV and JDG analyzed the data; HV and EF design and implementation of the statistical model; HV, JDG, CVG and JM interpreted the results of the experiments; HV and CVG prepared the figures; HV and CVG drafted manuscript; HV, JDG, CVG, JM, SDS, WW, CC, SVC and SP edited and revised the manuscript; CVG and JM approved the final version of the manuscript. All authors read and approved the final manuscript.

Competing interests
The authors declare that they have no competing interests.

Author details
[1]Laboratory of Applied Veterinary Morphology, Department of Veterinary Sciences, Faculty of Biomedical, Pharmaceutical and Veterinary Sciences, University of Antwerp, Campus Drie Eiken, Universiteitsplein 1, D.U.015, 2610 Wilrijk, Belgium. [2]Department of Applied Biosciences, Faculty of Bioscience Engineering, Ghent University, Ghent, Belgium. [3]Laboratory for Animal Nutrition and Animal Product Quality (LANUPRO), Department of Animal Production, Faculty of Bioscience Engineering, Ghent University, Melle, Belgium. [4]StatUa Center for Statistics, University of Antwerp, Antwerp, Belgium.

References
1. Lay Jr DC, Matteri RL, Carrol JA, Fangman TJ, Safranski TJ. Preweaning survival in swine. J Anim Sci. 2002;80(E. Suppl. 1):E74–86.
2. Tuchscherer M, Puppe B, Tuchscherer A, Tiemann U. Early identification of neonates at risk: traits of newborn piglets with respect to survival. Theriogenology. 2000;54(3):371–88. doi:10.1016/s0093-691x(00)00355-1.

3. Milligan BN, Dewey CE, de Grau AF. Neonatal-piglet weight variation and its relation to pre-weaning mortality and weight gain on commercial farms. Prev Vet Med. 2002;56(2):119–27.

4. Friel JK, Diehl-Jones B, Cockell KA, Chiu A, Rabanni R, Davies SS, et al. Evidence of oxidative stress in relation to feeding type during early life in premature infants. Pediatr Res. 2011;69(2):160–4. doi:10.1203/PDR.0b013e3182042a07.

5. Ahola T, Levonen AL, Fellman V, Lapatto R. Thiol metabolism in preterm infants during the first week of life. Scand J Clin Lab Invest. 2004;64(7):649–58. doi:10.1080/00365510410002959.

6. Rook D, Te Braake FW, Schierbeek H, Longini M, Buonocore G, Van Goudoever JB. Glutathione synthesis rates in early postnatal life. Pediatr Res. 2010;67(4):407–11. doi:10.1203/PDR.0b013e3181d22cf6.

7. Yin J, Ren W, Liu G, Duan J, Yang G, Wu L, et al. Birth oxidative stress and the development of an antioxidant system in newborn piglets. Free Radic Res. 2013;47(12):1027–35. doi:10.3109/10715762.2013.848277.

8. Tran H, Bundy JW, Li YS, Carney-Hinkle EE, Miller PS, Burkey TE. Effects of spray-dried porcine plasma on growth performance, immune response, total antioxidant capacity, and gut morphology of nursery pigs. J Anim Sci. 2014;92(10):4494–504. doi:10.2527/jas.2014-7620.

9. Michiels J, De Vos M, Missotten J, Ovyn A, De Smet S, Van Ginneken C. Maturation of digestive function is retarded and plasma antioxidant capacity lowered in fully weaned low birth weight piglets. Br J Nutr. 2013;109(1):65–75. doi:10.1017/s0007114512000670.

10. Krueger R, Derno M, Goers S, Metzler-Zebeli BU, Nuernberg G, Martens K, et al. Higher body fatness in intrauterine growth retarded juvenile pigs is associated with lower fat and higher carbohydrate oxidation during ad libitum and restricted feeding. Eur J Nutr. 2014;53(2):583–97. doi:10.1007/s00394-013-0567-x.

11. Aw TY. Intestinal glutathione: determinant of mucosal peroxide transport, metabolism, and oxidative susceptibility. Toxicol Appl Pharmacol. 2005;204(3):320–8. doi:10.1016/j.taap.2004.11.016.

12. Vergauwen H, Tambuyzer B, Jennes K, Degroote J, Wang W, De Smet S, et al. Trolox and ascorbic acid reduce direct and indirect oxidative stress in the IPEC-J2 cells, an in vitro model for the porcine gastrointestinal tract. PLoS One. 2015;10(3):e0120485. doi:10.1371/journal.pone.0120485.

13. Kelly FJ. Glutathione content of the small intestine: regulation and function. Br J Nutr. 1993;69(2):589–96.

14. Carrasco-Pozo C, Morales P, Gotteland M. Polyphenols Protect the Epithelial Barrier Function of Caco-2 Cells Exposed to Indomethacin through the Modulation of Occludin and Zonula Occludens-1 Expression. Journal of agricultural and food chemistry. 2013. doi:10.1021/jf400150p.

15. Bhattacharyya A, Chattopadhyay R, Mitra S, Crowe SE. Oxidative stress: an essential factor in the pathogenesis of gastrointestinal mucosal diseases. Physiol Rev. 2014;94(2):329–54. doi:10.1152/physrev.00040.2012.

16. Wang J, Chen L, Li D, Yin Y, Wang X, Li P, et al. Intrauterine growth restriction affects the proteomes of the small intestine, liver, and skeletal muscle in newborn pigs. J Nutr. 2008;138(1):60–6.

17. Wang X, Wu W, Lin G, Li D, Wu G, Wang J. Temporal proteomic analysis reveals continuous impairment of intestinal development in neonatal piglets with intrauterine growth restriction. J Proteome Res. 2010;9(2):924–35. doi:10.1021/pr900747d.

18. Mickiewicz M, Zabielski R, Grenier B, Le Normand L, Savary G, Holst JJ, et al. Structural and functional development of small intestine in intrauterine growth retarded porcine offspring born to gilts fed diets with differing protein ratios throughout pregnancy. J Physiol Pharmacol. 2012;63(3):225–39.

19. Ferenc K, Pietrzak P, Godlewski MM, Piwowarski J, Kilianczyk R, Guilloteau P, et al. Intrauterine growth retarded piglet as a model for humans–studies on the perinatal development of the gut structure and function. Reprod Biol. 2014;14(1):51–60. doi:10.1016/j.repbio.2014.01.005.

20. D'Inca R, Gras-Le Guen C, Che L, Sangild PT, Le Huerou-Luron I. Intrauterine growth restriction delays feeding-induced gut adaptation in term newborn pigs. Neonatology. 2011;99(3):208–16. doi:10.1159/000314919.

21. Wang W, Degroote J, Van Ginneken C, Van Poucke M, Vergauwen H, Dam TM et al. Intrauterine growth restriction in neonatal piglets affects small intestinal mucosal permeability and mRNA expression of redox-sensitive genes. FASEB journal : official publication of the Federation of American Societies for Experimental Biology. 2015. doi:10.1096/fj.15-274779.

22. Huygelen V, De Vos M, Willemen S, Tambuyzer B, Casteleyn C, Knapen D, et al. Increased intestinal barrier function in the small intestine of formula-fed neonatal piglets. J Anim Sci. 2012;90 Suppl 4:315–7. doi:10.2527/jas.53731.

23. Huygelen V, De Vos M, Willemen S, Fransen E, Casteleyn C, Van Cruchten S, et al. Age-related differences in mucosal barrier function and morphology of the small intestine in low and normal birth weight piglets. J Anim Sci. 2014;92(8):3398–406. doi:10.2527/jas.2014-7742.

24. Willemen S, Che L, De Vos M, Huygelen V, Tambuyzer B, Casteleyn C, et al. Perinatal growth restriction is not related to higher intestinal distribution and increased serum levels of 5-hydroxytryptamin in piglets. J Anim Sci. 2012;90 Suppl 4:305–7. doi:10.2527/jas.53730.

25. Ferrari CV, Sbardella PE, Bernardi ML, Coutinho ML, Vaz Jr IS, Wentz I, et al. Effect of birth weight and colostrum intake on mortality and performance of piglets after cross-fostering in sows of different parities. Prev Vet Med. 2014;114(3-4):259–66. doi:10.1016/j.prevetmed.2014.02.013.

26. De Vos M, Che L, Huygelen V, Willemen S, Michiels J, Van Cruchten S, et al. Nutritional interventions to prevent and rear low-birthweight piglets. J Anim Physiol Anim Nutr. 2014;98(4):609–19. doi:10.1111/jpn.12133.

27. Donovan TS, Dritz SS. Effect of split nursing on variation in pig growth from birth to weaning. J Am Vet Med Assoc. 2000;217(1):79–81.

28. De Vos M, Huygelen V, Willemen S, Fransen E, Casteleyn C, Van Cruchten S, et al. Artificial rearing of piglets: Effects on small intestinal morphology and digestion capacity. Livest Sci. 2014;159:165–73. doi:10.1016/j.livsci.2013.11.012.

29. Wedig J, Christian MS, Hoberman A, Diener RM, Thomas-Wedig R. A study to develop methodology for feeding 24-hour-old neonatal swine for 3 weeks. Int J Toxicol. 2002;21(5):361–70. doi:10.1080/10915810290096577.

30. Fiorotto ML, Reeds PJ, Cunningham JJ, Pond WG. A semiautomatic device for feeding liquid milk-replacer diets to infant pigs. J Anim Sci. 1993;71(1):78–85.

31. Degroote J, Michiels J, Claeys E, Ovyn A, De Smet S. Changes in the pig small intestinal mucosal glutathione kinetics after weaning. J Anim Sci. 2012;90 Suppl 4:359–61. doi:10.2527/jas.53809.

32. Zhu LH, Zhao KL, Chen XL, Xu JX. Impact of weaning and an antioxidant blend on intestinal barrier function and antioxidant status in pigs. J Anim Sci. 2012;90(8):2581–9. doi:10.2527/jas.2012-4444.

33. Sauerwein H, Schmitz S, Hiss S. The acute phase protein haptoglobin and its relation to oxidative status in piglets undergoing weaning-induced stress. Redox Report. 2005;10(6):295–302. doi:10.1179/135100005x83725.

34. Klobasa F, Werhahn E, Butler JE. Composition of sow milk during lactation. J Anim Sci. 1987;64(5):1458–66.

35. Csapo J, Martin TG, Csapo-Kiss ZS, Hazas Z. Protein, fats, vitamin and mineral concentration in porcine colostrum and milk from parturition to 60 days. Int Dairy J. 1996;6(8-9):881–902.

36. Ontsouka CE, Bruckmaier RM, Blum JW. Fractionized milk composition during removal of colostrum and mature milk. J Dairy Sci. 2003;86(6):2005–11. doi:10.3168/jds.S0022-0302(03)73789-8.

37. Buesa RJ, Peshkov MV. How much formalin is enough to fix tissues? Ann Diagn Pathol. 2012;16(3):202–9. doi:10.1016/j.anndiagpath.2011.12.003.

38. McKie AT, Zammit PS, Naftalin RJ. Comparison of cattle and sheep colonic permeabilities to horseradish peroxidase and hamster scrapie prion protein in vitro. Gut. 1999;45(6):879–88.

39. Yoshida T. Determination of reduced and oxidized glutathione in erythrocytes by high-performance liquid chromatography with ultraviolet absorbance detection. J Chromatogr B Biomed Appl. 1996;678(2):157–64.

40. Reed DJ, Babson JR, Beatty PW, Brodie AE, Ellis WW, Potter DW. High-performance liquid chromatography analysis of nanomole levels of glutathione, glutathione disulfide, and related thiols and disulfides. Anal Biochem. 1980;106(1):55–62.

41. Hernandez P, Zomeno L, Arino B, Blasco A. Antioxidant, lipolytic and proteolytic enzyme activities in pork meat from different genotypes. Meat Sci. 2004;66(3):525–9. doi:10.1016/s0309-1740(03)00155-4.

42. Grotto D, Santa Maria LD, Boeira S, Valentini J, Charao MF, Moro AM, et al. Rapid quantification of malondialdehyde in plasma by high performance liquid chromatography-visible detection. J Pharm Biomed Anal. 2007;43(2):619–24. doi:10.1016/j.jpba.2006.07.030.

43. Aguinaga MA, Gomez-Carballar F, Nieto R, Aguilera JF. Production and composition of Iberian sow's milk and use of milk nutrients by the suckling Iberian piglet. Animal. 2011;5(9):1390–7. doi:10.1017/s1751731111000474.

44. Han F, Hu L, Xuan Y, Ding X, Luo Y, Bai S, et al. Effects of high nutrient intake on the growth performance, intestinal morphology and immune

function of neonatal intra-uterine growth-retarded pigs. Br J Nutr. 2013;110(10):1819–27. doi:10.1017/s0007114513001232.

45. Ermer PM, Miller PS, Lewis AJ. Diet preference and meal patterns of weanling pigs offered diets containing either spray-dried porcine plasma or dried skim milk. J Anim Sci. 1994;72(6):1548–54.

46. Caspary WF. Physiology and pathophysiology of intestinal absorption. Am J Clin Nutr. 1992;55(1 Suppl):299s–308s.

47. Egberts HJ, de Groot EC, van Dijk JE, Vellenga L, Mouwen JM. Tight junctional structure and permeability of porcine jejunum after enterotoxic Escherichia coli infection. Res Vet Sci. 1993;55(1):10–4.

48. Van Ginneken C, Van Meir F, Weyns A. Stereologic characteristics of pig small intestine during normal development. Dig Dis Sci. 2002;47(4):868–78.

49. Cera KR, Mahan DC, Reinhart GA. Effect of weaning, week postweaning and diet composition on pancreatic and small intestinal luminal lipase response in young swine. Journal of animal science. 1990;68(2). doi:10.2527/1990.682384x.

50. Wang J, Zeng L, Tan B, Li G, Huang B, Xiong X, et al. Developmental changes in intercellular junctions and Kv channels in the intestine of piglets during the suckling and post-weaning periods. J Anim Sci Biotechnol. 2016;7:4. doi:10.1186/s40104-016-0063-2.

51. Huygelen V, De Vos M, Prims S, Vergauwen H, Fransen E, Casteleyn C, et al. Birth weight has no influence on the morphology, digestive capacity and motility of the small intestine in suckling pigs. Livest Sci. 2015;182:129–36. http://dx.doi.org/doi:10.1016/J.LIVSCI.2015.11.003.

52. Midtvedt AC, Carlstedt-Duke B, Midtvedt T. Establishment of a mucin-degrading intestinal microflora during the first two years of human life. J Pediatr Gastroenterol Nutr. 1994;18(3):321–6.

53. Phillips TE. Both crypt and villus intestinal goblet cells secrete mucin in response to cholinergic stimulation. Am J Physiol. 1992;262(2 Pt 1):G327–31.

54. Funderburke DW, Seerley RW. The effects of postweaning stressors on pig weight change, blood, liver and digestive tract characteristics. J Anim Sci. 1990;68(1):155–62.

55. van Beers-Schreurs HM, Nabuurs MJ, Vellenga L, Kalsbeek-van der Valk HJ, Wensing T, Breukink HJ. Weaning and the weanling diet influence the villous height and crypt depth in the small intestine of pigs and alter the concentrations of short-chain fatty acids in the large intestine and blood. J Nutr. 1998;128(6):947–53.

56. Kelly D, Smyth JA, McCracken KJ. Digestive development of the early-weaned pig. 1. Effect of continuous nutrient supply on the development of the digestive tract and on changes in digestive enzyme activity during the first week post-weaning. Br J Nutr. 1991;65(2):169–80.

57. Kelly D, Smyth JA, McCracken KJ. Digestive development of the early-weaned pig. 2. Effect of level of food intake on digestive enzyme activity during the immediate post-weaning period. Br J Nutr. 1991;65(2):181–8.

58. Spreeuwenberg MA, Verdonk JM, Gaskins HR, Verstegen MW. Small intestine epithelial barrier function is compromised in pigs with low feed intake at weaning. J Nutr. 2001;131(5):1520–7.

59. Hedemann MS, Hojsgaard S, Jensen BB. Small intestinal morphology and activity of intestinal peptidases in piglets around weaning. J Anim Physiol Anim Nutr. 2003;87(1-2):32–41.

60. Pierce JL, Cromwell GL, Lindemann MD, Russell LE, Weaver EM. Effects of spray-dried animal plasma and immunoglobulins on performance of early weaned pigs. J Anim Sci. 2005;83(12):2876–85.

61. Touchette KJ, Allee GL, Matteri RL, Dyer CJ, Carroll JA. Effect of spray-dried plasma and Escherichia coli on intestinal morphology and the hypothalamic–pituitary–adrenal (HPA) axis of the weaned pig. J Anim Sci. 1999;77(Supplement 1):56.

62. Jiang R, Chang X, Stoll B, Fan MZ, Arthington J, Weaver E, et al. Dietary plasma protein reduces small intestinal growth and lamina propria cell density in early weaned pigs. J Nutr. 2000;130(1):21–6.

63. van Dijk AJ, Niewold TA, Margry RJ, van den Hoven SG, Nabuurs MJ, Stockhofe-Zurwieden N, et al. Small intestinal morphology in weaned piglets fed a diet containing spray-dried porcine plasma. Res Vet Sci. 2001;71(1):17–22. doi:10.1053/rvsc.2001.0478.

64. Ramanan D, Cadwell K. Intrinsic Defense Mechanisms of the Intestinal Epithelium. Cell host & microbe. 2016. doi:10.1016/j.chom.2016.03.003.

65. Boudry G. The Ussing chamber technique to evaluate alternatives to in-feed antibiotics for young pigs. Animal Res. 2005;54:219–30.

66. Schneeberger EE, Lynch RD. The tight junction: a multifunctional complex. Am J Physiol Cell Physiol. 2004;286(6):C1213–28. doi:10.1152/ajpcell.00558.2003.

67. Lu Z, Ding L, Lu Q, Chen Y-H. Claudins in intestines: distribution and functional significance in health and diseases. Tissue Barriers. 2013;1(3):e24978. doi:10.4161/tisb.24978.

68. Anderson JM, Van Itallie CM, Fanning AS. Setting up a selective barrier at the apical junction complex. Curr Opin Cell Biol. 2004;16(2):140–5. doi:10.1016/j.ceb.2004.01.005.

69. Furuse M, Tsukita S. Claudins in occluding junctions of humans and flies. Trends Cell Biol. 2006;16(4):181–8. doi:10.1016/j.tcb.2006.02.006.

70. Yu AS, McCarthy KM, Francis SA, McCormack JM, Lai J, Rogers RA, et al. Knockdown of occludin expression leads to diverse phenotypic alterations in epithelial cells. Am J Physiol Cell Physiol. 2005;288(6):C1231–41. doi:10.1152/ajpcell.00581.2004.

71. Raleigh DR, Boe DM, Yu D, Weber CR, Marchiando AM, Bradford EM, et al. Occludin S408 phosphorylation regulates tight junction protein interactions and barrier function. J Cell Biol. 2011;193(3):565–82. doi:10.1083/jcb.201010065.

72. Patel RM, Myers LS, Kurundkar AR, Maheshwari A, Nusrat A, Lin PW. Probiotic bacteria induce maturation of intestinal claudin 3 expression and barrier function. Am J Pathol. 2012;180(2):626–35. doi:10.1016/j.ajpath.2011.10.025.

73. Holmes JL, Van Itallie CM, Rasmussen JE, Anderson JM. Claudin profiling in the mouse during postnatal intestinal development and along the gastrointestinal tract reveals complex expression patterns. Gene Expression Patterns. 2006;6(6):581–8. doi:10.1016/j.modgep.2005.12.001.

74. Mackie RI, Sghir A, Gaskins HR. Developmental microbial ecology of the neonatal gastrointestinal tract. Am J Clin Nutr. 1999;69(5):1035s–45s.

75. Milatz S, Krug SM, Rosenthal R, Gunzel D, Muller D, Schulzke JD, et al. Claudin-3 acts as a sealing component of the tight junction for ions of either charge and uncharged solutes. Biochim Biophys Acta. 2010;1798(11):2048–57. doi:10.1016/j.bbamem.2010.07.014.

76. Wang H, Zhang C, Wu G, Sun Y, Wang B, He B, et al. Glutamine enhances tight junction protein expression and modulates corticotropin-releasing factor signaling in the jejunum of weanling piglets. J Nutr. 2015;145(1):25–31. doi:10.3945/jn.114.202515.

77. Fukui A, Naito Y, Handa O, Kugai M, Tsuji T, Yoriki H et al. Acetyl salicylic acid induces damage to intestinal epithelial cells by oxidation-related modifications of ZO-1. American journal of physiology Gastrointestinal and liver physiology. 2012. doi:10.1152/ajpgi.00236.2012.

78. Lee TK, Li L, Ballatori N. Hepatic glutathione and glutathione S-conjugate transport mechanisms. Yale J Biol Med. 1997;70(4):287–300.

79. Aw TY. Biliary glutathione promotes the mucosal metabolism of luminal peroxidized lipids by rat small intestine in vivo. J Clin Investig. 1994;94(3):1218–25.

80. Ballatori N, Truong AT. Relation between biliary glutathione excretion and bile acid-independent bile flow. Am J Physiol. 1989;256(1 Pt 1):G22–30.

5

Disappearance and appearance of an indigestible marker in feces from growing pigs as affected by previous- and current-diet composition

Brandy M. Jacobs[1], John F. Patience[1], Merlin D. Lindemann[2], Kenneth J. Stalder[1] and Brian J. Kerr[3]*

Abstract

Background: Indigestible markers are commonly utilized in digestion studies, but the complete disappearance or maximum appearance of a marker in feces can be affected by diet composition, feed intake, or an animal's BW. The objectives of this study were to determine the impact of previous (Phase 1, P1) and current- (Phase 2, P2) diet composition on marker disappearance (Cr) and appearance (Ti) in pigs fed 3 diets differing in NDF content.

Results: When pigs were maintained on the 25.1, 72.5, and 125.0 g/kg NDF diets, it took 5.1, 4.1, and 2.5 d, respectively, for Cr levels to decrease below the limit of quantitation; or 4.6, 3.7, or 2.8 d, respectively, for Ti to be maximized. These effects were not, however, independent of the previous diet as indicated by the interaction between P1 and P2 diets on fecal marker concentrations ($P < 0.01$). When dietary NDF increased from P1 to P2, it took less time for fecal Cr to decrease or fecal Ti to be maximized (an average of 2.5 d), than if NDF decreased from P1 to P2 where it took longer for fecal Cr to decrease or fecal Ti to be maximized (an average of 3.4 d).

Conclusions: Because of the wide range in excretion times reported in the literature and improved laboratory methods for elemental detection, the data suggests that caution must be taken in considering dietary fiber concentrations of the past and currently fed diets so that no previous dietary marker addition remains in the digestive tract or feces such that a small amount of maker is present to confound subsequent experimental results, and that marker concentration have stabilized when these samples are collected.

Keywords: Adaptation, Digestibility, Fiber, Indigestible marker, Pig

Background

Indigestible markers are commonly used in animal nutrition studies to calculate digestibility coefficients, with chromic oxide, titanium dioxide, and acid insoluble ash being the most common in swine research [1]. Physiological aspects associated with gastric emptying or rate of passage are complex and affected by a variety of factors [2, 3]. Rate of passage can be affected by BW [4], feed intake level [5], dietary fiber type and level [6–8], particle size [9], and genetics [10]. In addition, rates of passage in the gastrointestinal tract are not consistent,

being pulsatile over time [11, 12]. The appearance of the first marker peak is relatively consistent at the terminal ileum of pigs, occurring approximately 6 h following a meal, dropping to minimum levels 24 h post-meal [13]. In contrast, digesta flow through the hind gut is longer and more variable, where mean transit times through the entire digestive tract have been reported to be less than 50 h [7] to over 100 h [6]. Imbeath et al. [13] reported that 4 d was needed before marker concentrations were near zero after marker withdrawal, while others [14, 15] have reported that their appearance in the feces is stabilized 4 to 5 d after feeding.

Currently, there is no standard time for pigs to be adapted to a diet, a specific number of days an animal should be sampled, or the number of days between

---

* Correspondence: brian.kerr@ars.usda.gov
[3]USDA-ARS-National Laboratory for Agricultural and the Environment, Ames 50010, IA, USA
Full list of author information is available at the end of the article

collection periods in swine research utilizing inert markers. As a consequence, the objectives of this study were to: 1) determine the impact of previous (P1) and currently-fed (P2) diet composition on the complete disappearance P2 marker (Cr) and 2) determine the impact of previous and currently-fed diet composition on the complete appearance of P2 marker (Ti) in growing pigs fed diets differing in fiber content.

Methods

The experiment was conducted under protocols approved by the University of Kentucky Institutional Animal Care and Use Committee.

Feeding management

Diets (Table 1) were formulated to contain varying levels of NDF through the utilization of dehulled, degermed corn (DDC), corn (C), soybean meal (S), and distillers dried grains with solubles (DDGS). Diets were formulated to meet requirements relative to NRC (1998) recommendations. The same diet composition was used in each of 2 phases, with Phase-1 (P1) diets utilizing chromic oxide and Phase-2 diets (P2) utilizing titanium dioxide, each added at 5.0 g/kg to the complete diet at the time of mixing to determine fecal marker concentrations. Two different inert markers were utilized to distinguish the feces originating from the diets consumed during P1 to the feces originating from the diet consumed in P2. This allows for the comparing the disappearance of the marker used in P1 and the appearance of the marker used in P2. This also prevents any potential contamination of the marker in the digestive tract in P1 with that of P2, which would have prevented the pre-planned comparisons of marker disappearance and appearance during the P2 period relative to a diet change. Pigs were provided ad libitum access to feed and water throughout the experiment.

Pig management and collections

Seventy two crossbred barrows [(Yorkshire × Landrace × Duroc) × Chester White] were individually penned and randomly assigned to 1 of 3 dietary treatments. Pigs were initially separated into 3 treatment groupings of 24 pigs (d-0; 59.2 kg BW, 4.81 kg SD) and fed ad libitum P1 diets for 14 d (d-14; 75.4 kg BW, 5.71 kg SD) and then randomly reassigned within P1 dietary treatment into 1 of 3 P2 dietary treatments, and fed ad libitum an additional 14 days (d-28; 88.6 kg BW, 5.46 kg SD), resulting in 9 treatment groups of 8 pigs each (Fig. 1). For each pig and each day during P2 (d-14 through d-28), freshly excreted fecal samples (samples either from the anus or after just dropping on the floor—but not contaminated with feed or existing feces) were collected into plastic containers and placed into a −20 °C freezer until analyzed. Samples were collected from 0700 to 1200 h on

Table 1 Composition of Phase-1 and Phase-2 diets, as-fed basis[a]

	DDC	CS	DDGS
Ingredient, g/kg			
Corn	–	784.0	567.0
Soybean meal	180.0	180.0	150.0
Dehulled, degermed corn	781.9	–	–
Dried distillers grains with solubles	–	–	250.0
Soybean oil	5.0	5.0	5.0
L-Lysine · HCl	1.1	–	–
Dicalcium phosphate	8.5	7.0	1.5
Limestone	7.0	7.5	10.0
Sodium chloride	5.0	5.0	5.0
Vitamin premix[b]	0.5	0.5	0.5
Trace mineral premix[c]	0.5	0.5	0.5
Marker[d]	5.0	5.0	5.0
Clay[e]	5.0	5.0	5.0
Antibiotic[f]	0.5	0.5	0.5
Calculated composition, g/kg unless otherwise noted			
Calcium	5.0	5.0	5.0
Crude fat,	12.0	41.0	53.0
Crude protein	142.7	150.6	187.6
Lysine	7.5	7.5	7.5
Metabolizable energy, kcal/kg	3,293	3,332	3,193
NDF	45.0	91.0	154.0
Phosphorus	3.4	4.7	4.8
Sulfur	1.0	1.8	2.1
Analyzed composition, g/kg unless otherwise noted[g]			
Crude fat	12.8	37.2	47.7
Crude protein	132.5	160.0	197.5
Gross energy, kcal/kg	3,770	3,973	4,131
NDF	25.1	72.5	125.0
Phosphorus	2.8	4.6	4.9
Sulfur	1.7	2.0	3.2

[a]Abbreviations: DDC dehulled, degermed corn, CS corn, soybean meal, DDGS distillers dried grains with solubles
[b]Supplied per kilogram of diet: vitamin A, 6,600 IU; vitamin D$_3$, 880 IU; vitamin E, 44 IU; vitamin K (menadione sodium bisulfate complex), 6.4 mg; thiamin, 4.0 mg; riboflavin, 8.8 mg; pyridoxine, 4.4 mg; vitamin B$_{12}$, 33 µg; folic acid, 1.3 mg; niacin, 44 mg; pantothenic acid, 22 mg; and D-biotin, 0.22 mg
[c]Supplied per kilogram of diet: Zn, 131 mg as ZnO; Fe, 131 mg as FeSO$_4$ · H$_2$O; Mn 45 mg, as MnO; Cu, 13 mg as CuSO$_4$ · 5H$_2$O; I, 1.5 mg as CaI$_2$O$_6$; Co, 0.23 mg as CoCO$_3$; and Se, 0.28 mg as Na$_2$O$_3$Se
[d]The addition of 0.5%, Cr$_2$O$_3$ (≥98% purity; Elementis Chromium LP, Corpus Christi, TX) represents an addition of 3.35 mg Cr/g diet; averaged across diets, the analyzed content equaled 2.76 mg Cr/kg diet (Phase-1). The addition of 0.5% TiO2 (99% purity, Tronox Pigments GmBH, Krefield, Germany) represents an addition of 2.97 mg titanium/g diet; averaged across diets, the analyzed content equaled 2.89 mg titanium/kg diet (Phase-2)
[e]AB-20 (Prince Agriproducts, Quincy, IL)
[f]Tylan-40 supplied 44 mg/kg of diet (Elanco, Greenfield, IN)
[g]Diets were analyzed at the USDA-ARS (Ames, IA), except for phosphorus which was analyzed by SDK Labs (Hutchison, KS)

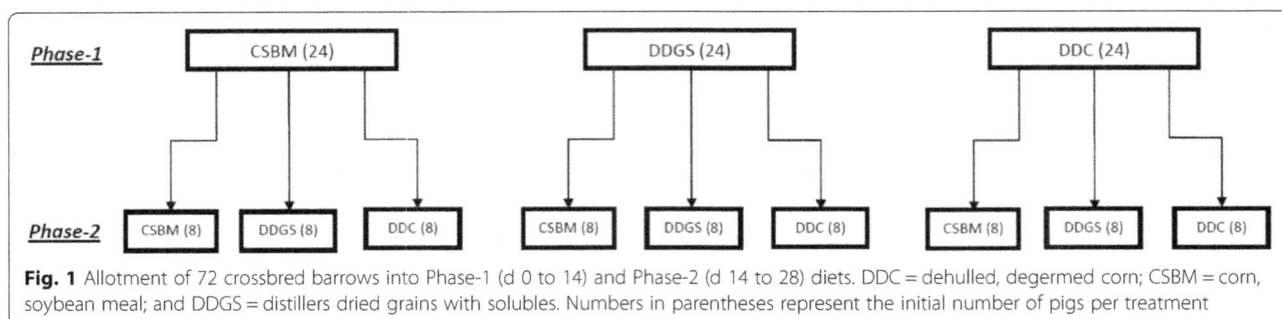

Fig. 1 Allotment of 72 crossbred barrows into Phase-1 (d 0 to 14) and Phase-2 (d 14 to 28) diets. DDC = dehulled, degermed corn; CSBM = corn, soybean meal; and DDGS = distillers dried grains with solubles. Numbers in parentheses represent the initial number of pigs per treatment

each collection day to be consistent in sample collection during the 14 d and to ensure an adequate sample size for subsequent analysis.

Chemical analysis

Prior to analysis, fecal samples were dried in a forced-air oven at 70 °C for 48 h prior to grinding. Feed and fecal samples were ground through a 1-mm screen before composition was determined. Chromic oxide in feces was analyzed for Cr at a commercial laboratory (SDK Labs, Hutchinson, KS) by inductively coupled plasma spectroscopy (Ultima 2; Horiba Jobin-Yvon Inc., Edison, NJ) according to standard method (3120B; American Public Health Association, 1992) with a limit of quantitation (LOQ) of 0.3 mg Cr/kg sample. Titanium dioxide in feces was analyzed for Ti by digesting the samples in sulfuric acid and hydrogen peroxide and subsequent absorbance was measured using a UV spectrophotometer (Method 988.05; [16]), with a LOQ of 6 mg Ti/kg sample (USDA-ARS, Ames, IA). Because reporting a zero (0) for data below the LOQ artificially skews analytical values to 0, any value analyzed below the LOQ but above the limit of detection (values above the blank value used in standard curve assays), was assumed to be 50% of the LOQ, which is common in the chemical analysis industry.

Calculations and statistical analysis

All data were analyzed using mixed model methods using PROC MIXED (SAS Inst., Cary, NC). The model included P1 dietary treatment, P2 dietary treatment, and P1 × P2 dietary interaction as fixed effects. For fecal Cr disappearance or Ti appearance during P2 as affected by the P1 diet, both d-14 BW and d-8 to 14 ADFI were used as model linear covariates [17]. However, for fecal Cr disappearance or Ti appearance during P2 as affected by the P2 diet, both d-14 BW and d-15 to 21 ADFI were used as model linear covariates. Only BW was utilized as a model linear covariate for the interaction between P1 and P2 diet. Regardless of significance, BW (which was often significant) and ADFI (which was often not significant) were retained in the model. Pig within treatment

was included as a random effect in all models. Means are reported as least square means with fecal Cr or Ti concentrations plotted over time to show the disappearance of Cr and appearance of Ti during P2, relative to P1 or P2 diet composition. Fecal Cr or Ti concentrations for the final 7 d in P2, are not shown because there were

Table 2 Marker concentrations in corn-soybean meal diet containing either titanium dioxide, chromic oxide, or both titanium dioxide and chromic oxide[a]

	Titanium, mg/kg diet	Chromium, mg/kg diet
Diet 1-Cr_2O_3		
Sample 1	240	2,700
Sample 2	173	2,400
Sample 3	246	2,900
Sample 4	161	2,400
Mean	205	2,600
SD	44	245
CV	21.6	9.4
Diet 2-TiO_2		
Sample 1	2,491	<0.01
Sample 2	2,529	<0.01
Sample 3	2,635	<0.01
Sample 4	2,465	<0.01
Mean	2,530	–
SD	75	–
CV	3.0	–
Diet 3-Cr_2O_3 and TiO_2		
Sample 1	2,942	2,400
Sample 2	2,831	2,500
Sample 3	2,995	2,800
Sample 4	2,768	2,700
Mean	2,884	2,600
SD	103	183
CV	3.6	7.0

[a]The addition of 5,000 mg Cr_2O_3 added × 99.3% purity × 684 g/kg Cr would result in an expected level of 3,395 mg Cr/kg diet. The addition of 5,000 mg TiO_2 added × 99.0% purity × 600 g/kg Cr would result in an expected level of 2,970 mg Ti/kg diet

no changes in fecal Cr or Ti during that time period or the levels were below LOQ. Estimates of the number of days for fecal Cr to decrease to the LOQ or for fecal Ti to reach 95% of its maximum value for each P1 × P2 combination was determined fitting a 4 parameter sigmoidal logistic function [$\left(y = D + \frac{(A-D)}{1+\left(\frac{x}{C}\right)^B}\right)$; where x = collection time, y = the response value (Cr or Ti concentration), A = minimum point in the line B = slope in the middle of the curve, C = point of inflection, D = maximum of the line; Microsoft Excel 2010] to the overall treatment means.

Results

Dual marker recovery

A critical factor for the present study was that analysis of Cr and Ti in the same diet would not interfere with the analysis of either element. To evaluate this, 3 separate corn-soybean meal diets were mixed which contained either 5 g chromic oxide/kg diet (Diet 1), 5 g titanium dioxide/kg diet (Diet 2), or both 5 g chromic oxide and 5 g titanium dioxide/kg diet (Diet 3). Although Cr analysis was lower than expected averaging 2,600 mg Cr/kg diet versus an expected level of 3,395 mg Cr/kg diet (5,000 mg Cr_2O_3 added × 99.3% purity × 684 g/kg Cr), it did not differ whether added either alone (Diet 1) or with titanium dioxide (Diet 3), Table 2. Titanium in Diets 2 and 3 averaged 2,502 mg/kg diet after subtracting out the apparent background Ti level noted in Diet 1. This too was lower than the expected value of 2,970 mg Ti/kg diet (5,000 mg TiO_2 added × 99.0% purity × 600 g/kg Ti). There were slight differences in Ti levels between Diet 2 (2,530 mg Ti/kg diet) with only TiO_2 added, and Diet 3 (2,884 mg Ti/kg diet) when both and Cr_2O_3 and TiO_2.

Table 3 Fecal chromium (mg/g fecal DM) of growing pigs during Phase-2 when fed different diets during Phase-1 and Phase-2

Phase × diet combinations		Collection day[b]							
Phase-1	Phase-2	14	15	16	17	18	19	20	21
CS[a]	CS	21.6	20.2	16.2	3.2	0.6	LOQ[c]	LOQ	LOQ
CS	DDGS	17.8	20.6	17.7	1.1	0.4	LOQ	LOQ	LOQ
CS	DDC	19.5	21.4	18.0	5.7	1.8	0.8	0.3	LOQ
DDGS	CS	15.3	15.4	10.7	0.4	LOQ	LOQ	LOQ	LOQ
DDGS	DDGS	14.2	15.8	11.3	0.3	LOQ	LOQ	LOQ	LOQ
DDGS	DDC	15.3	17.5	18.2	9.7	2.4	0.4	0.3	LOQ
DDC	CS	60.4	60.8	32.8	0.6	LOQ	LOQ	LOQ	LOQ
DDC	DDGS	64.0	55.8	26.3	3.0	LOQ	LOQ	LOQ	LOQ
DDC	DDC	63.5	60.4	32.3	8.3	1.5	0.4	LOQ	LOQ
SE		2.47	2.85	4.92	2.21	0.35	0.11	0.04	0.02
Interaction P value		0.01	0.01	0.01	0.02	0.01	0.01	0.01	0.01
Main effect of Phase-1 diet									
Phase-1	*Phase-2*								
CS	CS/DDGS/DDC	20.1	20.1	17.4	3.5	0.9	0.3	LOQ	LOQ
DDGS	CS/DDGS/DDC	15.5	16.2	13.8	3.6	0.9	LOQ	LOQ	LOQ
DDC	CS/DDGS/DDC	62.5	58.6	30.1	3.8	0.6	LOQ	LOQ	LOQ
SE		1.00	1.45	2.82	1.37	0.36	0.10	0.04	0.03
P value		0.01	0.01	0.01	0.98	0.76	0.46	0.28	0.60
Main effect of Phase-2 diet									
Phase-1	*Phase-2*								
CS/DDGS/DDC	CS	29.9	26.4	20.0	1.6	0.3	LOQ	LOQ	LOQ
CS/DDGS/DDC	DDGS	31.9	29.2	18.9	2.0	0.3	LOQ	LOQ	LOQ
CS/DDGS/DDC	DDC	31.8	34.3	22.4	7.7	1.9	0.5	0.3	LOQ
SE		5.75	6.08	3.85	1.06	0.21	0.01	0.02	0.01
P value		0.95	0.57	0.81	0.01	0.01	0.01	0.01	0.01

[a]*Abbreviations*: *CS* corn-soybean meal based diet, *DDGS* corn-soybean meal-distillers dried grains based diet, *DDC* dehulled, degermed corn-soybean meal based diet. For Phase-1, ADFI from d 1 to 14 was 2.94, 2.79, and 2.66 kg (SE = 0.07 kg) for pigs fed the DDC, CS, and DDGS diets, respectively. For Phase-2, ADFI from d 14 to 28 was 2.91, 2.84, and 2.63 kg (SE = 0.21 kg) for pigs fed the DDC, CS, and DDGS diets, respectively

[b]Collection day following change from Phase-1 to Phase-2 diet, with d 14 representing the last day of the diet containing the Cr marker was fed

[c]LOQ = limit of quantitation; 0.3 mg Cr/kg; with ½ LOQ used for statistical analysis

Fecal Cr disappearance

Interactions occurred between P1 and P2 diets on fecal Cr disappearance ($P < 0.01$) during P2, with specific values and significance levels listed in Table 3 and graphically depicted in Fig. 2a. Averaged across diet changes, when dietary NDF was increased in the diets fed to pigs from P1 to P2 (i.e., pigs fed the CS diet

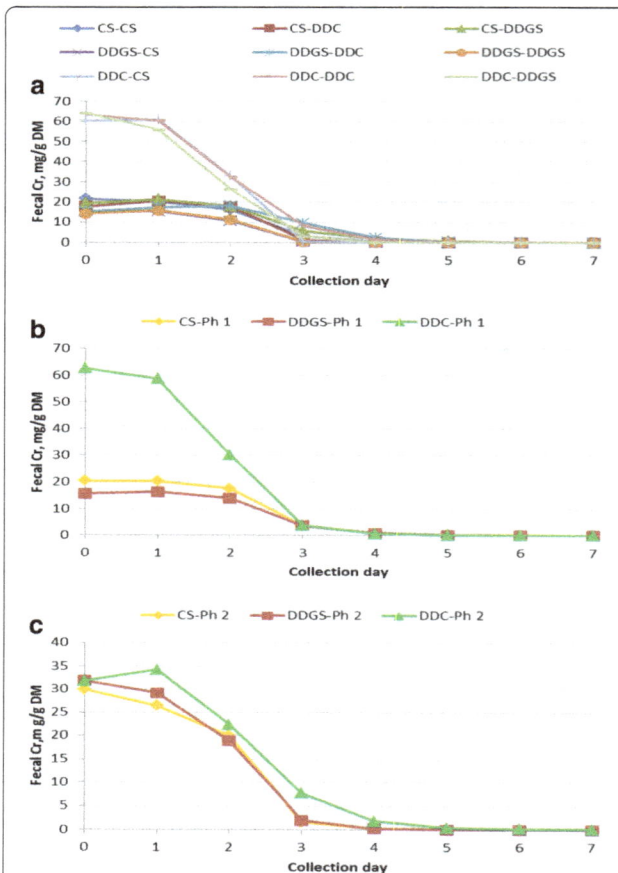

Fig. 2 a Fecal Cr concentration of growing pigs during Phase 2 as affected by the combination of Phase 1 and Phase 2 diets. Abbreviations: CS, corn-soybean meal based diet; DDGS, corn-soybean meal-distillers dried grains based diet; DDC, dehulled, degermed corn-soybean meal based diet. First abbreviation in legend represents the Phase-1 diet and the second abbreviation in the legend represents the Phase-2 diet. Collection day following change from Phase-1 to Phase-2 diet with d-0 being the day of diet change. b Fecal Cr concentration of growing pigs during Phase 2 as affected by Phase 1 diet. Legend abbreviations: CS, corn-soybean meal based diet; DDGS, corn-soybean meal-distillers dried grains based diet; DDC, dehulled, degermed corn-soybean meal based diet. Collection day following change from Phase-1 to Phase-2 diet with d-0 being the day of diet change. c Fecal Cr concentration of growing pigs during Phase 2 as affected by Phase 2 diet. Legend abbreviations: CS, corn-soybean meal based diet; DDGS, corn-soybean meal-distillers dried grains based diet; DDC, dehulled, degermed corn-soybean meal based diet. Collection day following change from Phase-1 to Phase-2 diet with d-0 being the day of diet change

switched to the DDGS diet and pigs fed the DDC diet switched to either the CS or DDGS diet), it took 2.6 d for each 5 percentage unit increase in NDF for P2 fecal Cr to decrease below the LOQ of 0.3 mg/kg fecal DM. In contrast, when dietary NDF was decreased in the diets fed to pigs from P1 to P2 (i.e., pigs fed the CS diet switched to the DDC diet and pigs fed the DDGS diet switched to either the CS or DDC diet), it took 3.5 d for each 5 percentage unit decrease in NDF for P2 fecal Cr to decrease below the LOQ (Tables 3 and 5). When pigs remained on the same diets from P1 to P2, pigs continually fed the DDC diet containing 25.1 g/kg NDF took 5.1 d for P2 fecal Cr to decrease below the LOQ, while pigs fed the CS diet containing 72.5 g/kg NDF and the DDGS diet containing 125.0 g/kg NDF took 4.1 d and 2.5 d, respectively, for P2 fecal Cr to decrease below the LOQ (Tables 3 and 5).

The main effect of the P1 diet on P2 fecal Cr concentration is reported in Table 3 and graphically depicted in Fig. 2b. For d-14 and the following 2 d, pigs fed the DDC diet in P1 had a greater P2 fecal Cr concentration of than for pigs fed either the CS or DDGS diet, with pigs fed the CS diet having a higher P2 fecal Cr than pigs fed the DDGS diet for d-14 and d-15, but equal on d-16. No dietary differences were noted thereafter. The main effect of P2 diet on P2 fecal Cr concentration is additionally reported in Table 3 and graphically depicted in Fig. 2c. Phase 2 diets had no impact on P2 fecal Cr concentration among pigs fed the diets for d-14 through d-16, with pigs fed the DDC diet having a higher P2 fecal Cr concentration than pigs fed either the CS or DDGS diets on d-17 and d-18, with no differences in P2 fecal Cr concentration between pigs fed the CS or DDGS diets. Subsequent to d-18, P2 fecal Cr fell below the LOQ for pigs fed the CS or the DDGS diet, but did not decrease below the LOQ in pigs fed the DDC diet until d-21.

Fecal Ti appearance

Similar to that observed for fecal Cr disappearance, interactions were noted between P1 and P2 diets on fecal Ti appearance during P2, with specific values and significance levels listed in Table 4, and graphically depicted in Fig. 3a. Averaged across diet changes, when dietary NDF was increased in the diets fed to pigs from P1 to P2 (i.e., pigs fed the CS diet switched to the DDGS diet and pigs fed the DDC diet switched to either the CS or DDGS diet), it took 2.4 d for each 5 percentage units increase in NDF for P2 fecal Ti to approach its maximum level. In contrast, when dietary NDF was decreased in the diets fed to pigs from P1 to P2 (i.e., pigs fed the CS diet switched to the DDC diet and pigs fed the DDGS diet switched to either the CS or DDC diet), it took 3.2 d for each 5 percentage units decrease in NDF for P2 fecal Ti to approach its maximum level (Tables 4 and 5). When

Table 4 Fecal titanium (mg/g fecal DM) of growing pigs during Phase-2 when fed different diets during Phase-1 and Phase-2

Phase × diet combinations		Collection day[b]							
Phase-1	Phase-2	14	15	16	17	18	19	20	21
CS[a]	CS	LOQ[c]	LOQ	7.3	22.0	25.9	28.1	27.4	24.8
CS	DDGS	LOQ	LOQ	LOQ	15.4	15.9	17.9	18.3	16.1
CS	DDC	LOQ	LOQ	10.1	45.6	56.7	60.6	58.1	58.8
DDGS	CS	LOQ	LOQ	9.2	22.9	25.6	25.5	26.2	25.7
DDGS	DDGS	LOQ	LOQ	6.5	16.8	16.1	17.2	18.7	16.4
DDGS	DDC	LOQ	LOQ	LOQ	34.9	55.6	59.8	61.4	62.1
DDC	CS	LOQ	LOQ	10.0	24.2	22.2	23.2	23.8	23.4
DDC	DDGS	LOQ	LOQ	9.7	17.3	17.9	17.6	18.9	19.1
DDC	DDC	LOQ	LOQ	26.5	50.4	64.0	65.3	65.2	68.3
SE		0.11	0.58	2.82	1.87	2.06	2.16	1.81	2.06
Interaction P value		0.01	0.83	0.01	0.01	0.01	0.01	0.01	0.01
Main effect of Phase-1 diet									
Phase-1	*Phase-2*								
CS	CS/DDGS/DDC	LOQ	LOQ	8.2	26.6	31.4	33.1	32.2	31.1
DDGS	CS/DDGS/DDC	LOQ	LOQ	6.1	23.5	32.1	33.2	34.2	34.0
DDC	CS/DDGS/DDC	LOQ	LOQ	15.7	31.8	33.6	35.7	34.8	37.9
SE		0.06	0.20	2.62	3.67	4.74	3.99	5.08	5.31
P value		0.01	0.01	0.03	0.22	0.93	0.88	0.91	0.61
Main effect of Phase-2 diet									
Phase-1	*Phase-2*								
CS/DDGS/DDC	CS	LOQ	LOQ	9.9	23.1	24.8	25.5	26.1	25.3
CS/DDGS/DDC	DDGS	LOQ	LOQ	10.0	17.4	18.9	19.8	19.9	19.7
CS/DDGS/DDC	DDC	LOQ	LOQ	10.9	42.6	56.3	60.1	60.8	61.3
SE		0.17	0.36	2.86	1.52	0.70	0.81	0.78	0.76
P value		0.79	0.48	0.96	0.01	0.01	0.01	0.01	0.01

[a]*Abbreviations*: *CS* corn-soybean meal based diet, *DDGS* corn-soybean meal-distillers dried grains based diet, *DDC* dehulled, degermed corn-soybean meal based diet. For Phase-1, ADFI from d 1 to 14 was 2.94, 2.79, and 2.66 kg (SE = 0.07 kg) for pigs fed the DDC, CS, and DDGS diets, respectively. For Phase-2, ADFI from d 14 to 28 was 2.91, 2.84, and 2.63 kg (SE = 0.21 kg) for pigs fed the DDC, CS, and DDGS diets, respectively

[b]Collection day following change from Phase-1 to Phase-2 diet, with d 14 representing the first day of the diet containing the Ti marker was fed

[c]LOQ = limit of quantitation; 6 mg Ti/kg. with ½ LOQ used for statistical analysis

pigs remained on the same diets from P1 to P2, pigs continually fed the DDC diet containing 25.1 g/kg NDF took 4.6 d for P2 fecal Ti to reach 95% of the maximum level, while pigs fed the CS diet containing 72.5 g/kg NDF and the DDGS diet containing 125.0 g/kg NDF took 3.7 d and 2.8 d, respectively, for P2 fecal Ti to reach 95% of its maximum level (Tables 4 and 5).th=tlb=

The main effect of the P1 diet on P2 fecal Ti concentration is reported in Table 4 and graphically depicted in Fig. 3b. Prior to d-16, fecal Ti was below the laboratory LOQ of 6 mg/kg fecal DM. On d-16, P2 fecal Ti for pigs fed the DDC diet in P1 was greater than for pigs fed the CS or DDGS diets, with no difference observed in P2 fecal Ti between pigs fed the CS and DDGS diets. After d-16, diets fed during P1 had no effect on P2 fecal Ti

concentrations. The main effect of P2 diet on P2 fecal Ti concentration is reported in Table 4 and graphically depicted in Fig. 3c. There were no differences observed between P2 fecal Ti concentrations among pigs fed the diets for d-14 through d-16. From d-17 through d-21, pigs fed the DDC diet during P2 had a higher P2 fecal Ti concentration than pigs fed either the CS or DDGS diets, and pigs fed the CS diet had a higher P2 fecal Ti concentration when compared to pigs fed the DDGS diets.

Discussion

Others [18–20] have reviewed criteria necessary for the use of markers in digestibility studies, but in addition to these, a critical factor for the present study was that analysis of Cr and Ti in the same diet would not interfere

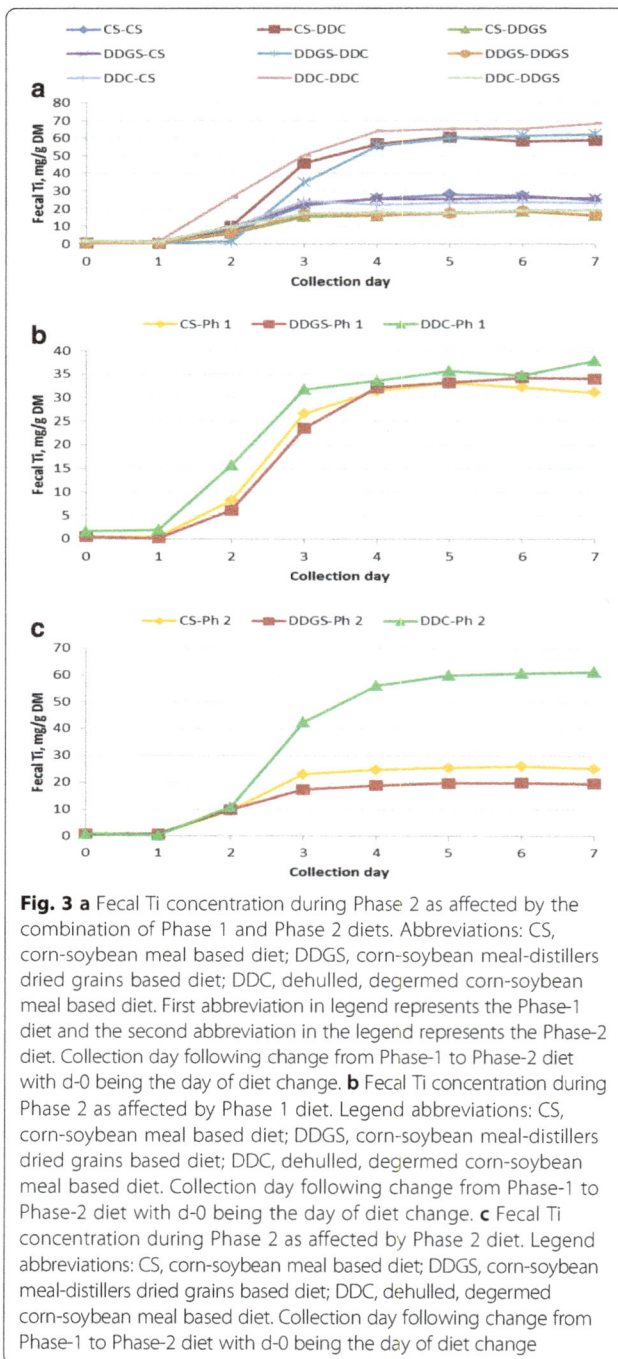

Fig. 3 a Fecal Ti concentration during Phase 2 as affected by the combination of Phase 1 and Phase 2 diets. Abbreviations: CS, corn-soybean meal based diet; DDGS, corn-soybean meal-distillers dried grains based diet; DDC, dehulled, degermed corn-soybean meal based diet. First abbreviation in legend represents the Phase-1 diet and the second abbreviation in the legend represents the Phase-2 diet. Collection day following change from Phase-1 to Phase-2 diet with d-0 being the day of diet change. **b** Fecal Ti concentration during Phase 2 as affected by Phase 1 diet. Legend abbreviations: CS, corn-soybean meal based diet; DDGS, corn-soybean meal-distillers dried grains based diet; DDC, dehulled, degermed corn-soybean meal based diet. Collection day following change from Phase-1 to Phase-2 diet with d-0 being the day of diet change. **c** Fecal Ti concentration during Phase 2 as affected by Phase 2 diet. Legend abbreviations: CS, corn-soybean meal based diet; DDGS, corn-soybean meal-distillers dried grains based diet; DDC, dehulled, degermed corn-soybean meal based diet. Collection day following change from Phase-1 to Phase-2 diet with d-0 being the day of diet change

Table 5 Sigmoidal response parameters for P2 fecal Cr disappearance and P2 fecal Ti appearance for growing pigs fed different diets during Phase-1 and Phase-2

Phase × diet combinations		NDF Δ	Fecal Cr disappearance,	Fecal Ti appearance,
Phase-1	Phase-2	% Units[b]	d reach limit of quantitation[c]	d to reach 95% maximum[d]
CS[a]	CS	0	4.1	3.7
CS	DDGS	+5.25	3.3	3.5
CS	DDC	−4.74	4.9	3.9
DDGS	CS	−5.25	3.1	3.5
DDGS	DDGS	0	2.5	2.8
DDGS	DDC	−9.99	4.8	4.2
DDC	CS	+4.74	2.6	2.2
DDC	DDGS	+9.99	3.8	3.2
DDC	DDC	0	5.1	4.6

[a]*Abbreviations*: *CS* corn-soybean meal based diet, *DDGS* corn-soybean meal-distillers dried grains based diet, *DDC* dehulled, degermed corn-soybean meal based diet

[b]Change in analyzed dietary NDF, percentage units

[c]As determined by sigmoidal response of phase × diet treatment means obtained from Table 3. The Cr limit of quantitation was 0.3 mg Cr/kg

[d]As determined by sigmoidal response of phase × diet treatment means obtained from Table 4. The Ti limit of quantitation was 6 mg Ti/kg

recovery. Nonetheless, despite any potential differences in marker recovery, we believe that the data obtained in our animal experiment is valid in determining the time from which a new collection period could begin without the previous marker interfering with the results obtained in the subsequent collection period. Taken together, the literature and our data suggest that use of two markers within the digestive tract does not compromise or confound the results that we obtained in our animal experiment. We also chose to sample pigs at the same time each day to eliminate any confounding effects relative to diurnal variation in fecal composition that has been previously reported [12, 26, 27].

Numerous experiments have been conducted to describe the time of first or 5% marker appearance [4, 5, 11, 28], mean transit rate [6, 8, 28–31] or 25, 50, 80, or 95% of the marker excreted [4, 5, 7]; values which are useful in mathematical modeling of digestion [3]. This was not the focus of our experiment as we chose to only determine when P2 fecal Cr reached its minimum LOQ and when P2 fecal Ti reached 95% of its maximum because we were interested in if the previous or present diet affected when a dietary marker was completely excreted (Cr, Table 3) or stabilized (Ti, Table 4).

It is well known that the dietary fiber type and level affects rate of passage [6–8]. These effects were not, however, independent from the previous diet fed as indicated by the interaction between diets fed during P1 and P2. The current data indicate that as dietary NDF increased from P1 to P2, it took less time for P2 fecal Cr to decrease

with the analysis of either element. Full recovery of Cr [20, 21] and Ti [22, 23] has been shown to be problematic, which was the case in our dual marker recovery experiment (Cr recovery of 77%, Ti recovery of 84%; Table 2) and animal experiment (Cr recovery of 82%, Ti recovery of 97%; Table 1) as well. The lack of any major differences in recovery of dual markers in our experiment is supported by others [24, 25] who have noted little impact of dual markers on individual marker

(2.6 d) or P2 fecal Ti to be maximized (2.4 d), than if NDF was decreased from P1 to P2, where it took 3.5 d for P2 fecal Cr to decrease or 3.2 d for P2 fecal Ti to be maximized. These effects were independent from feed intake in the current study because in most instances ADFI was not a significant covariate (although ADFI was still retained in the model to eliminate even minimal feed intake differences). Differences in P2 fecal marker concentration (Cr or Ti in the current study) by diet type were expected due to digestibility differences among ingredients utilized in diet formulations. With components in the diets digested to different degrees but the marker remaining undigested, subsequent Cr or Ti concentration should have changed proportionally. In the current experiment, pigs fed the diet having the greatest digestibility (DDC) resulted in the greatest fecal marker concentration, followed by pigs fed the CS diet, and lastly, by pigs fed the DDGS diet.

Conclusions

Overall, data from the present study indicate that as the digestibility of the diets increased (i.e., decreasing dietary NDF), it took progressively longer for P2 fecal Cr to be excreted or P2 fecal Ti to be maximized (approximately 2.5 d) than if diets that were decreasing in digestibility (i.e., increasing dietary NDF) were fed (approximately 3.4 d), a difference of approximately 1 d. For pigs fed diets containing a moderate amount of DDGS or only corn and soybean meal, the complete elimination of Cr in P2 feces or maximum appearance of Ti in P2 feces was approximately 3 and 4 d, respectively. In contrast, pigs fed diets containing highly digestible ingredients such as DDC (or semi-purified diets as are used in some experimental methodologies) took longer for clearance or equilibrium, approximately 5 d in the current experiment. This information is critical to know when pigs are utilized either once or for multiple times in digestibility experiments containing inert makers, and suggest that caution must be taken so as to not have previous dietary marker addition remain in the digestive tract or feces to confound subsequent experimental results.

Abbreviations
ADFI: Average daily feed intake; BW: Body weight; C: Corn; Cr: Chromium; CV: Coefficient of variation; d: Day; DDC: Dehulled ddgermed corn; DDGS: Distillers dried grains with solubles; h: Hour; LOQ: Limit of quantitation; NDF: Neutral detergent fiber; P1: Phase 1; P2: Phase 2; S: Soybean meal; SD: Standard deviation; Se: Standard error; Ti: Titanium

Acknowledgements
The authors express thanks to J. Cook at the National Laboratory for Agriculture and the Environment (Ames, Iowa) for laboratory assistance. Appreciation is also expressed to H. J. Monegue and W. Patton (University of Kentucky, Lexington) for assistance in the care of pigs and to D. Higginbotham (University of Kentucky, Lexington) for help in diet preparation; and to Akey Inc. (Lewisburg, OH) and DSM Nutritional Products Inc. (Parsippany, NJ) for ingredients used in the diets. Mention of a trade name, proprietary product, or specific equipment does not constitute a guarantee or warranty by the USDA, Iowa State University, or the University of Kentucky and does not imply approval to the exclusion of other products that may be suitable. The USDA is an equal opportunity provider and employer.

Funding
No external funds were used in the design, collection, analysis, interpretation, or writing of this manuscript.

Authors' contributions
BMJ, MDL, and BJK conceived and performed the experiment. All authors assisted in data analysis, interpreting and proofing the manuscript, and approving the final version of the manuscript.

Competing interests
The authors declare that they have no conflicts of interests that could be perceived as prejudicing the impartiality of this paper.

Author details
[1]Department of Animal Science, Iowa State University, Ames 50010, USA. [2]Department of Animal and Food Sciences, University of Kentucky, Lexington 40546, KY, USA. [3]USDA-ARS-National Laboratory for Agricultural and the Environment, Ames 50010, IA, USA.

References
1. Adeola A. Digestion and balance techniques in pigs. In: Lewis AJ, Southern LL, editors. Swine nutrition. Boca Raton: CRC Press; 2001. p. 903–16.
2. Low AG. Nutritional regulation of gastric secretion, digestion and emptying. Nutr Res Rev. 1990;3:229–52.
3. Bastianelli D, Sauvant D, Rerat A. Mathematical modeling of digestion and nutrient absorption in pigs. J Anim Sci. 1996;74:1879–87.
4. Castle EJ, Castle ME. The rate of passage of food through the alimentary tract of pigs. J Agric Sci. 1956;47:196–203.
5. Castle EJ, Castle ME. Further studies of the rate of passage of food through the alimentary tract of pigs. J Agric Sci. 1957;49:106–12.
6. Cherbut C, Barry JJL, Wyers M, Delort-Laval J. Effect of the nature of dietary fibre on transit time and faecal excretion in the growing pig. Anim Feed Sci Technol. 1988;20:327–33.
7. Freire JPB, Guerreiro AJG, Cunha LF, Aumaitre A. Effect of dietary fibre source on total tract digestibility, caecum volatile fatty acids and digestive transit time in the weaned piglet. Anim Feed Sci Technol. 2000;87:71–83.
8. Le Goff G, van Milgen J, Noblet J. Influence of dietary fibre on digestive utilization and rate of passage in growing pigs, finishing pigs and adult sow. Anim Sci. 2002;74:503–15.
9. Potkins ZV, Lawrence TLJ, Thomlinson JR. Effects of structural and non-structural polysaccharides in the diet of the growing pig on gastric emptying rate and rate of passage of digesta to the terminal ileum and through the total gastrointestinal tract. Br J Nutr. 1991;65:391–413.
10. Varel VH, Jung HG, Pond WG. Effects of dietary fiber of young adult genetically lean, obese and contemporary pigs: rate of passage, digestibility and microbiological data. J Anim Sci. 1988;66:707–12.
11. Holzgraefe DP, Fahey Jr GC, Jensen AH. Influence of dietary alfalfa: orchardgrass hay and lasalocid on in vitro estimates of dry matter digestibility and volatile fatty acid concentrations of cecal contents and rate of digesta passage in sows. J Anim Sci. 1985;60:1235–46.
12. Jorgensen H, Lindberg JE, Andersson C. Diurnal variation in the composition of ileal digesta and the ileal digestibilities of nutrients in growing pigs. J Sci Food Agric. 1997;74:244–50.
13. Imbeah M, Sauer WC, Caine WR. Comparison of the single dose and withdrawal method for measuring the rate of passage of two digestibility markers in digesta collected form the distal ileum and feces in growing pigs. Anim Feed Sci Technol. 1995;52:41–50.
14. Clawson AJ, Reid JT, Sheffy BE, Willman JP. Use of chromium oxide in digestion studies with swine. J Anim Sci. 1955;14:700–9.
15. Agudelo JH, Lindemann MD, Cromwell GK. A comparison of two methods to assess nutrient digestibility in pigs. Livest Sci. 2010;133:74–7.
16. Association of Official Analytical Chemists (AOAC). Official methods of analysis. 9th ed. Washington, DC: AOAC; 1978.
17. Jacobs BM, Patience JF, Lindemann MD, Stalder KJ, Kerr BJ. The use of covariate reduces experimental error in nutrient digestion studies in growing pigs. J Anim Sci. 2013;91:804–10.
18. Kotb AR, Luckey TD. Markers in nutrition. Nutr Abstr Rev. 1972;42:813–45.

19. Saha DC, Gilbreath RL. Analytical recovery of chromium from the diet and faeces determined by colorimetry and atomic absorption spectrophotometry. J Sci Food Agric. 1991;55:433–46.

20. Jagger S, Wiseman J, Cole DJA, Craigon J. Evaluation of inert markers for the determination of ileal and faecal apparent digestibility values in the pig. Br J Nutr. 1992;68:729–39.

21. Carciofi AC, Vasconcellos RS, de Oliveira LD, Brunetto MA, Valerio AG, Baxolli RS, et al. Chromic oxide as a digestibility marker for dogs—a comparison of methods of analysis. Anim Feed Sci Technol. 2007;134:273–82.

22. Yin JL, McEvoy DG, Schulze H, McCracken KJ. Studies on cannulation method and alternative indigestible markers and the effects of food enzyme supplementation in barley-based diets on ileal and overall apparent digestibility in growing pigs. Anim Sci. 2000;70:63–72.

23. Yin JL, McEvoy JDG, Schulze H, Henning U, Souffrant WB, McCracken KJ. Apparent digestibility (ileal and overall) of nutrients as evaluated with PVTC-cannulated or ileo-rectal anastomised pigs fed diets containing two indigestible markers. Livest Prod Sci. 2000;62:133–41.

24. Kavanagh S, Lynch PB, O'Mara F, Caffrey PJ. A comparison of total collection and marker technique for the measurement of apparent digestibility of diets for growing pigs. Anim Feed Sci Technol. 2001;89:49–58.

25. Olukosi OA, Bolarinwa OA, Cowieson AJ, Adeola O. Marker type but not concentration influenced apparent ileal amino acid digestibility in phytase-supplemented diets for broiler chickens and pigs. J Anim Sci. 2012;90:4414–20.

26. Moore JH. Diurnal variations in the composition of the faeces of pigs on diets containing chromium oxide. Br J Nutr. 1957;11:273–88.

27. Horvath DJ, Peterson ME, Clawson AJ, Sheffy BE, Loosli JK. Diurnal variations in the composition of swine feces. J Anim Sci. 1958;17:869–74.

28. Ehle FR, Jeraci JL, Robertson JB, Van Soest PJ. The influence of dietary fiber on digestibility, rate of passage and gastrointestinal fermentation in pigs. J Anim Sci. 1982;55:1071–81.

29. Furuya S, Sakamoto K, Asano T, Takahashi S, Kameoka K. Effects of added dietary sodium polyacrylate on passage rate of markers and apparent digestibility by growing swine. J Anim Sci. 1978;47:159–65.

30. Pond WG, Pond KR, Ellis WC, Matis JH. Markers for estimating digesta flow in pigs and the effects of dietary fiber. J Anim Sci. 1986;63:1140–9.

31. Wilfart A, Montagne L, Simmins H, Noblet J, Van Milgen J. Digesta transit in different segments of the gastrointestinal tract of pigs as affected by insoluble fibre supplied by wheat bran. Br J Nutr. 2007;98:54–62.

6

Effect of *Bacillus subtilis* and *Bacillus licheniformis* supplementation in diets with low- and high-protein content on ileal crude protein and amino acid digestibility and intestinal microbiota composition of growing pigs

Chanwit Kaewtapee[1,2], Katharina Burbach[1], Georgina Tomforde[1], Thomas Hartinger[1,3], Amélia Camarinha-Silva[1], Sonja Heinritz[1], Jana Seifert[1], Markus Wiltafsky[4], Rainer Mosenthin[1] and Pia Rosenfelder-Kuon[1*]

Abstract

Background: *Bacillus* spp. seem to be an alternative to antimicrobial growth promoters for improving animals' health and performance. However, there is little information on the effect of *Bacillus* spp. in combination with different dietary crude protein (CP) levels on the ileal digestibility and microbiota composition. Therefore, the objective of this study was to determine the effect of *Bacillus* spp. supplementation to low- (LP) and high-protein diets (HP) on ileal CP and amino acid (AA) digestibility and intestinal microbiota composition.

Methods: Eight ileally cannulated pigs with an initial body weight of 28.5 kg were randomly allocated to a row-column design with 8 pigs and 3 periods of 16 d each. The assay diets were based on wheat-barley-soybean meal with two protein levels: LP (14% CP, as-fed) and HP diet (18% CP, as-fed). The LP and HP diets were supplemented with or without *Bacillus* spp. at a level of 0.04% (as-fed). The apparent ileal digestibility (AID) and standardized ileal digestibility (SID) of CP and AA was determined. Bacterial community composition from ileal digesta was analyzed by Illumina amplicon sequencing and quantitative real-time PCR. Data were analyzed as a 2×2 factorial design using the GLIMMIX procedures of SAS.

Results: The supplementation with *Bacillus* spp. did not affect both AID and SID of CP and AA in growing pigs. Moreover, there was no difference in AID of CP and AA between HP and LP diets, but SID of cystine, glutamic acid, glycine, and proline was lower ($P < 0.05$) in pigs fed the HP diets. The HP diets increased abundance of *Bifidobacterium* spp. and *Lactobacillus* spp., ($P < 0.05$) and by amplicon sequencing the latter was identified as predominant genus in microbiota from HP with *Bacillus* spp., whereas dietary supplementation of *Bacillus* spp. increased ($P < 0.05$) abundance of *Roseburia* spp..

Conclusions: The HP diet increased abundance of *Lactobacillus* spp. and *Bifidobacterium* spp.. The supplementation of *Bacillus* spp. resulted in a higher abundance of healthy gut associated bacteria without affecting ileal CP and AA digestibility, whereas LP diet may reduce the flow of undigested protein to the large intestine of pigs.

Keywords: *Bacillus* spp., Growing pigs, Ileal digestibility, Microbiota, Protein levels

* Correspondence: pia.rosenfelder@uni-hohenheim.de
[1]University of Hohenheim, Institute of Animal Science, Emil-Wolff-Strasse 10, 70599 Stuttgart, Germany
Full list of author information is available at the end of the article

Background

Due to the ban of antimicrobial growth promoters in livestock feeding by the European Union in 2006 [1], probiotics are considered as an alternative for improving animals' health and performance [2, 3]. Within this regard, *Bacillus* spp. have the ability to sporulate, thereby making them stable during thermal treatment of feed, and resistant to enzymatic digestion along the gastro-intestinal tract (GIT) [4]. Thus, *Bacillus* spp. such as *Bacillus subtilis* (*B. subtilis*) and *Bacillus licheniformis* (*B. licheniformis*) are frequently supplemented to pig diets [4–6] as these two species have been listed to be added as non-toxigenic, biological supplements to livestock diets [7], and additionally, they are widely used for the large-scale industrial production of proteins including extracellular enzymes [8]. Positive effects of dietary supplementation of *B. subtilis* and *B. licheniformis* on pigs' growth performance have been reported before [9, 10].

Activity of probiotics is influenced by diet composition [11] and variations in dietary protein supply, thereby possibly affecting microbial composition in the gut [12, 13]. Accordingly, reducing the dietary crude protein (CP) level has been reported to markedly reduce the production of potentially harmful microbial metabolites such as ammonia and amines due to the lower availability of undigested protein for microbial fermentation [14]. Thus, excessive nitrogen (N) excretion by pigs is mitigated, resulting in a decrease of environmental pollutants [15, 16]. In contrast, increasing the dietary CP intake may stimulate the proliferation of almost all bacteria groups over the entire GIT including beneficial bacteria, such as *Bifidobacterium* spp., and potentially pathogenic bacteria, such as *Bacteroides* groups [17]. Furthermore, there is increasing evidence that interactions of supplemental probiotics with dietary CP level affect the intestinal microbiome at the ileal level [18].

According to the results of previous studies [19, 20] *Bacillus* spp. enhanced the development and activities of digestive enzymes in the GIT, which was associated with a numerical increase in apparent ileal digestibility (AID) and standardized ileal digestibility (SID) of some amino acids (AA) in weaning pigs [21]. However, studies with growing pigs in which *Bacillus* spp. were supplemented to diets varying in CP content are still lacking. Therefore, the objective of the present study was to test the hypothesis, if *B. subtilis* and *B. licheniformis* supplementation to low- and high-protein diets will affect ileal CP and AA digestibility and intestinal microbiota composition in growing pigs.

Methods

The research protocol was reviewed and approved by the German Ethical Commission for Animal Welfare, and care of the animals throughout this experiment was in accordance with guidelines issued by the Council Directive [22].

Animals, housing, and surgical procedures

Eight pigs were obtained from the University of Hohenheim Research Station. The average initial and final body weight (BW) of the experimental animals were 28.5 ± 0.8 and 64.3 ± 1.5 kg, respectively. The pigs were housed individually in stainless steel metabolic crates (0.8 m × 1.5 m). Each metabolic crate was equipped with an infrared heating lamp and a low pressure drinking nipple which allowed free access to water. The research unit was equipped with an automated temperature control system kept at 20 °C. Until the beginning of the experiment, the pigs were fed a commercial starter diet at a daily level of 4% (as-fed)/kg of average BW (Porcigold® SMA 134, Raiffeisen Kraftfutterwerke Süd GmbH, Würzburg, Germany; 17.5% CP and 13.4 MJ metabolizable energy (ME)/kg, as-fed). After arrival at the research unit, the pigs were surgically fitted with a simple T-cannula at the distal ileum as described by Li et al. [23]. The pigs were allowed a recovery period of at least 7 d. During this period, the feed allowance was gradually increased, starting from 50 g/d the day after surgery until 1000 g/d (as-fed) were consumed.

Experimental design, diets, and procedures

The experiment was arranged as a row-column design with 8 pigs and 3 experimental periods of 16 d each. Pigs were fed assay diets twice daily at 0700 and 1900 h at a level of 4% (as-fed)/kg of their average BW corresponding to 3 times their energy requirement for maintenance (i.e. 0.44 MJ ME/kg $BW^{0.75}$) [24]. Pigs' BW was determined at the beginning of each experimental period.

The assay diets were based on wheat, barley, and soybean meal with 2 protein levels resulting in a low-protein (14% CP, as-fed; LP) and a high-protein diet (18% CP, as-fed; HP). The LP diet was accomplished by blending the HP diet with 25% of native cornstarch. The contents of oil, minerals, vitamins, and titanium dioxide were the same for all diets. The *Bacillus* spp. product is comprised of a mixture of spray-dried spores of *B. licheniformis* and *B. subtilis*. The LP and HP diets were supplemented with (+) or without (-) *Bacillus* spp. at a level of 0.04% (as-fed). All assay diets were formulated (Table 1) to meet or exceed the dietary threshold levels for CP and AA according to Fan et al. [25] and NRC [26] nutrient recommendations for pigs from 25 to 50 kg BW. Vitamins and minerals were supplemented to all diets to meet or exceed NRC [26] standard, and all diets contained titanium dioxide at a level of 0.4% (as-fed basis) as an indigestible marker.

The assay diets were in a mash form mixed with water (1/1, w/v). During each of the 3 experimental periods, the pigs were allowed to adapt to their assay diets for 14 d before ileal digesta was collected for a total of 24 h from 0700 to 1900 h on d 15 and from 1900 on d 16 to 0700 h on d 17. Digesta collection procedure was

Table 1 Ingredient composition of assay diets, % as-fed basis

Item	High-protein	Low-protein
Barley	20.00	15.00
Wheat	51.00	38.24
Soybean meal	21.51	16.13
Oil[a]	1.50	1.50
Cornstarch[b]	2.08	25.55
Vitamins and minerals premix[c]	0.76	0.76
Sodium chloride	0.07	0.07
Monocalcium phosphate	0.66	0.66
Calcium carbonate	0.65	0.65
Vitamin E[d]	0.03	0.03
L-Lysine-HCl[e]	0.61	0.46
DL-Methionine[e]	0.22	0.16
L-Isoleucine[e]	0.03	0.02
L-Leucine[e]	0.13	0.10
L-Threonine[e]	0.22	0.17
L-Tryptophan[e]	0.01	0.01
L-Valine[e]	0.12	0.09
Titanium dioxide	0.40	0.40
Bacillus spp.[f]	-	-
Calculated chemical composition[g]		
Metabolizable energy, MJ/kg	13.43	14.25
Crude protein, %	18.00	14.00
Calcium, %	0.66	0.63
Available Phosphorus, %	0.27	0.25
SID[g] Lysine, %	1.20	0.92
SID[g] Methionine, %	0.43	0.33
SID[g] Threonine, %	0.73	0.56

[a]Blend of rapeseed oil (75%) and soybean oil (25%)
[b]Roquette, Lestrem, France
[c]Vilomin® 18950, Deutsche VilomixTierernährung GmbH, Neuenkirchen-Vörden, Germany; provided the following quantities of minerals and vitamins per kg of diet: Ca, 1.86 g; P, 0.38 g; Na, 0.42 g; Mg, 76.00 mg; Fe, 30.40 mg (FeSO$_4$·H$_2$O); Cu, 3.80 mg (CuSO$_4$·5H$_2$O); Mn, 20.29 mg (MnO); Zn, 25.38 mg (ZnO); I, 0.51 mg (Ca(IO$_3$)$_2$); Se, 0.10 mg (Na$_2$SeO$_3$); Co, 0.06 mg (2CoCO$_3$·3Co(OH)$_2$·H$_2$O); vitamin A, 3,040 IU; vitamin D$_3$, 456 IU; vitamin E, 19.00 mg; vitamin B$_1$, 0.38 mg; vitamin B$_2$, 1.18 mg; vitamin B$_6$, 0.95 mg; vitamin B$_{12}$, 7.60 µg; vitamin K$_3$, 0.76 mg; niacin, 4.75 mg; calcium pantothenate, 2.85 mg; folic acid, 0.19 mg; choline chloride, 57.00 mg
[d]LutavitE 50, BASF, Ludwigshafen, Germany
[e]All crystalline amino acids (AA) were supplied by Evonik Industries AG (Hanau-Wolfgang, Germany). The purity of all crystalline AA was 99%, with the exception of L-Lysine-HCl (78%)
[f]High- and low-protein diets were supplemented with or without 0.04% (as-fed) of Bacillus spp. product at the expense of cornstarch
[g]SID standardized ileal digestibility

adapted from Li et al. [23] using soft plastic bags attached to the barrel of the cannula by elastic bands. The bags were changed at least every 20 min. To minimize further bacterial fermentation 4 mL of 2.5 mol/L formic acid were added to the sampling bags and then immediately frozen at -18 °C. The individual digesta samples of each pig were pooled

for each sampling period, freeze-dried, and ground to 0.5 mm prior to analyses. For analyses of intestinal microbiota composition, ileal digesta and feces samples were taken prior to the first experimental period (starter period) and on d 15 once for each experimental period. Ileal digesta and feces samples for microbial community analysis were immediately put on ice before being stored in a freezer at -80 °C for subsequent treatment and analyses.

Chemical analyses

Official standard methods [27] were used to determine contents of proximate nutrients, neutral detergent fiber (NDF), acid detergent fiber (ADF), acid detergent lignin (ADL), and microbial numbers of B. subtilis and B. licheniformis in assay diets. The assay diets and digesta samples were analyzed for DM (method 3.1). In addition, assay diets were analyzed for ash (method 8.1); ether extract (EE; method 5.1.1 using petroleum ether), NDF assayed with a heat-stable amylase and expressed inclusive of residual ash (method 6.5.1), ADF expressed inclusive of residual ash (method 6.5.2), and ADL determined by solubilization of cellulose with sulphuric acid (method 6.5.3). Moreover, microbial numbers of B. subtilis and B. licheniformis in assay diets were determined by method 28.2.2 [27]. Nitrogen contents in assay diets and ileal digesta samples were analyzed using a gas combustion method according to official method 990.03 of the AOAC International [28] (FP-2000, Leco Corp., St Joseph, MI, US). Ethylenediaminetetraacetic acid was used as a reference standard before and after all N analyses. Crude protein contents were calculated by multiplying the content of N with 6.25. Amino acid contents in assay diets and ileal digesta samples were determined by using ion-exchange chromatography with postcolumn derivatization with ninhydrin [29]. Tryptophan was determined by HPLC with fluorescence detection (extinction 280 nm, emission 356 nm), after alkaline hydrolysis with barium hydroxide octahydrate for 20 h at 110 °C according to the procedure as outlined by Commission Directive [30]. The titanium dioxide content in the assay diets and ileal digesta samples was performed according to the procedure described by Brandt and Allam [31].

DNA extraction of ileal digesta and feces samples

Genomic DNA was extracted from 250 mg ileal digesta and feces using Fast DNA Spin Kit for Soil (MP Biomedicals GmbH, Heidelberg, Germany). Extraction procedure was performed with slight modifications to manufacturer's instructions as described by Burbach et al. [32].

Amplicon sequencing analysis

Illumina amplicon sequencing libraries of the V1-2 region of the 16S rRNA gene was performed similar to procedures described previously [33]. Library preparation, however, was modified as follows: the V1-2 region was amplified with a 27 F-modified forward primer (AGRGTTHGATYMTGGCTCAG) in a 20 µL reaction. 1 µL of this first PCR was used as template in a second PCR using multiplexing and indexing primers as described previously [33]. Amplicons were verified by agarose gel electrophoresis and normalized using SequalPrep™ Normalization Plate Kit (Invitrogen, Thermo Fisher Scientific, Waltham, USA). Libraries were pooled by index, purified with MinElute PCR Purification Kit (Qiagen, Hilden, Germany), quantified with Qubit® 2.0 Fluorometer (Invitrogen) and sequenced on Illumina MiSeq platform using 250 bp paired end sequencing chemistry. All analyzed samples comprised around 2.8 million reads, with an average of 43,646 reads per sample. Reads were quality filtered, assembled and aligned using Mothur pipeline [34]. UCHIME was used to find possible chimeras and reads were clustered at 97% identity into 2601 operative taxonomic units (OTU). The closest representative was manually identified with seqmatch from RDP [35]. Sequences classified as Chloroplast/Cyanobacteria were removed from OTU dataset as it was assumed that they represent undigested plant material. Sequences were submitted to European Nucleotide Archive under the accession number PRJEB14413 (http://www.ebi.ac.uk/ena/data/view/PRJEB14413).

Quantitative real time PCR

Quantitative real-time PCR (qPCR) was used to analyze the following bacteria groups in the ileal digesta samples:

Total bacteria, *Lactobacillus* spp., *Bifidobacterium* spp., *Roseburia* spp., *Enterobacteriaceae*, *Bacteroides-Prevotella-Porphyromonas* group, *Clostridium* Cluster IV, and *Bacillus* spp.. All used primers were selected from literature and are listed in Table 2. Optimization of primer conditions was done in order to determine optimal annealing temperatures and primer concentrations by running a standard PCR with diverse primer concentrations (200 nmol/L, 400 nmol/L, 600 nmol/L) and a temperature gradient from 55.0 °C to 65.0 °C. According to melt curves on standard PCR and the agarose gel electrophoresis results, optimal primer concentration and annealing temperature was set for each primer.

Standard curves for each primer were designed using serial dilutions of the purified and quantified PCR products generated by standard PCR and genomic DNA from pig feces. The PCR products were checked by agarose gel electrophoresis (2% agarose) to ensure correct primer specific products. Quantity of purified PCR amplification products was determined using Qubit® 2.0 Fluorometer (Invitrogen).

Quantification was carried out using the CFX Connect™ Real-Time System (Bio-Rad Laboratories GmbH, Munich, Germany), associated with the Bio Rad CFX Manager™ Software 3.1 (Bio-Rad Laboratories GmbH, Munich, Germany). All samples were determined in duplicate and all standards were pipetted in triplicate on each plate. The order of samples and standards on the plates was randomized. The reaction mixture for each bacterial group consisted of 10 µL of KAPA SYBR FAST (PEQLAB Biotechnologie GmbH, Erlangen, Germany),

Table 2 Oligonucleotide primers used for real-time PCR

Target group	Item	Oligonucleotide sequence (5'→3')	Primer conc., nmol/L	Annealing temp., °C	Product size, bp	Reference
Total bacteria	Forward	GTGSTGCAYGGYYGTCGTCA	600	52	147	Fuller et al. [80]
	Reverse	ACGTCRTCCMCNCCTTCCTC				
Lactobacillus spp.	Forward	AGAGGTAGTAACTGGCCTTTA	400	59	391	Malinen et al. [81]
	Reverse	GCGGAAACCTCCCAACA				
Bifidobacterium spp.	Forward	TCGCGTCYGGTGTGAAAG	400	59	243	Rinttilä et al. [82]
	Reverse	CCACATCCAGCRTCCAC				
Roseburia spp.	Forward	AGGCGGTACGGCAAGTCT	400	59	353	Veiga et al. [83]
	Reverse	AGTTTYATTCTTGCGAACG				Rinttilä et al. [82]
Enterobacteriaceae	Forward	CATTGACGTTACCCGCAGAAGAAGC	200	59	195	Bartosch et al. [84]
	Reverse	CTCTACGAGACTCAAGCTTGC				
Clostridium Cluster IV	Rflbr730F	GGCGGCYTRCTGGGCTTT	400	65	147	Ramirez-Farias et al. [85]
	Clep866mR§	CCAGGTGGATWACTTATTGTGTTAA				Lay et al. [86]
Bacteroides-Prevotella-Porphyromonas	Forward	GGTGTCGGCTTAAGTGCCAT	600	58	140	Rinttilä et al. [82]
	Reverse	CGGAYGTAAGGGCCGTGC				
Bacillus spp.	Forward	CCTACGGGAGGCAGCAGTAG	600	59	78	Fernández-No et al. [87]
	Reverse	GCGTTGCTCCGTCAGACTTT				

1 μL template DNA (ileal digesta samples and standards), the optimized primer concentrations of forward and reverse primers (Table 2), and was filled up to a total volume of 20 μL with PCR grade water (Carl Roth GmbH, Karlsruhe, Germany). Amplification conditions were: activation of polymerase at 95.0 °C for 3 min, followed by 40 cycles consisting of denaturation at 95.0 °C for 5 s, primer annealing for 20 s (at optimized temperatures, Table 2), and extension at 72.0 °C for 1 s. Subsequently, a final elongation step at 72.0 °C for 1 min followed. The melt curve was obtained by stepwise (0.5 °C) increase of temperature from 55 °C to 95 °C. Results were reported as $\log_{10}$ 16S rRNA gene copies/g digesta.

Calculations

The AID of CP and AA in the assay diets was calculated according to the following equation:

$$AID_D = [1 - (I_D \times A_I) / (A_D \times I_I)] \times 100\,\%$$

where AID_D = AID of CP or AA in the assay diet (%), I_D = marker content in the assay diet (g/kg DM), A_I = CP or AA content in ileal digesta (g/kg DM), A_D = CP or AA content in the assay diet (g/kg DM), and I_I = marker content in ileal digesta (g/kg DM).

According to Stein et al. [36] and Jansman et al. [37], the basal ileal endogenous loss of CP and AA (IAA_{end}) is considered to be constant among groups of pigs, and therefore, mean values for IAA_{end} [37] can be used for transformation of AID into their SID values.

The SID of CP and AA in assay diets was estimated according to the following equation:

$$SID_D = AID_D + (IAA_{end} / A_D) \times 100\,\%$$

where SID_D = SID of CP or AA in the assay diet (%).

Statistical analyses

Homogeneity of variances and normal distribution of the data were confirmed using the UNIVARIATE procedure of SAS (SAS Inst., Inc., Cary, NC). Data were analyzed as a 2×2 factorial using the GLIMMIX procedures of SAS. The model included the protein level, probiotic supplementation, and the interactive effects of protein level and probiotic supplementation as the fixed effects, and pig and period as the random effects. In case of interaction, the significant differences between treatments based on a t-test were set at $\alpha = 0.05$ using the algorithm for letter-based representation of all pair-wise comparisons according to Piepho [38]. For microbiota analyses, bacterial 16S rRNA gene copy numbers in pre-treatment period was considered as covariate. Least squares means and standard error of the means are presented, and a probability level of $P < 0.05$

was considered to be statistically significant, whereas a $P < 0.10$ was considered to constitute a tendency.

Illumina amplicon sequencing data were analyzed using statistic software PRIMER (v.6.1.16, PRIMER-E; Plymouth Marine Laboratory, Plymouth, UK) [39]. Samples were standardized by total and resemblance matrix was calculated using Bray-Curtis coefficient. Overall community structures were explored by nonmetric multidimensional scaling (MDS). One way analysis of similarity (ANOSIM) and permutational multivariate analysis of variance (PERMANOVA) were used to evaluate similarity between different dietary groups, different protein levels and probiotic treatments, and a probability level of $P \leq 0.05$ was considered to be significant different. The ANOSIM R values range from -1 to 1; the farer from zero the more distinct and the closer to zero the more similar are the compared groups. Variables contributing to observed differences were identified by similarity percentages routine. The bacterial families contributing to overall 70% of dissimilarities among treatment groups were considered to be the most important and their abundance data were graphically plotted according to a color key from zero to maximal abundance. Shannon index was used to measure diversity in bacterial communities from different sample groups, taking into account the number of OTUs and the proportion of each OTU. A Mantel-type test (RELATE) on Bray-Curtis matrices was used to quantify the correlation between results from bacterial community analysis. To enable comparison between amplicon sequencing and qPCR approaches, RELATE routine was run on untransformed datasets, restricted to bacteria groups targeted by qPCR primers and the generated Spearman Rho was considered to be significant if $P \leq 0.05$.

Results

All pigs remained healthy throughout the experiment and readily consumed their daily feed allowances. The analyzed CP and AA contents of the assay diets and microbial numbers of *B. subtilis* and *B. licheniformis* in assay diets are presented in Table 3. As expected, CP and AA contents in LP were approximately 76.5 and 76.6% that of HP, respectively. The contents of ash, EE, NDF, ADF, and ADL in the HP diets were also greater than in the LP diets. The *Bacillus* spores determined in the experimental diets amounted to 1.54×10^9 CFU/kg feed for HP + and LP + diets, whereas HP - and LP - diets contained 0.02×10^9 and 0.04×10^9 CFU/kg feed, respectively.

The AID and SID of CP and AA in the assay diets are shown in Tables 4 and 5, respectively. The supplementation with *Bacillus* spp. did not affect both AID and SID of CP and AA. Furthermore, there was no difference in AID of CP and AA between HP and LP diets,

Table 3 Analyzed chemical composition and *Bacillus* cell numbers in assay diets

Item	High-protein		Low-protein	
	-	+	-	+
Dry matter, %	88.6	88.7	88.3	88.6
Crude protein, % DM	20.6	20.3	15.2	16.1
Ash, % DM	6.1	6.0	5.2	5.3
Ether extract, % DM	3.7	3.6	3.2	3.3
Neutral detergent fiber, % DM	12.7	13.1	10.1	10.5
Acid detergent fiber, % DM	7.0	6.6	5.4	5.1
Acid detergent lignin, % DM	1.1	0.9	0.8	0.8
Indispensable amino acids, % DM				
Arginine	1.26	1.25	0.93	0.99
Histidine	0.46	0.46	0.35	0.36
Isoleucine	0.80	0.80	0.61	0.63
Leucine	1.53	1.53	1.15	1.19
Lysine	1.48	1.50	1.12	1.12
Methionine	0.50	0.51	0.38	0.37
Phenylalanine	0.95	0.95	0.69	0.74
Threonine	0.91	0.91	0.68	0.70
Tryptophan	0.27	0.27	0.20	0.21
Valine	1.02	1.01	0.76	0.79
Dispensable amino acids, % DM				
Alanine	0.80	0.79	0.60	0.63
Aspartic acid	1.73	1.71	1.29	1.37
Cystine	0.33	0.33	0.25	0.26
Glutamic acid	4.08	4.03	3.05	3.20
Glycine	0.81	0.80	0.61	0.64
Proline	1.32	1.31	0.99	1.04
Serine	0.93	0.91	0.69	0.73
Bacillus cell numbers, CFU/kg feed				
Bacillus subtilis	0.022×10^9	0.860×10^9	0.038×10^9	0.970×10^9
Bacillus licheniformis	$<0.002 \times 10^9$	0.680×10^9	0.006×10^9	0.570×10^9

but SID of cystine, glutamic acid, glycine, and proline was lower ($P < 0.05$) in the HP diets than in the LP diets. Moreover, SID of CP, alanine, aspartic acid, and serine also tended to be lower ($P < 0.10$) in the HP diets. However, no interactions between CP level and *Bacillus* spp. supplementation could be observed for AID and SID of CP and AA in the present study.

The overall structure in bacterial communities from ileal digesta was evaluated by 16S rRNA gene amplicon sequencing. Analysis of similarity revealed significant differences in microbiota composition due to different dietary treatments ($P = 0.05$), but a statistic R value close to zero ($R = 0.176$) suggests a weak separation of the different treatment groups (Fig. 1a).

When compared with the starter period, bacterial communities were different ($P < 0.01$) between dietary

treatments. Within assay diets, however, there were no effects (Table 6).

Taxonomical composition of ileal digesta samples demonstrated some variation among dietary treatments. At phylum level, the bacterial communities were dominated by *Firmicutes* and *Bacteroidetes*. Within the assay diets from periods 1 to 3, the relative abundance of *Firmicutes* was higher than *Bacteroidetes* when compared to the starter period. The reduction of *Bacteroidetes* was mainly due to lower abundance of *Prevotellaceae*, with an average abundance of 27% in the starter diet compared to 5% in the HP diets, 4% in LP - and 11% in LP +. Nine bacterial families contributed to the overall dissimilarities among microbiota structure in ileal digesta samples of different dietary treatments (Fig. 1b). Ileal microbiota from dietary

Table 4 Apparent ileal digestibility of crude protein and amino acids of the assay diets[a]

Item	High-protein		Low-protein		SEM	P-value		
	-	+	-	+		P[1]	B[2]	P × B[3]
Crude protein	76.4	75.4	80.0	76.6	2.09	0.273	0.310	0.573
Indispensable amino acids								
Arginine	85.4	84.8	87.1	84.7	1.35	0.563	0.286	0.534
Histidine	80.4	79.2	82.9	79.7	1.70	0.398	0.211	0.551
Isoleucine	79.3	78.6	82.5	79.1	1.96	0.356	0.313	0.513
Leucine	81.1	80.3	83.8	80.6	1.73	0.389	0.269	0.501
Lysine	84.8	84.3	87.5	84.2	1.38	0.351	0.194	0.327
Methionine	89.7	89.3	91.4	89.0	1.03	0.492	0.200	0.349
Phenylalanine	78.6	78.1	82.0	79.0	1.99	0.295	0.386	0.522
Threonine	76.6	75.7	79.5	75.3	2.09	0.546	0.235	0.446
Tryptophan	75.1	73.2	78.1	73.8	2.47	0.470	0.226	0.628
Valine	78.8	78.9	81.9	78.1	1.93	0.403	0.232	0.471
Dispensable amino acids								
Alanine	70.6	68.9	75.5	71.0	2.66	0.197	0.257	0.605
Aspartic acid	74.6	73.4	78.6	74.7	2.32	0.263	0.284	0.566
Cystine	74.3	72.6	78.9	74.5	2.33	0.176	0.200	0.568
Glutamic acid	86.4	85.7	88.9	86.9	1.22	0.148	0.267	0.601
Glycine	65.5	63.4	70.1	64.3	2.91	0.352	0.188	0.521
Proline	82.7	81.4	85.3	82.2	1.64	0.312	0.197	0.611
Serine	76.7	75.2	79.7	76.3	2.10	0.341	0.244	0.665

[1]P-value of protein level
[2]P-value of probiotic supplementation with Bacillus spp.
[3]P-value of interaction between protein level and probiotic supplementation with Bacillus spp.
[a]LS means and standard error of the means, %

Table 5 Standardized ileal digestibility of crude protein and amino acids of the assay diets[a]

Item	High-protein		Low-protein		SEM	P-value		
	-	+	-	+		P[1]	B[2]	P × B[3]
Crude protein	82.1	81.3	87.8	83.9	2.09	0.063	0.274	0.488
Indispensable amino acids								
Arginine	88.5	87.9	91.3	88.7	1.35	0.210	0.254	0.474
Histidine	84.5	83.4	88.4	84.9	1.70	0.128	0.187	0.492
Isoleucine	84.0	83.3	88.8	85.2	1.96	0.110	0.289	0.474
Leucine	84.3	83.5	88.0	84.7	1.73	0.164	0.255	0.472
Lysine	87.5	87.0	91.1	87.8	1.38	0.126	0.192	0.333
Methionine	91.9	91.5	94.3	92.0	1.03	0.166	0.201	0.363
Phenylalanine	82.2	81.7	86.9	83.6	1.99	0.112	0.348	0.474
Threonine	83.3	82.4	88.5	84.1	2.09	0.114	0.212	0.410
Tryptophan	80.3	78.4	85.0	80.3	2.47	0.195	0.202	0.584
Valine	84.2	83.2	89.0	85.0	1.93	0.106	0.218	0.436
Dispensable amino acids								
Alanine	76.9	75.2	83.9	79.0	2.66	0.056	0.232	0.549
Aspartic acid	79.3	78.2	85.0	80.7	2.32	0.094	0.259	0.509
Cystine	80.7	78.9	87.5	82.5	2.33	0.037	0.166	0.499
Glutamic acid	89.5	88.8	93.1	90.8	1.22	0.034	0.242	0.539
Glycine	76.8	74.8	85.3	78.7	2.91	0.045	0.159	0.436
Proline	91.4	90.1	96.8	93.2	1.64	0.018	0.158	0.486
Serine	84.1	82.6	89.6	85.6	2.10	0.055	0.209	0.547

[1]P-value of protein level
[2]P-value of probiotic supplementation with Bacillus spp.
[3]P-value of interaction between protein level and probiotic supplementation with Bacillus spp.
[a]LS means and standard error of the means, %

treatments without probiotic supplementation consisted mainly of *Peptostreptococcaceae*, *Clostridiaceae* 1, *Streptococcaceae*, *Lactobacillaceae* and *Erysipelotrichaceae* with even proportions, except for *Peptostreptococcaceae* and *Streptococcaceae* being the predominant family in the HP and LP treatment, respectively. *Streptococcus alactolyticus* accounted for 15% of total microbiota in samples of LP - treatment. Compared to this, ileal digesta samples from LP + were enhanced in *Clostridiaceae* 1, *Erysipelotrichaceae* and *Prevotellaceae*. In HP +, the bacterial composition was dominated by *Lactobacillaceae*, with an average abundance of 40%. Here, an uncultured *Lactobacillus* from porcine intestine (relative abundance of 21.5%) and *Lactobacillus amylovorus* (14.2%) were the predominant species.

Lactobacillus spp. and other bacteria groups of interest were quantified in ileal digesta by qPCR (Table 7). Mantel test showed a significant correlation between the two approaches, sequencing and qPCR (Rho = 0.852, P < 0.01), thus confirming that both methodological approaches resulted in comparable results. The HP diets increased abundance of *Lactobacillus* spp.

and *Bifidobacterium* spp. (P < 0.05). No effects of CP content on ileal gene copy numbers of total bacteria, *Roseburia* spp., *Enterobacteriaceae*, *Bacteroides-Prevotella-Porphyromonas*, *Clostridium* cluster IV and *Bacillus* spp. were found. Likewise, no significant effect of supplementation of *Bacillus* spp. was observed for ileal gene copy numbers of total bacteria, *Lactobacillus* spp., *Bifidobacterium* spp., *Enterobacteriaceae*, *Clostridium* cluster IV and *Bacillus* spp.. However, dietary supplementation of *Bacillus* spp. increased (P < 0.05) abundance of *Roseburia* spp., while it tended (P < 0.10) to promote *Bacillus* spp. and total bacteria. Furthermore, there was an interaction (P < 0.05) of protein level and *Bacillus* spp. supplementation for ileal gene copy numbers of *Bacteroides-Prevotella-Porphyromonas*. The LP + resulted in higher (P < 0.05) abundance of *Bacteroides-Prevotella-Porphyromonas* than the LP -, but did not differ from the HP diets.

The analysis of fecal microbiota by 16S rRNA gene amplicon sequencing showed no statistical effect on overall community structure. However, feces microbiota from each assay diet revealed to be significant different

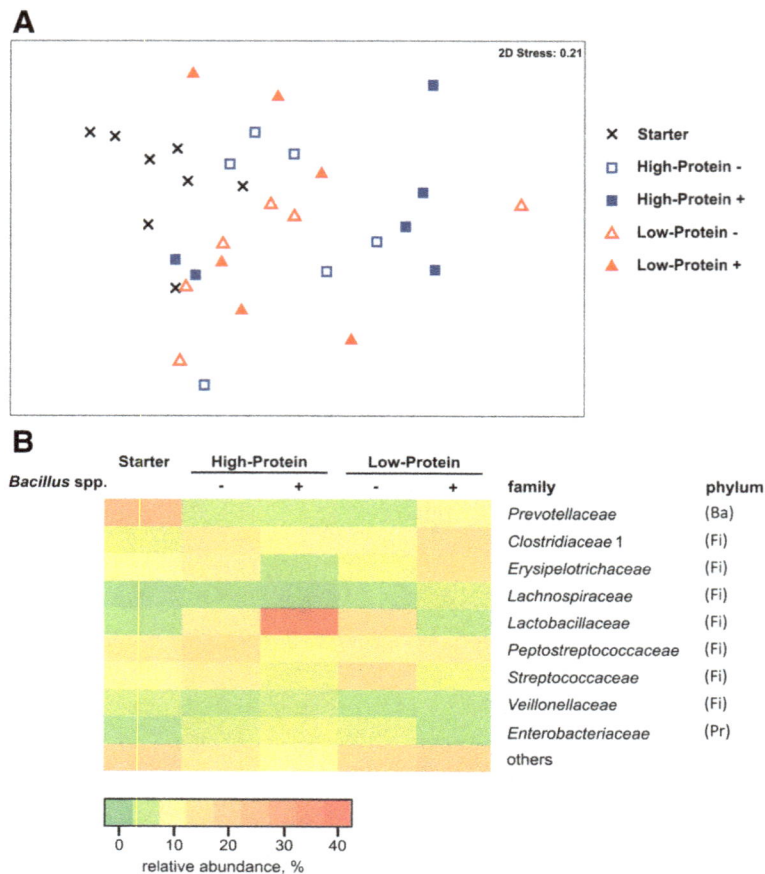

Fig. 1 Microbiota composition in ileal digesta samples from pigs fed starter diet and assay diets. **a** Multidimensional scaling plot based on Bray Curtis similarity matrix of 16S rDNA sequence data from ileal digesta. **b** Abundance plot of most important bacterial families in overall microbiota structure of ileal digesta. Phyla: *Firmicutes* (Fi), *Bacteroidetes* (Ba), *Proteobacteria* (Pr)

to that from the starter period ($P < 0.01$; Fig. 2a). At family level, *Prevotellaceae* exhibited the strongest impact on these dissimilarities (Fig. 2b), with *Prevotella* being the predominant genus. The average abundances of *Prevotella* showed slight variations for treatment groups with different protein levels; starter (15%), LP diets (13%), and HP diets (19%).

Comparing sequencing results from porcine ileal digesta and feces revealed distinct differences in bacterial communities structure ($R = 0.924$, $P < 0.01$) (Fig. 3a).

Samples from ileal digesta showed a lower diversity compared to feces (Shannon index in average 2.9 vs. 4.7) (Fig. 3b and c). Mainly *Streptococcus alactolyticus* contributed to the dissimilarity with an average abundance of 9.7% in ileal digesta compared to 2.0% in feces. At family level differences were mainly due to *Lactobacillaceae* and *Ruminococcaceae*. The abundance of *Lactobacillaceae* was higher in ileal digesta (16%) than in feces (2%), and contrary the abundance of *Ruminococcaceae* was higher in feces (23%) than in ileal digesta (0.5%).

Table 6 Results from PERMANOVA test for dietary effect on 16S rRNA sequencing data from ileal digesta

Source	Degrees of freedom	Sum of squares	Mean square	Pseudo-F	P(perm)	Unique perms
P[1]	1	2022.4	2022.4	0.770	0.638	998
B[2]	1	1340.1	1340.1	0.511	0.901	998
P × B[3]	1	2691.9	2691.9	10.255	0.424	999
Res	20	52,497	2624.8			
Total	23	58,551				

[1]P(perm)-value of protein level
[2]P(perm)-value of probiotic supplementation with *Bacillus* spp.
[3]P(perm)-value of interaction between protein level and probiotic supplementation with *Bacillus* spp.

Table 7 Ileal gene copy numbers[a] in ileal digesta of growing pigs

Item	High-protein		Low-protein		SEM	P-value		
	-	+	-	+		P[1]	B[2]	P × B[3]
Total bacteria	8.9	9.1	8.4	9.1	0.30	0.286	0.070	0.226
Lactobacillus spp.	7.9	8.8	6.9	7.1	0.44	0.002	0.109	0.279
Bifidobacterium spp.	6.2	6.4	5.3	6.0	0.32	0.024	0.179	0.354
Roseburia spp.	7.1	7.3	6.5	7.7	0.33	0.834	0.033	0.111
Enterobacteriaceae	7.7	7.9	7.4	8.3	0.43	0.836	0.139	0.274
Bacteroides-Prevotella-Porphyromonas	8.1[b,c]	8.2[b,c]	7.6[c]	8.6[b]	0.26	0.968	0.013	0.042
Clostridium cluster IV	5.6	5.7	5.2	5.9	0.30	0.735	0.185	0.324
Bacillus spp.	8.0	8.3	7.5	8.1	0.23	0.100	0.054	0.498

[1]P-value of protein level
[2]P-value of probiotic supplementation with Bacillus spp.
[3]P-value of interaction between protein level and probiotic supplementation with Bacillus spp.
[a]$\log_{10}$ 16S rRNA gene copies/g digesta (LS means and standard error of the means)
[b,c]Within a row, LS means with a common superscript are not different at $\alpha = 0.05$

Discussion

According to previous studies, *B. subtilis* and *B. licheniformis* produce extracellular enzymes including proteases and α-amylase [19, 20], which may enhance nutrient digestibility resulting in improved feed conversion in finisher pigs [40]. In addition, *B. subtilis* exceeds *B. licheniformis* in production of glycosyl hydrolases [4], which assist in the hydrolysis of glycosidic bonds in complex sugars. However concerning antibiotic resistance, which is considered to be an important key requirement for probiotics, a higher concentration of antibiotics is tolerated by *B. licheniformis* than by *B. subtilis* [4]. Recently, probiotic characteristics were described for spores of *B. subtilis*, although interactions with porcine epithelial cells are not understood so far [4]. For example, the supplementation of *B. subtilis* to a soybean meal diet showed slight improvements in AID and SID of some AA in weaning pigs as described by Kim et al. [21]. However, in the present study, there was no improvement in AID and SID of CP and AA in growing pigs fed diets supplemented with *B. subtilis* and *B. licheniformis*. Similarly, previous studies [6, 41] failed to demonstrate that the inclusion of *B. subtilis* and *B. licheniformis* in diets would affect apparent total tract digestibility of CP in growing-finishing pigs. The lack of probiotic treatment effects may be due to low quantity of the supplemented bacterial species in porcine intestine, as in treatments with probiotic supplementation the *Bacillus* spp. numbers were not significantly higher compared to numbers in treatments without probiotic supplementation. The gene copy numbers of *Bacillus* spp. in the treatments without probiotic supplementation correspond to results of a study by Dowd et al. [42] on *Bacillus* spp. in the ileum of piglets using 16S rRNA gene sequencing. In addition to the qPCR results, further *Bacillus* species (*B. pumilus* and *B. cereus*) were identified by amplicon sequencing. Operative taxonomic units corresponding to *Bacillus* genus appeared in very low abundance (<1%), and were present in samples with and without *Bacillus* spp. supplementation. These results are in accordance with previous studies demonstrating the ability of germinated *Bacillus* spores to proliferate in mammal GIT, even if only at a low rate [5], and therefore might not be persistent [43].

Positive effects of diets supplemented with *B. subtilis* and *B. licheniformis* on feed conversion in pigs have been reported before [40, 44], however, the underlying mechanisms of *Bacillus*' probiotic activity are little understood, and may be attributed to competitive adhesion and immunomodulation by *Bacillus* spores or to enzymes and other substances produced by the germinated, vegetative cells of *Bacillus* [5]. Notably, probiotic supplements may be more effective under stress such as practical field conditions [45, 46]. This might be one reason for the missing effect of *Bacillus* spp. supplementation on digestibility values in the present study, as pigs were individually housed and kept in a clean environment under optimal temperature and minimal stress conditions. Furthermore, the age of pigs may be associated with probiotic efficacy [47]. The use of probiotics tended to be more effective in early age of pigs rather than the growing period [48, 49]. In the present study, grower pigs (13- and 20-week old at the initial and final BW, respectively) fed diets supplemented with *Bacillus* spp. did not show any differences in ileal digestibility of CP and AA. It has been suggested that increasing age may be a contributing factor in building up the complexity of the microbial community [50] with growing pigs being more resistant to intestinal disorders than young pigs [51].

Dietary content of CP has been reported to be associated with AID due to the variation in endogenous CP and AA losses in ileal digesta [52]. Previous research [53] suggests that AID shows segmented quadratic with plateau relationships as the level of CP and AA in the diet

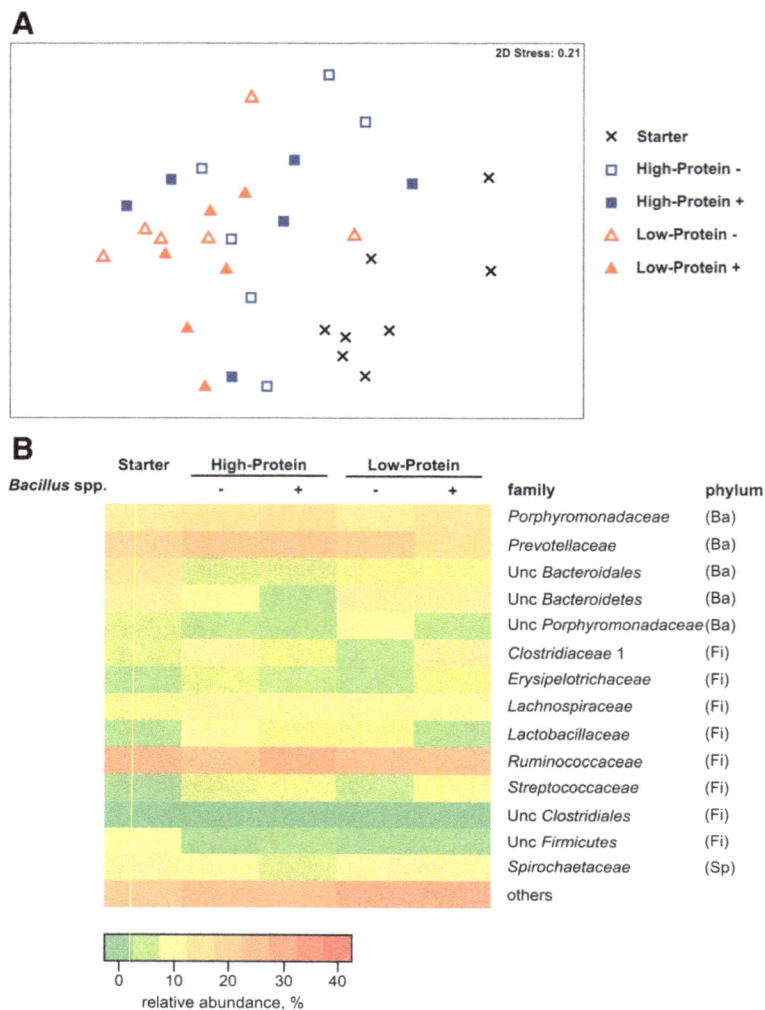

Fig. 2 Microbiota composition in fecal samples from pigs fed starter diet and assay diets. **a** Multidimensional scaling plot based on Bray Curtis similarity matrix of 16S rDNA sequence data from fecal samples. **b** Abundance plot of most important bacterial families in overall microbiota structure of feces. Phyla: *Firmicutes* (Fi), *Bacteroidetes* (Ba), *Spirochaetes* (Sp)

increased from 4 to 24% (as-fed). Alternatively, SID has been widely accepted to overcome this problem by correcting AID values for basal endogenous losses of CP and AA [54]. In general, SID values are higher in comparison to their corresponding AID values as the basal endogenous losses of CP and AA are subtracted from ileal CP and AA outflow [36]. In the present study, SID of some AA was lower in HP diets than in LP diets. Apparently, higher fiber contents in HP diets, associated with enhanced secretion of endogenous AA [55, 56], may have contributed to higher rate of digesta passage in the digestive tract of pigs [57], thereby, decreasing SID values. This is confirmed by the results of a recent study [58], where SID of CP and most AA decreased linearly with increasing dietary CP from 6.8 to 21.4% (as-fed) due to the greater NDF and ADF contents.

The higher numbers of *Lactobacillus* spp. and *Bifidobacterium* spp. in ileal digesta of HP treatments are in agreement with a recent study by Rist et al. [17], where piglets fed high dietary CP levels showed an increased growth and proliferation of lactic-acid bacteria in ileal digesta. As content of soybean meal in the present study was greater in HP than in LP diets, enhanced availability of fermentable carbohydrates in the small intestine can be suggested, thereby stimulating ileal growth of lactobacilli and bifidobacteria [17]. Furthermore, HP diets could increase the availability of free AA in the small intestine [17], contributing much more preformed AA of dietary and endogenous origin to bacterial growth in the upper part of the digestive tract than microbial *de novo* synthesis of AA [13]. Furthermore, analysis of overall microbiota composition in ileal digesta by amplicon sequencing

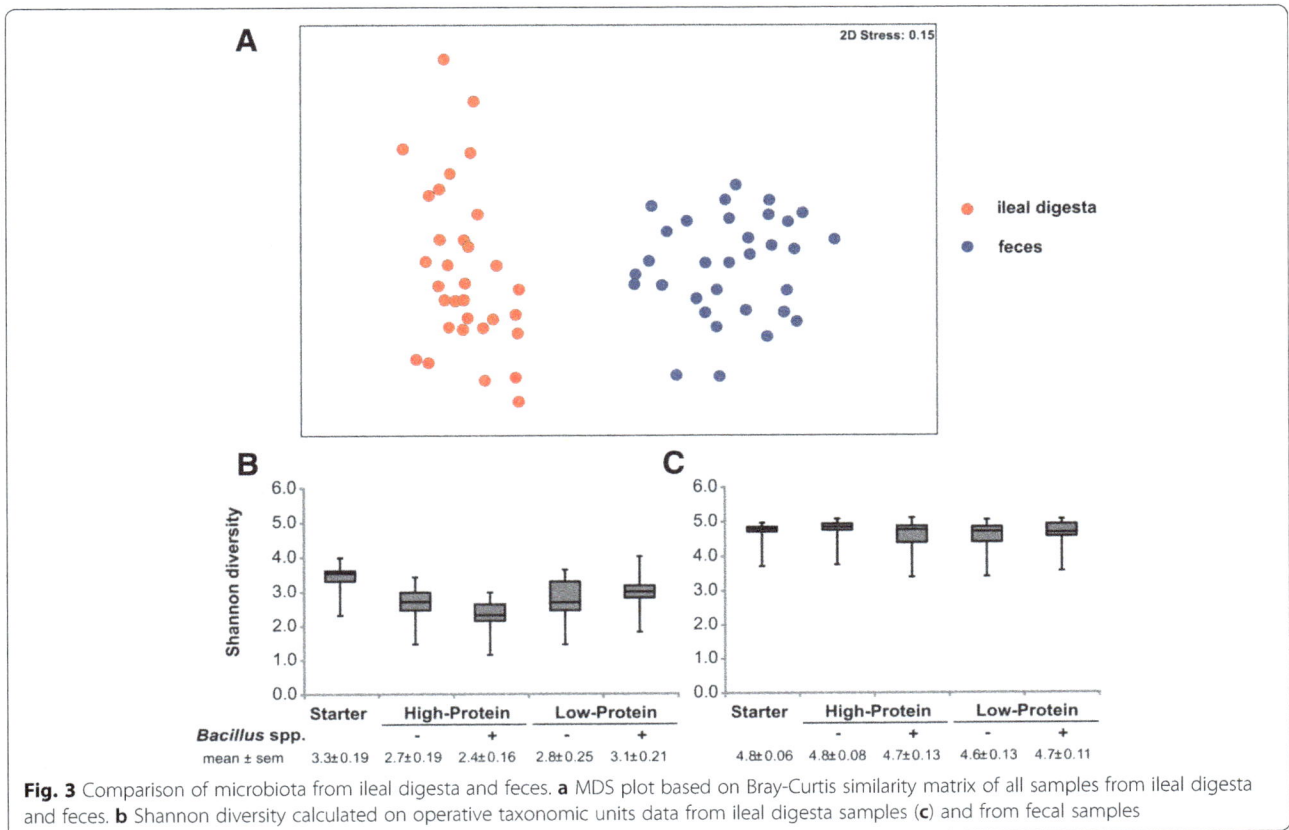

Fig. 3 Comparison of microbiota from ileal digesta and feces. **a** MDS plot based on Bray-Curtis similarity matrix of all samples from ileal digesta and feces. **b** Shannon diversity calculated on operative taxonomic units data from ileal digesta samples (**c**) and from fecal samples

supported an increasing effect on *Lactobacillus* proportion upon feeding of HP + diets. The presence of *Lactobacillus* spp. and *Bifidobacterium* spp. in the GIT has been reported to be beneficial for the host animal [17] due to their ability for bacteriocin production [59]. Moreover, proliferation of pathogenic bacteria may be inhibited through the production of short-chain fatty acids (SCFA) and lactic acid, being associated with a lower pH, causing a hostile environment for some acid-sensitive bacteria strains [60]. The presented sequencing results for *Lactobacillus* spp. are supported by qPCR results, which revealed a higher number of *Lactobacillus* gene copies in HP diets. The identified *Lactobacillus* spp. were dominated by an uncultured bacterium, previously isolated from porcine intestine [61], and the species *L. amylovorus*. *L. amylovorus* is a synonym expression for *Lactobacillus sobrius*, which is characterized by amylolytic activity, and being previously identified with high prevalence in porcine intestine [62–65]. Application of an oral probiotic mixture including a *L. amylovorus* strain has been shown to promote growth performance of pigs [66]. In general, the enhancement of potential beneficial *Lactobacillus* spp. is considered to promote gut health. However, the above described supporting effect of HP + diet on abundance of *Lactobacillus* caused a reduced community diversity compared to microbiota from ileal digesta of other dietary

treatments. A high diversity in intestinal microbiota might be preferable to cope effectively with potential challenging conditions [67].

Regardless of dietary protein level, the supplementation of *B. subtilis* and *B. licheniformis* had a stimulating effect on targeted quantity of *Roseburia* spp., known as an important butyrate producer [68]. Butyrate represents the most preferential energy source of colonocytes [69], resulting in the stimulation of epithelial cell proliferation and mucus secretion [70]. Therefore, the supplementation of *B. subtilis* and *B. licheniformis* may contribute to an improved gut health of pigs.

Assay diets did not significantly impact overall microbiota, but influence was demonstrated for bacterial copy numbers of *Bacteroides-Prevotella-Porphyromonas*. In the present study, the dietary CP level and the supplementation of *B. subtilis* and *B. licheniformis* showed an interaction, as supplementation of *B. subtilis* and *B. licheniformis* increased *Bacteroides-Prevotella-Porphyromonas* in the LP diets when compared to HP diets. The *Bacteroides-Prevotella-Porphyromonas* group includes phylogenetic related species from *Bacteroidetes* phylum that commonly inhabit GIT. Sequencing results confirmed an increased abundance of *Prevotella* in ileal digesta from LP + treatment when compared to the other assay diets. This finding is in agreement with other studies, which showed an enhancing effect of low protein diets on gene copy numbers of *Bacteroides-Prevotella-Porphyromonas* group in ileal digesta [17], and a significant increase in the abundance of *Prevotella* genus in cecum [71] when compared to samples of treatments with a higher protein level [17, 71]. *Prevotella* dominate the porcine fecal metagenome [72], play an important role in intestinal carbohydrate fermentation [73] and also show proteolytic activity [74]. Sequencing results also revealed members of *Prevotella* as main discriminators of community structure from ileal microbiota of starter and experimental periods. The observed decrease over experimental time is in agreement with a longitudinal study of Kim et al. [75]. Thus, the observations on relative proportion of *Prevotella* represent the general impact of diet and age on porcine intestinal microbiota. Contrary to ileal digesta, where abundance of *Prevotella* was highest in LP +, the fecal proportion of *Prevotella* was higher in HP than LP treatment and slightly increased over experimental time. This variation along sampling sites is in agreement with a previous study, investigating as well ileal digesta and fecal samples from pigs [17], where abundance patterns of *Prevotella* species in the GIT of pigs were different between ileal digesta and fecal samples.

The results of this study demonstrate an overall lower bacterial diversity for ileal digesta compared with fecal samples. Metagenome studies on porcine microbiota collected from different intestine sites revealed different contributions of bacterial species and activities along the GIT [76, 77]. The fecal collection is an easy accessible sampling site with samples showing high similarity to microbiota composition from proximal intestine. However, microbiota composition from fecal samples is not identical representatives to those from ileal digesta. Therefore, collecting samples of different sites of the GIT, where close interactions between the microbiome and the digestive processes occur, will improve understanding of probable functional changes and the effects of dietary treatments such as the addition of probiotics.

Undigested dietary components passing into the large intestine are subjected to fermentation by the intestinal microbiota [17]. As a result, fermentation products such as SCFA are rapidly absorbed across the gut wall, contributing up to 30% of growing pigs' maintenance requirement for energy [78]. On the other hand, increasing protein fermentation may result in the formation of detrimental fermentation products such as ammonia and amines in the colon [79]. A lower dietary protein level may reduce ammonia production, as observed by Htoo et al. [14] in cecal samples of pigs, while supplementation of diets with *B. subtilis* and *B. licheniformis* showed similar results in slurry samples from pigs due to a lowering effect on the pH [6]. Therefore, LP diets

supplemented with *B. subtilis* and *B. licheniformis* might be used to reduce the production of harmful microbial metabolites in the large intestine of pigs.

Conclusions

Supplementation with *Bacillus* spp. did not affect both AID and SID of CP and AA in growing pigs. The higher SID of some AA in the LP diets when compared to HP diets hints towards the possibility of reducing N excretion through diet manipulation. Regarding microbiota, the assay diets had no significant effect on overall community structure, neither in ileal digesta nor feces. Nevertheless, dietary protein content and *Bacillus* spp. supplementation may enhance various community members in ileal digesta. Within this regard, feeding of the HP diet resulted in a higher abundance of *Lactobacillus* spp. and *Bifidobacterium* spp., whereas LP diet may support bacteria important for carbohydrate degradation such as *Prevotella*. Furthermore, relative proportion of *Prevotella* was altered during pig's age. The supplementation of *Bacillus* spp. promoted gene copy numbers of *Roseburia* spp., which may be beneficial due to ascribed health promoting properties of this butyrate producer, and this phenomenon may be more effective under stress condition. The LP diet supplemented with *B. subtilis* and *B. licheniformis* may be used as an alternative feeding strategy to support gut health in pigs.

Abbreviations

AA: Amino acid; ADF: Acid detergent fiber; ADL: Acid detergent lignin; AID: Apparent ileal digestibility; ANOSIM: Analysis of similarity; BW: Body weight; CFU: Colony forming units; CP: Crude protein; DM: Dry matter; GIT: Gastrointestinal tract; HP: High-protein diet; LP: Low-protein diet; ME: Metabolizable energy; N: Nitrogen; NDF: Neutral detergent fiber; OTU: Operativ taxonomic unit; PERMANOVA: Permutational multivariate analysis of variance; qPCR: Quantitative real-time PCR; RELATE: Mantel type test; SCFA: Short chain fatty acids; SID: Standardized ileal digestibility

Acknowledgements

The authors would like to thank S. Stabenow for taking care of pigs and M. Steffl and N. Nautscher for their excellent work with animal surgery. The support of M. Eklund in designing this study is acknowledged.

Funding

This research was financially supported by Chr. Hansen A/S (Hørsholm, Denmark) and the Foundation fiat panis (Ulm, Germany). The receipt of a scholarship for Chanwit Kaewtapee granted by Food Security Center (FSC), University of Hohenheim under the German Academic Exchange Service (DAAD) program exceed with funds of the Federal Ministry for Economic Cooperation and Development (BMZ) of Germany is gratefully acknowledged.

Authors' contributions

CK, KB, PRK, SH, JS and RM conceived the study, performed the statistics and drafted the manuscript. KB, GT and TH participated in the animal care and digesta collection. KB, ACS, TH and SH conducted the DNA extraction and quantitative real-time PCR. MW analyzed contents of CP and AA in diets and ileal digesta. All authors read and approved the final manuscript.

Competing interests

The authors declare that they have no competing interests.

Author details

[1]University of Hohenheim, Institute of Animal Science, Emil-Wolff-Strasse 10, 70599 Stuttgart, Germany. [2]Present address: Department of Animal Science, Faculty of Agriculture, Kasetsart University, 50 Ngam Wong Wan Rd, Chatuchak, Bangkok 10900, Thailand. [3]Present address: University of Bonn, Institute of Animal Science, Endenicher Allee 15, 53115 Bonn, Germany. [4]Evonik Nutrition & Care GmbH, Rodenbacher Chaussee 4, 63457 Hanau-Wolfgang, Germany.

References

1. Barton MD. Impact of antibiotic use in the swine industry. Curr Opin Microbiol. 2014;19:9–15. doi:10.1016/j.mib.2014.05.017 .
2. Chen YJ, Min BJ, Cho JH, Kwon OS, Son KS, Kim HJ, et al. Effects of dietary Bacillus-based probiotic on growth performance, nutrients digestibility, blood characteristics and fecal noxious gas content in finishing pigs. Asian Australas J Anim Sci. 2006;19:587–92. doi:10.5713/ajas.2006.587 .
3. Meng QW, Yan L, Ao X, Zhou TX, Wang JP, Lee JH, et al. Influence of probiotics in different energy and nutrient density diets on growth performance, nutrient digestibility, meat quality, and blood characteristics in growing-finishing pigs. J Anim Sci. 2010;88:3320–6. doi:10.2527/jas.2009-2308 .
4. Larsen N, Thorsen L, Kpikpi EN, Stuer-Lauridsen B, Cantor MD, Nielsen B, et al. Characterization of Bacillus spp. strains for use as probiotic additives in pig feed. Appl Microbiol Biotechnol. 2014;98:1105–18. doi:10.1007/s00253-013-5343-6 .
5. Leser TD, Knarreborg A, Worm J. Germination and outgrowth of Bacillus subtilis and Bacillus licheniformis spores in the gastrointestinal tract of pigs. J Appl Microbiol. 2008;104:1025–33. doi:10.1111/j.1365-2672.2007.03633.x .
6. Wang Y, Cho JH, Chen YJ, Yoo JS, Huang Y, Kim HJ, et al. The effect of probiotic BioPlus 2B® on growth performance, dry matter and nitrogen digestibility and slurry noxious gas emission in growing pigs. Livest Sci. 2009;120:35–42. doi:10.1016/j.livsci.2008.04.018 .
7. EFSA (European Food Safety Authority). Scientific Opinion on the maintenance of the list of QPS biological agents intentionally added to food and feed (2010 update). EFSA J. 2010;8:1944. doi:10.2903/j.efsa.2010.1944 .
8. Degering C, Eggert T, Puls M, Bongaerts J, Evers S, Maurer KH, et al. Optimization of protease secretion in Bacillus subtilis and Bacillus licheniformis by screening of homologous and heterologous signal peptides. Appl Environ Microbiol. 2010;76:6370–6. doi:10.1128/AEM.01146-10 .
9. Upadhaya SD, Kim SC, Valientes RA, Kim IH. The effect of Bacillus-based feed additive on growth performance, nutrient digestibility, fecal gas emission, and pen cleanup characteristics of growing-finishing pigs. Asian Australas J Anim Sci. 2015;28:999–1005. doi:10.5713/ajas.15.0066 .
10. Jørgensen JN, Laguna JS, Millán C, Casabuena O, Gracia MI. Effects of a Bacillus-based probiotic and dietary energy content on the performance and nutrient digestibility of wean to finish pigs. Anim Feed Sci Technol. 2016;221:54–61. doi:10.1016/j.anifeedsci.2016.08.008 .
11. Mosenthin R, Hambrecht E, Sauer WC. Utilisation of different fibres in piglet feeds. In: Garnsworthy PC, Wiseman J, editors. Recent advances in animal nutrition. Nottingham: Nottingham University Press; 1999. p. 227–56.
12. Wellock IJ, Fortomaris PD, Houdijk JGM, Kyriazakis I. The effect of dietary protein supply on the performance and risk of post-weaning enteric disorders in newly weaned pigs. Anim Sci. 2006;82:327–35. doi:10.1079/ASC200643 .
13. Libao-Mercado AJO, Zhu CL, Cant JP, Lapierre H, Thibault JN, Sève B, et al. Dietary and endogenous amino acids are the main contributors to microbial protein in the upper gut of normally nourished pigs. J Nutr. 2009;139:1088–94. doi:10.3945/jn.108.103267 .
14. Htoo JK, Araiza BA, Sauer WC, Rademacher M, Zhang Y, Cervantes M, et al. Effect of dietary protein content on ileal amino acid digestibility, growth performance, and formation of microbial metabolites in ileal and cecal digesta of early-weaned pigs. J Anim Sci. 2007;85:3303–12. doi:10.2527/jas.2007-0105 .
15. Ferket PR, van Heugten E, van Kempen TATG, Angel R. Nutritional strategies to reduce environmental emissions from nonruminants. J Anim Sci. 2002;80(E-Suppl 2):E168–82. doi:10.2527/animalsci2002.80E-Suppl_2E168x .
16. Lee JH, Kim JH, Kim JD, Kim SW, Han IK. Effects of low crude protein diets supplemented with synthetic amino acids on performance, nutrient

17. Rist VTS, Weiss E, Sauer N, Mosenthin R, Eklund M. Effect of dietary protein supply originating from soybean meal or casein on the intestinal microbiota of piglets. Anaerobe. 2014;25:72–9. doi:10.1016/j.anaerobe.2013.10.003 .
18. Bhandari SK, Opapeju FO, Krause DO, Nyachoti CM. Dietary protein level and probiotic supplementation effects on piglet response to Escherichia coli K88 challenge: Performance and gut microbial population. Livest Sci. 2010;133:185–8. doi:10.1016/j.livsci.2010.06.060 .
19. Priest FG. Extracellular enzyme synthesis in the genus Bacillus. Bacteriol Rev. 1977;41:711–53.
20. Carlisle GE, Falkinham III JO. Enzyme activities and antibiotic susceptibility of colonial variants of Bacillus subtilis and Bacillus licheniformis. Appl Environ Microbiol. 1989;55:3026–8.
21. Kim DH, Heo PS, Jang JC, Jin SS, Hong JS, Kim YY. Effect of different soybean meal type on ileal digestibility of amino acid in weaning pigs. J Anim Sci Technol. 2015;57:11. doi:10.1186/s40781-015-0041-9 .
22. Council Directive. Council Directive of 24 November 1986 on the approximation of laws, regulations and administrative provisions of the Member States regarding the protection of animals used for experimental and other scientific purposes (86/609/EEC). Off J Eur Union. 1986;L 358:1–28.
23. Li S, Sauer WC, Fan MZ. The effect of dietary crude protein level on ileal and fecal amino acid digestibility in early-weaned pigs. J Anim Physiol Anim Nutr. 1993;70:117–28. doi:10.1111/j.1439-0396.1993.tb00314.x .
24. NRC. Nutrient requirements of swine. 10th ed. Washington: National Academies Press; 1998.
25. Fan MZ, Sauer WC, Hardin RT, Lien KA. Determination of apparent ileal amino acid digestibility in pigs: effect of dietary amino acid level. J Anim Sci. 1994;72:2851–9. doi:10.2527/1994.72112851x .
26. NRC. Nutrient requirements of swine. 11th ed. Washington: National Academies Press; 2012.
27. Verband Deutscher Landwirtschaftlicher Untersuchungs- und Forschungsanstalten (VDLUFA). Handbuch der Landwirtschaftlichen Versuchs- und Untersuchungsmethodik (VDLUFA-Methodenbuch), Bd. III Die chemische Untersuchung von Futtermitteln mit 1. – 8. Ergänzungslieferung (1983 – 2012). (In German). 3rd ed. Darmstadt: VDLUFA-Verlag; 1976.
28. AOAC. Official methods of analysis. 17th ed. Gaithersberg: AOAC International; 2000.
29. Llames CR, Fontaine J. Determination of amino acids in feeds: collaborative study. J Assoc Off Anal Chem. 1994;77:1362–402.
30. Commission Directive. Commission Directive 2000/45/EC of 6 July 2000 on establishing community methods of analysis for the determination of vitamin A, vitamin E and tryptophan in feedingstuffs. Annex part C. Determination of tryptophan. Off J Eur Union. 2000;L 174:32–50.
31. Brandt M, Allam SM. Analytik von TiO_2 im Darminhalt und Kot nach Kjeldahlaufschluß. Arch Anim Nutr. 1987;37:453–4.
32. Burbach K, Seifert J, Pieper DH, Camarinha-Silva A. Evaluation of DAN extraction kits and phylogenetic diversity of the porcine gastrointestinal tract based on Illumina sequencing of two hypervariable regions. Microbiologyopen. 2016;5:70–82. doi:10.1002/mbo3.312 .
33. Camarinha-Silva A, Jáuregui R, Chaves-Moreno D, Oxley APA, Schaumburg F, Becker K, et al. Comparing the anterior nare bacterial community of two discrete human populations using Illumina amplicon sequencing. Environ Microbiol. 2014;16:2939–52. doi:10.1111/1462-2920.12362 .
34. Schloss PD, Westcott SL, Ryabin T, Hall JR, Hartmann M, Hollister EB, et al. Introducing mothur: open-source, platform-independent, community-supported software for describing and comparing microbial communities. Appl Environ Microbiol. 2009;75:7537–41. doi:10.1128/AEM.01541-09 .
35. Wang Q, Garrity GM, Tiedje JM, Cole JR. Naïve Bayesian classifier for rapid assignment of rRNA sequences into the new bacterial taxonomy. Appl Environ Microbiol. 2007;73:5261–7. doi:10.1128/AEM.00062-07 .
36. Stein HH, Sève B, Fuller MF, Moughan PJ, de Lange CFM. Invited review: amino acid bioavailability and digestibility in pig feed ingredients: terminology and application. J Anim Sci. 2007;85:172–80. doi:10.2527/jas.2005-742 .
37. Jansman AJM, Smink W, van Leeuwen P, Rademacher M. Evaluation through literature data of the amount and amino acid composition of basal endogenous crude protein at the terminal ileum of pigs. Anim Feed Sci Technol. 2002;98:49–60. doi:10.1016/S0377-8401(02)00015-9 .

38. Piepho H-P. A SAS macro for generating letter displays of pairwise mean comparisons. CBCS. 2012;7:4–13.

39. Clarke KR, Warwick RM. Change in marine communities: an approach to statistical analysis and interpretation. Plymouth: PRIMER-E Ltd; 2001.

40. Davis ME, Parrott T, Brown DC, de Rodas BZ, Johnson ZB, Maxwell CV, et al. Effect of a *Bacillus*-based direct-fed microbial feed supplement on growth performance and pen cleaning characteristics of growing-finishing pigs. J Anim Sci. 2008;86:1459–67. doi:10.2527/jas.2007-0603 .

41. Kornegay ET, Risley CR. Nutrient digestibilities of a corn-soybean meal diet as influenced by *Bacillus* products fed to finishing swine. J Anim Sci. 1996; 74:799–805. doi:10.2527/1996.744799x .

42. Dowd SE, Sun Y, Wolcott RD, Domingo A, Carroll JA. Bacterial tag-encoded FLX amplicon pyrosequencing (bTEFAP) for microbiome studies: bacterial diversity in the ileum of newly weaned *Salmonella*-infected pigs. Foodborne Pathog Dis. 2008;5:459–72. doi:10.1089/fpd.2008.0107 .

43. Tam NKM, Uyen NQ, Hong HA, Le Duc H, Hoa TT, Serra CR, et al. The intestinal life cycle of *Bacillus subtilis* and close relatives. J Bacteriol. 2006;188:2692–700. doi:10.1128/JB.188.7.2692-2700.2006 .

44. Alexopoulos C, Georgoulakis IE, Tzivara A, Kyriakis CS, Govaris A, Kyriakis SC. Field evaluation of the effect of a probiotic-containing *Bacillus licheniformis* and *Bacillus subtilis* spores on the health status, performance, and carcass quality of grower and finisher pigs. J Vet Med. 2004;A 51:306–12. doi:10.1111/j.1439-0442.2004.00637.x .

45. Shon KS, Hong JW, Kwon OS, Min BJ, Lee WB, Kim IH, et al. Effects of *Lactobacillus reuteri*-based direct-fed microbial supplementation for growing-finishing pigs. Asian Australas J Anim Sci. 2005;18:370–4. doi:10.5713/ajas.2005.370 .

46. Weiss E, Eklund M, Semaskaite A, Urbaityte R, Metzler-Zebeli B, Sauer N, et al. Combinations of feed additives affect ileal fibre digestibility and bacterial numbers in ileal digesta of piglets. Czech J Anim Sci. 2013;58:351–9.

47. Lessard M, Brisson GJ. Effect of a *Lactobacillus* fermentation product on growth, immune response and fecal enzyme activity in weaned pigs. Can J Anim Sci. 1987;67:509–16. doi:10.4141/cjas87-049 .

48. Pollman DS, Danielson DM, Peo Jr ER. Effects of microbial feed additives on performance of starter and growing-finishing pigs. J Anim Sci. 1980;51:557–81.

49. Link R, Kováč G. The effect of probiotic BioPlus 2B on feed efficiency and metabolic parameters in swine. Biologia. 2006;61:783–7. doi:10.2478/s11756-006-0158-x .

50. Baker AA, Davis E, Spencer JD, Moser R, Rehberger T. The effect of a *Bacillus*-based direct-fed microbial supplemented to sows on the gastrointestinal microbiota of their neonatal piglets. J Anim Sci. 2013;91:3390–9. doi:10.2527/jas2012-5821.

51. Nousiainen J, Setälä J. Lactic acid bacteria as animal probiotics. In: Salminen S, von Wright A, editors. Lactic acid bacteria. New York: Marcel Dekker; 1993. p. 315–56.

52. Eklund M, Mosenthin R, Piepho H-P, Rademacher M. Estimates of dietary threshold levels for crude protein and amino acids to obtain plateau values of apparent ileal crude protein and amino acid digestibilities in newly weaned pigs. Arch Anim Nutr. 2010;64:357–72. doi:10.1080/1745039X.2010.492139 .

53. Fan MZ, Sauer WC, McBurney MI. Estimation by regression analysis of endogenous amino acid levels in digesta collected from the distal ileum of pigs. J Anim Sci. 1995;73:2319–28. doi:10.2527/1995.7382319x .

54. Eklund M, Mosenthin R, Piepho H-P, Rademacher M. Effect of dietary crude protein level on basal ileal endogenous losses and standardized ileal digestibilities of crude protein and amino acids in newly weaned pigs. J Anim Physiol Anim Nutr. 2008;92:578–90. doi:10.1111/j.1439-0396.2007.00751.x.

55. Grala W, Verstegen MWA, Jansman AJM, Huisman J, van Leeusen P. Ileal apparent protein and amino acid digestibilities and endogenous nitrogen losses in pigs fed soybean and rapeseed products. J Anim Sci. 1998;76:557–68. doi:10.2527/1998.762557x.

56. Jondreville C, van den Broecke J, Gâtel F, Grosjean F, Van Cauwenberghe S, Sève B. Ileal amino acid digestibility and estimates of endogenous amino acid losses in pigs fed rapeseed meal, sunflower meal and soybean meal. Can J Anim Sci. 2000;80:495–506. doi:10.4141/A99-104 .

57. Calvert CC. Fiber utilization by swine. In: Miller ER, Ullrey DE, Lewis AJ, editors. Swine nutrition. Massachusetts: Butterworth-Heinemann; 1991. p. 285–96.

58. Zhai H, Adeola O. Apparent and standardized ileal digestibilities of amino acids for pigs fed corn- and soybean meal-based diets at varying crude protein levels. J Anim Sci. 2011;89:3626–33. doi:10.2527/jas.2010-3732 .

59. Vuotto C, Longo F, Donelli G. Probiotics to counteract biofilm-associated

infections: promising and conflicting data. Int J Oral Sci. 2014;6:189–94. doi:10.1038/ijos.2014.52 .

60. Kleerebezem M, Vaughan EE. Probiotic and gut lactobacilli and bifidobacteria: molecular approaches to study diversity and activity. Annu Rev Microbiol. 2009;63:269–90. doi:10.1146/annurev.micro.091208.073341 .

61. Leser TD, Amenuvor JZ, Jensen TK, Lindecrona RH, Boye M, Møller K. Culture-independent analysis of gut bacteria: the pig gastrointestinal tract microbiota revisited. Appl Environ Microbiol. 2002;68:673–90. doi:10.1128/AEM.68.2.673-690.2002 .

62. Jakava-Viljanen M, Murros A, Palva A, Björkroth KJ. *Lactobacillus sobrius* Konstantinov et al. 2006 is a later synonym of *Lactobacillus amylovorus* Nakamura 1981. Int J Syst Evol Microbiol. 2008;58:910–3. doi:10.1099/ijs.0.65432-0 .

63. Nakamula LK. *Lactobacillus amylovorus*, a new starch-hydrolyzing species from cattle waste-corn fermentations. Int J Syst Evol Microbiol. 1981;31:56–63. doi:10.1099/00207713-31-1-56 .

64. Mann E, Schmitz-Esser S, Zebeli Q, Wagner M, Ritzmann M, Metzler-Zebeli BU. Mucosa-associated bacterial microbiome of the gastrointestinal tract of weaned pigs and dynamics linked to dietary calcium-phosphorus. PLoS One. 2014;9:e86950. doi:10.1371/journal.pone.0086950 .

65. Konstantinov SR, Awati AA, Williams BA, Miller BG, Jones P, Stokes CR, et al. Post-natal development of the porcine microbiota composition and activities. Environ Microbiol. 2006;8:1191–9. doi:10.1111/j.1462-2920.2006.01009.x .

66. Ross GR, Gusils C, Oliszewski R, de Holgado SC, González SN. Effects of probiotic administration in swine. J Biosci Bioeng. 2010;109:545–9. doi:10.1016/j.jbiosc.2009.11.007 .

67. Lozupone CA, Stombaugh JI, Gordon JI, Jansson JK, Knight R. Diversity, stability and resilience of the human gut microbiota. Nature. 2012;489:220–30. doi:10.1038/nature11550 .

68. Flint HJ, Scott KP, Louis P, Duncan SH. The role of the gut microbiota in nutrition and health. Nat Rev Gastroenterol Hepatol. 2012;9:577–89. doi:10.1038/nrgastro.2012.156 .

69. Roediger WEW. Utilization of nutrients by isolated epithelial cells of the rat colon. Gastroenterology. 1982;83:424–9.

70. Tsukahara T, Hashizume K, Koyama H, Ushida K. Stimulation of butyrate production through the metabolic interaction among lactic acid bacteria, *Lactobacillus acidophilus*, and lactic acid-utilizing bacteria, *Megasphaera elsdenii*, in porcine cecal digesta. Anim Sci J. 2006;77:454–61. doi:10.1111/j.1740-0929.2006.00372.x .

71. Zhou L, Fang L, Sun Y, Su Y, Zhu W. Effects of the dietary protein level on the microbial composition and metabolomic profile in the hindgut of the pig. Anaerobe. 2016;38:61–9. doi:10.1016/j.anaerobe.2015.12.009 .

72. Lamendella R, Domingo JWS, Ghosh S, Martinson J, Oerther DB. Comparative fecal metagenomics unveils unique functional capacity of the swine gut. BMC Microbiol. 2011;11:103. doi:10.1186/1471-2180-11-103 .

73. De Filippo C, Cavalieri D, Di Paola M, Ramazzotti M, Poullet JB, Massart S, et al. Impact of diet in shaping gut microbiota revealed by a comparative study in children from Europe and rural Africa. Proc Natl Acad Sci U S A. 2010;107:14691–6. doi:10.1073/pnas.1005963107 .

74. Purushe J, Fouts DE, Morrison M, White BA, Mackie RI, the North American Consortium for Rumen Bacteria, et al. Comparative genome analysis of *Prevotella ruminicola* and *Prevotella bryantii*: insights into their environmental niche. Microb Ecol. 2010;60:721–9. doi:10.1007/s00248-010-9692-8 .

75. Kim HB, Borewicz K, White BA, Singer RS, Sreevatsan S, Tu ZJ, et al. Longitudinal investigation of the age-related bacterial diversity in the feces of commercial pigs. Vet Microbiol. 2011;153:124–33. doi:10.1016/j.vetmic.2011.05.021 .

76. Looft T, Allen HK, Cantarel BL, Levine UY, Bayles DO, Alt DP, et al. Bacteria, phages and pigs: the effects of in-feed antibiotics on the microbiome at different gut locations. ISME J. 2014;8:1566–76. doi:10.1038/ismej.2014.12 .

77. Zhao W, Wang Y, Liu S, Huang J, Zhai Z, He C, et al. The dynamic distribution of porcine microbiota across different ages and gastrointestinal tract segments. PLoS One. 2015;10:e0117441. doi:10.1371/journal.pone.0117441 .

78. Varel VH, Yen JT. Microbial perspective on fiber utilization by swine. J Anim Sci. 1997;75:2715–22. doi:10.2527/1997.75102715x .

79. Macfarlane GT, Gibson GR, Beatty E, Cummings JH. Estimation of short-chain fatty acid production from protein by human intestinal bacteria based on branched-chain fatty acid measurements. FEMS Microbiol Lett. 1992;101:81–8. doi:10.1111/j.1574-6968.1992.tb05764.x .

80. Fuller Z, Louis P, Mihajlovski A, Rungapamestry V, Ratcliffe B, Duncan AJ. Influence of cabbage processing methods and prebiotic manipulation of

colonic microflora on glucosinolate breakdown in man. Br J Nutr. 2007;98: 364–72. doi:10.1017/S0007114507709091 .

81. Malinen E, Kassinen A, Rinttilä T, Palva A. Comparison of real-time PCR with SYBR Green I or 5'-nuclease assays and dot-blot hybridization with rDNA-targeted oligonucleotide probes in quantification of selected faecal bacteria. Microbiology. 2003;149:269–77. doi:10.1099/mic.0.25975-0 .

82. Rinttilä T, Kassinen A, Malinen E, Krogius L, Palva A. Development of an extensive set of 16S rDNA-targeted primers for quantification of pathogenic and indigenous bacteria in faecal samples by real-time PCR. J Appl Microbiol. 2004;97:1166–77. doi:10.1111/j.1365-2672.2004.02409.x .

83. Veiga P, Gallini CA, Beal C, Michaud M, Delaney ML, DuBois A, et al. Bifidobacterium animalis subsp. lactis fermented milk product reduces inflammation by altering a niche for colitogenic microbes. Proc Natl Acad Sci U S A. 2010;107:18132–7. doi:10.1073/pnas.1011737107 .

84. Bartosch S, Fite A, Macfarlane GT, McMurdo MET. Characterization of bacterial communities in feces from healthy elderly volunteers and hospitalized elderly patients by using real-time PCR and effects of antibiotic treatment on the fecal microbiota. Appl Environ Microbiol. 2004;70:3575–81. doi:10.1128/AEM.70.6. 3575-3581.2004 .

85. Ramirez-Farias C, Slezak K, Fuller Z, Duncan A, Holtrop G, Louis P. Effect of inulin on the human gut microbiota: stimulation of Bifidobacterium adolescentis and Faecalibacterium prausnitzii. Br J Nutr. 2009;101:541–50. doi:10.1017/S0007114508019880 .

86. Lay C, Sutren M, Rochet V, Saunier K, Doré J, Rigottier-Gois L. Design and validation of 16S rRNA probes to enumerate members of the Clostridium leptum subgroup in human faecal microbiota. Environ Microbiol. 2005;7: 933–46. doi:10.1111/j.1462-2920.2005.00763.x .

87. Fernández-No IC, Guarddon M, Böhme K, Cepeda A, Calo-Mata P, Barros-Velázquez J. Detection and quantification of spoilage and pathogenic Bacillus cereus, Bacillus subtilis and Bacillus licheniformis by real-time PCR. Food Microbiol. 2011;28:605–10. doi:10.1016/j.fm.2010.10.014.

7

Net energy of corn, soybean meal and rapeseed meal in growing pigs

Zhongchao Li[1], Yakui Li[1], Zhiqian Lv[1], Hu Liu[1], Jinbiao Zhao[1], Jean Noblet[2], Fenglai Wang[1], Changhua Lai[1] and Defa Li[1*]

Abstract

Background: Two experiments were conducted to estimate the net energy (NE) of corn, soybean meal, expeller-pressed rapeseed meal (EP-RSM) and solvent-extracted rapeseed meal (SE-RSM) using indirect calorimetry and to validate the NE of these four ingredients using pig growth performance.

Methods: In Exp.1, 24 barrows (initial BW = 36.4 ± 1.6 kg) were allotted to 1 of 4 diets which included a corn basal diet, a corn-soybean meal basal diet and two rapeseed meal diets containing 20% EP-RSM (9.5% ether extract) or SE-RSM (1.1% ether extract) substituted for corn and soybean meal. The design allowed the calculation of NE values of corn, soybean meal and rapeseed meals according to the difference method. In Exp.2, 175 growing pigs (initial BW = 36.0 ± 5.2 kg) were fed 1 of 5 diets for 28 d, with five pigs per pen and seven replications (pens) per treatment in order to validate the measured energy values. Diets were a corn-soybean meal diet and four diets including 10% or 20% EP-RSM and 10% or 20% SE-RSM.

Results: The NE of corn, soybean meal, EP-RSM and SE-RSM were 12.46, 11.34, 11.71 and 8.83 MJ/kg DM, respectively. The NE to ME ratio of corn (78%) was similar to tabular values, however, the NE to ME ratios of soybean meal (70%) and rapeseed meal (76%) were greater than tabular values. The greater NE value in EP-RSM than in SE-RSM is consistent with its higher EE content. Increasing EP-RSM or SE-RSM did not affect the growth performance of pigs and the caloric efficiency of NE was comparable for all diets.

Conclusions: The NE of EP-RSM was similar to soybean meal, and both were greater than SE-RSM. The DE, ME and NE values measured in Exp.1 are confirmed by results of Exp. 2 with comparable caloric efficiencies of DE, ME or NE for all diets.

Keywords: Caloric efficiency, Growing pig, Heat production, Net energy, Rapeseed meal

Background

Following soybean, the most abundant oilseed produced in the world is rapeseed, with a production of about 70 million tons in 2015 [1]. There are two primary rapeseed co-products associated with oil extraction processing methods: expeller-pressed rapeseed meal (EP-RSM) containing 10% to 15% residual oil and solvent-extraction rapeseed meal (SE-RSM) containing much less oil (1–2%) than EP-RSM. In a previous work, the DE, ME and standardized ileal digestible (SID) amino acid (AA) of SE-RSM [2] and EP-RSM [3] were evaluated. Little et al. [4] found high-protein or conventional rapeseed meal may fully replace soybean meal as protein supplements in growing-finishing pig diets without impairing pig performance or carcass quality. However, there are limited reports that estimate and compare the NE of EP-RSM and SE-RSM or that have validated the NE value of rapeseed meals using a growth trial. In addition, there is no clear agreement on the proper method of validation of energy values using growth trials. Some researchers have proposed the concept of caloric efficiency [5–8] meaning that, if the assigned energy value is correct, regardless of the test ingredient inclusion level, a similar caloric efficiency (ie, dietary energy per kg of BW gain) will be calculated among the dietary treatments. However, in these experiments, the NE value of corn and soybean meal, which are the two main

* Correspondence: defali@cau.edu.cn
[1]State Key Laboratory of Animal Nutrition, Ministry of Agriculture Feed Industry Centre, China Agricultural University, No. 2, Yuanminyuan west road, Haidian district, Beijing 100193, China
Full list of author information is available at the end of the article

ingredients in the formula, were referenced from tabular values. However, in the current experiment, we have measured the NE of all ingredients in a first experiment (Exp. 1) and used the obtained values in a second validation experiment (Exp. 2). Practically, the NE of corn, soybean meal, EP-RSM and SE-RSM were determined by indirect calorimetry and these four ingredients were used subsequently to prepare five diets fed in a growth trial.

Methods

The two experimental protocols used in this study were approved by the Institutional Animal Care and Use Committee of China Agricultural University (Beijing, China).

Exp.1: Net energy experiment

Twenty-four barrows (Duroc × Large White × Landrace, initial BW = 36.4 ± 1.6 kg) were used in this experiment conducted at the FengNing Swine Research Unit of China Agricultural University (Hebei Province, China). The 24 barrows were allotted to 1 of 4 dietary treatments with six pigs per diet and four periods. Diets included a corn basal diet supplemented with free amino acids in order to improve the amino acids balance of corn and maximize protein gain, a corn-soybean meal basal diet (to calculate the energy of soybean meal using the difference method), and two rapeseed meal diets containing 20% EP-RSM or SE-RSM substituted for corn and soybean meal (to calculate the energy of rapeseed meals using the difference method). The analyzed nutrient composition of ingredients are shown in Table 1. The ratios between SID essential amino acids and SID Lys met or exceeded the recommended amino acid ratios (NRC 2012) in all diets; but the amino acids content in the corn diet was below the standard recommendations for growing pigs (Table 2).

During each period, pigs were individually housed in metabolism crates for 16 d, which included seven initial days to adapt to the feed, metabolism crate and environmental conditions. On d 8, the pigs were transferred to the open-circuit respiration chambers [9, 10] for measurement of daily O_2 consumption and CO_2 and CH_4 productions. During this time, pigs were fed one of the four diets at 2,200 kJ ME/(kg $BW^{0.6}$·d). Total feces and urine were collected from d 9 to d 13; gas exchanges were measured over the same period for calculating heat production (HP). On d 14 and 15, pigs were fed at their maintenance requirement level (890 kJ ME/(kg $BW^{0.6}$·d) [9] in order to adapt them from the fed to the fasted state. HP was also measured at this low feed level, but the results are not included in the present paper. On the last day of each period (d 16), pigs were fasted; the HP measured during the last 8 h of d 16 from 2230 h (d 16) to 0630 h (d 17) was considered as fasting heat production (FHP). The FHP period started then 31 h after the last meal and always in the dark with expected minimal physical activity.

Table 1 Analyzed nutrient composition of ingredients used in the experiment (%, DM basis)[a]

Item	Corn	SBM	EP-RSM	SE-RSM
GE, MJ/kg	18.44	19.50	21.33	19.54
Crude protein	9.2	48.5	39.2	42.4
Ether extract	3.0	1.2	9.5	1.1
NDF	14.0	17.1	37.0	30.8
ADF	2.5	10.0	22.8	20.7
Ash	1.2	6.2	6.4	7.0
Indispensable AA				
Arg	0.39	3.81	2.75	2.64
His	0.24	1.46	1.27	1.27
Leu	1.13	3.82	2.84	2.93
Ile	0.27	2.14	1.50	1.59
Lys	0.28	3.23	2.70	2.57
Met	0.16	0.57	0.75	0.77
Phe	0.41	2.46	1.52	1.41
Thr	0.36	2.12	1.77	1.89
Trp	0.06	0.61	0.61	0.65
Val	0.42	2.34	2.03	2.14
Dispensable AA				
Ala	0.65	2.25	1.91	1.98
Asp	0.62	5.74	2.82	2.97
Cys	0.17	0.63	0.95	0.95
Glu	1.69	8.77	7.18	7.31
Gly	0.32	2.21	2.14	2.22
Pro	0.86	2.72	2.86	2.92
Ser	0.45	2.54	1.72	1.83
Tyr	0.31	1.66	1.02	0.93

[a]All samples were analyzed in duplicate, *SBM* soybean meal, *EP-RSM* expelled press rapeseed meal, *SE-RSM* solvent extracted rapeseed meal

Pigs were fed equal size meals twice daily at 15÷30 and 1530 h and had free access to water via a low-pressure nipple drinker throughout the trial. The chambers were opened for approximately 1 h per day at 0830 and 1530 h to feed the pigs and collect the feces. The O_2 consumption and CO_2 and CH_4 productions during this time were not included in the calculation of daily HP. The concentration of CO_2 in the chamber increased when the door was closed. The calculation of HP began when the concentration of CO_2 in the chamber was above 2,000 ppm [10].

During d 9 to d 13, feed refusals and spillage were collected twice daily and dried and weighed. Total but separate collections of feces and urine were conducted according to the methods described by Li et al. [10]. Feces were collected twice daily at 0830 and 1530 h when the chamber door was opened and immediately stored at −20 ° C. Urine was collected each morning at 0830 h for each pig from plastic buckets containing 50 mL of 6 N HCl and

Table 2 Ingredients and chemical composition of the Exp.1 diets[a]

Diet	Corn	SBM	EP-RSM	SE-RSM
Ingredients, %				
Corn	97.03	77.13	61.31	61.31
Soybean meal	-	20.00	15.90	15.90
EP-RSM	-	-	20.00	0.00
SE-RSM	-	-	-	20.00
Dicalcium phosphate	0.90	0.90	0.90	0.90
Limestone	0.75	0.75	0.75	0.75
Salt	0.35	0.35	0.35	0.35
Vitamins and minerals premix[b]	0.50	0.50	0.50	0.50
Lys-HCl	0.26	0.21	0.16	0.16
DL-Met	0.14	0.11	0.09	0.09
L-Thr	0.03	0.02	0.02	0.02
L-Trp	0.04	0.03	0.03	0.03
DM, %	86.69	86.93	87.95	87.27
Analyzed composition, DM basis				
CP, %	9.3	17.2	21.4	22.5
EE, %	2.9	2.8	4.0	2.3
NDF, %	12.8	16.7	21.8	18.7
ADF, %	3.0	4.6	7.9	7.3
Ash, %	3.4	4.4	5.2	5.3
GE, MJ/kg	18.12	18.23	18.73	18.36
Calculated composition				
SID Lys	0.38	0.80	0.99	0.98
SID Thr/SID Lys	0.70	0.65	0.65	0.67
SID Trp/SID Lys	0.22	0.20	0.21	0.22
SID Val/SID Lys	0.78	0.72	0.74	0.76
SID Leu/SID Lys	2.21	1.55	1.41	1.43
SID Ile/SID Lys	0.50	0.60	0.61	0.61
SID His/SID Lys	0.45	0.44	0.47	0.47
SID Met + Cys/SID lys	0.98	0.59	0.63	0.64
SID Phe + Tyr/SID Lys	1.34	1.29	1.20	1.17

[a] All samples were analyzed in duplicate, SBM, soybean meal; EP-RSM, expelled press rapeseed meal; SE-RSM, solvent extracted rapeseed meal
[b]Vitamin-mineral premix supplied the following per kg of diet: vitamin A, 5,512 IU; vitamin D_3, 2,200 IU; vitamin E, 30 IU; vitamin K_3, 2.2 mg; vitamin B_{12}, 27.6 µg; riboflavin, 4 mg; pantothenic acid, 14 mg; niacin, 30 mg; choline chloride, 400 mg; folic acid, 0.7 mg; thiamine, 1.5 mg; pyridoxine, 3 mg; biotin, 44 µg; Mn (MnO), 40 mg; Fe ($FeSO_4 \cdot H_2O$), 75 mg; Zn (ZnO), 75 mg; Cu ($CuSO_4 \cdot 5H_2O$), 100 mg; I (KI), 0.3 mg; Se (Na_2SeO_3), 0.3 mg

filtered through cotton gauze. The total urinary volume produced by each pig was measured and 5% of the daily urinary excretion was stored at −20 °C. At the end of the urinary collection, urine samples were thawed, and thoroughly mixed, and a sub-sample was saved for analysis. Urine was collected separately during the 24 h fasting state to calculate urinary nitrogen (N) losses for the calculation of FHP.

At the end of the experiment, fecal samples were thawed, mixed, weighed, and sub-samples were oven-dried for 72 h at 65 °C for measurement of dry matter fecal excretion. The feed and fecal samples were ground through a 1-mm screen prior to chemical analysis. Six chambers were available for the current experiment. Therefore, Exp.1 was conducted at four consecutive periods; the experimental design is detailed in the Table 7 in Appendix.

Exp.2: Growth performance experiment

A total of 175 growing pigs (initial BW = 36.0 ± 5.2 kg) were fed 1 of 5 diets for 28 d, with five pigs per pen and seven replications (pens) per treatment. Diets were a corn-soybean meal based diet and diets including 10% or 20% EP-RSM and 10 or 20% SE-RSM (Table 3). All ingredients in Exp. 2 (corn, SBM, EP-RSM and SE-RSM) were the same as in Exp. 1; the NE values of diets in Exp. 2 were calculated according to the results of Exp. 1 and the ratio between SID Lys and NE value was the same among the five diets. The ratio between SID essential amino acids and SID Lys met the ideal protein profile. Pigs had free access to feed and water throughout the experiment; room temperature was maintained between 22 and 24 °C.

Chemical analyses and calculations

Ingredients, diets and feces were analyzed for dry matter (DM) (AOAC 2007, 930.15), crude protein (CP) (AOAC 2007, 984.13), ash (AOAC 2007, 942.05) [11], and ether extract (EE) [12]. The organic matter (OM) was calculated as DM minus ash content. Neutral detergent fiber (NDF) and acid detergent fiber (ADF) were determined using filter bags and fiber analyzer equipment (Fiber Analyzer, Ankom Technology, Macedon, NY) following a modification of the procedure of Van Soest et al. [13]. The gross energy (GE) was analyzed using an isoperibol calorimeter (Parr 6300 Calorimeter, Moline, IL) with benzoic acid as a standard.

The ingredients were hydrolyzed with 6 N HCl at 110 °C for 24 h and analyzed for 15 amino acids using an Amino Acid Analyzer (Hitachi L-8900, Tokyo, Japan). Methionine and cysteine were determined as methionine sulfone and cysteic acid after cold performic acid oxidation overnight and hydrolyzing with 7.5 N HCl at 110 °C for 24 h using an Amino Acid Analyzer (Hitachi L-8800, Tokyo, Japan). Tryptophan was determined after LiOH hydrolysis for 22 h at 110 °C using HPLC (Agilent 1200 Series, Santa Clara, CA, USA).

In Exp.1, the DM intake from d 9 to d 13 in each period was calculated as the product of feed intake and DM content of diets. GE intake was calculated as the product of the GE content of the diet and the actual feed

Table 3 Ingredients and chemical composition of the Exp.2 diets[a]

Diets	Basal	EP-RSM		SE-RSM	
		10%	20%	10%	20%
Ingredients, %					
Corn	76.64	69.24	61.82	69.35	62.01
Soybean meal	20.00	18.00	16.00	18.00	16.00
EP-RSM	-	10.00	20.00	-	-
SE-RSM	-	-	-	10.00	20.00
Dicalcium phosphate	0.60	0.40	0.35	0.45	0.35
Limestone	0.75	0.75	0.60	0.70	0.60
Salt	0.35	0.35	0.35	0.35	0.35
Vitamin and mineral premix[b]	0.50	0.50	0.50	0.50	0.50
Lys-HCl	0.50	0.36	0.22	0.33	0.16
Met	0.20	0.12	0.05	0.10	0.03
Thr	0.18	0.11	0.05	0.09	0.00
Trp	0.07	0.04	0.01	0.03	0.00
Val	0.21	0.13	0.05	0.10	0.00
Calculated composition					
CP	15.67	17.51	19.35	17.60	19.57
Ca	0.53	0.53	0.52	0.53	0.52
STTD P	0.26	0.24	0.26	0.25	0.25
DE[c]	14.25	14.28	14.32	13.92	13.59
ME[c]	13.75	13.76	13.77	13.37	12.99
NE[c]	10.49	10.48	10.47	10.20	9.92
SID Lys[d]	1.03	1.03	1.03	1.00	0.98
SID Lys/NE, g/MJ	0.98	0.98	0.98	0.98	0.98
SID Thr/SID Lys	0.65	0.65	0.65	0.65	0.65
SID Trp/SID Lys	0.19	0.19	0.19	0.19	0.19
SID Val/SID Lys	0.76	0.76	0.76	0.76	0.76
SID Leu/SID Lys	1.20	1.28	1.35	1.32	1.44
SID Ile/SID Lys	0.50	0.53	0.58	0.54	0.62
SID His/SID Lys	0.35	0.40	0.45	0.41	0.47
SID Met + Cys/SID lys	0.55	0.55	0.57	0.55	0.58
SID Phe + Tyr/SID Lys	1.00	1.08	1.15	1.09	1.18

[a]A total of 175 pigs were used in an 28-d study. There were five pigs per pen and seven pens per treatment; SBM, soybean meal; EP-RSM, expelled press rapeseed meal; SE-RSM, solvent extracted rapeseed meal

[b]Vitamin-mineral premix supplied the following per kg of diet: vitamin A, 5,512 IU; vitamin D_3, 2,200 IU; vitamin E, 30 IU; vitamin K_3, 2.2 mg; vitamin B_{12}, 27.6 µg; riboflavin, 4 mg; pantothenic acid, 14 mg; niacin, 30 mg; choline chloride, 400 mg; folic acid, 0.7 mg; thiamine, 1.5 mg; pyridoxine, 3 mg; biotin, 44 µg; Mn (MnO), 40 mg; Fe ($FeSO_4 \cdot H_2O$), 75 mg; Zn (ZnO), 75 mg; Cu ($CuSO_4 \cdot 5H_2O$), 100 mg; I (KI), 0.3 mg; Se (Na_2SeO_3), 0.3 mg

[c]DE, ME and NE calculated from values measured in Exp. 1

[d]SID values were referenced from NRC (2012)

DM intake over the 5-d collection period from d 9 to 13. The energy lost in feces, urine and methane were measured for each animal on a given diet. The ME included energy lost as urine and methane. Energy lost as methane was calculated using a 39.54 kJ/L conversion factor [14].

During d 9 to d 13 of each period, O_2, CO_2, and CH_4 concentrations in both ingoing and outgoing air, and outgoing air flow rates were measured at 5 min intervals. These concentrations were then used to calculate O_2 consumption and CO_2 and CH_4 productions during each 5 min interval and these values were averaged and extrapolated to a 24 h period. Total heat production (THP) was then calculated for each day from gas exchanges and urinary loss of N according to Brouwer [14] using the following equation:

$$THP(kJ) = 16.18 \times O_2(L) + 5.02 \times CO_2(L)\text{-}2.17$$
$$\times CH_4(L)\text{-}5.99 \times urinaryN(g)$$

Retention of energy (RE) was calculated according to the following equation:

$$RE\,(MJ/d) = [ME\ intake\,(MJ/d) - THP\,(MJ/d)]$$

Retention of energy as protein (RE_P) was calculated from nitrogen balance and the retention of energy as lipid (RE_L) corresponded to the difference between RE and RE_P [10].

FHP was calculated using the equation used for THP with gas concentrations and air flow obtained from only the last 8-h heat production measurement on d 16 (ie, from 2230 to 0630 h or started 31 h after the last maintenance meal) [9]. In order to base production using the same time span as used for THP, the 8-h heat production was extrapolated to a 24-h period. Net energy of each diet was calculated according to Noblet et al. [15]

$$NE(kJ/kg\ DM) = [RE(kJ/d) + FHP(kJ/d)]/DM\ intake(kg/d)$$

The FHP value obtained on the fasting day [kJ/(kg $BW^{0.6}$·d)] was used during the fed days according to the estimated average BW over the fed days.

The DM of ingredients was measured prior to the preparation of diets in order to calculate the DM ratio of each test ingredient in the diet. The DE, ME and NE of the corn was calculated using direct method. The percentage minerals and vitamins in the corn diet was 2.79% (DM basis) and was not a source of energy; therefore, the energy values of the corn and amino acids were divided by 0.9721; the energy of corn per se was calculated by subtracting the energy of amino acids (Evapig). The difference method was used to calculate the average GE, DE, ME and NE contributions of soybean meal assuming that the average GE, DE, ME and NE values of the corn and amino acids mixture estimated from the corn diet were applicable to the corn-soybean meal diet [16]. The DE/GE, ME/DE and NE/ME ratios could then be calculated for soybean meal from these calculated GE, DE, ME and NE values and used to estimate the final DE, ME and NE values as the product of measured GE and DE/GE for DE, measured GE and DE/GE and ME/DE for ME and measured GE and DE/GE and ME/DE and NE/ME for NE. All calculations were done on a DM basis. Similarly, the corn-soybean meal diet was used as the basal diet when the energy of rapeseed meal was calculated according to the difference method. The apparent total tract digestibility (ATTD) of nutrients in diets was calculated according to the methods of Noblet et al. [15]. The respiratory quotient (RQ) corresponds to the ratio between CO_2 production and O_2 consumption.

In Exp. 2, all pigs were weighed on d 1and d 29 to determine average daily gain (ADG), average daily feed intake (ADFI) and gain to feed ratio (G:F). The caloric efficiency of dietary energy was calculated as ADFI (kg/d) × Diet energy content (MJ/kg)/ADG (kg/d) with energy content expressed as DE, ME or NE.

Statistical analysis

Data from Exp. 1 were analyzed using PROC MIXED (SAS Inst. Inc., Cary, NC, USA) with diet as fixed effects and period and chamber as random effects. When the F-test for diet treatment was significant, differences between dietary treatments were tested among least square means using Tukey's least significant difference. Data in Exp. 2 were analyzed using MIXED procedure of SAS for a completely randomized design with fixed effects of treatment and pen as the experiment unit. In all analyses, the differences were considered significant if $P < 0.05$, and considered as a trend at $P < 0.10$.

Results

Chemical composition of rapeseed meal

The percentage of EE in EP-RSM was greater than in SE-RSM (9.5% vs. 1.1%), while the CP content in the EP-RSM was lower than in the SE-RSM (39.2% vs. 42.4%). The NDF content in the EP-RSM was greater than in SE-RSM (37.0% vs. 30.8%). These two rapeseed meals contained similar proportions of amino acids. The composition of the four diets is in agreement with the ingredients percentages and their composition with the lowest CP content in the corn diet and the highest CP contents in the rapeseed meal diets. Finally, the EE content was higher in the EP-RSM diet.

Exp.1: Net energy experiment

The ATTD of DM, OM and GE was similar in the corn and corn-soybean meal diets, which values were greater than in the two rapeseed meal diets (Table 4, 90.2% vs. 86.6% for ATTD of DM, 90.3% vs. 87.0% for ATTD of OM, 88.3% vs. 85.1% for ATTD of GE; $P < 0.01$). However, the ATTD of CP was similar in the corn-soybean meal diet and the two rapeseed meal diets, these values being greater than in the corn diet in connection with its lower CP content (86.4% vs. 79.9%; $P < 0.01$).

The nitrogen intake increased with the CP content in the diet ($P < 0.01$). The nitrogen output from feces and urine was similar in the corn-soybean meal diet and the corn diet but lower than in the two rapeseed meal diets ($P < 0.01$). Consequently, the urinary energy as a percentage of DE was greater in the SE-RSM diet than in the corn-soybean meal diet, and the lowest in the corn diet ($P < 0.01$). These led to ME to DE ratio greater in the corn diet than in the corn-soybean meal diet, and the lowest in the SE-RSM diet ($P < 0.01$).

The retention of nitrogen was similar in the two rapeseed meal diets and the corn-soybean diet, which was

Table 4 Effect of diet characteristics on energy and nitrogen balances of growing pigs (Exp. 1)[1]

Item	Corn	SBM	EP-RSM	SE-RSM	SEM	P - value
BW, kg	40.2[b]	44.0[a]	44.1[a]	44.1[a]	0.9	0.01
DM intake, kg/d	1.33	1.34	1.37	1.40	0.02	0.07
Digestibility coefficients, %						
DM	90.4[a]	90.0[a]	87.3[b]	85.8[b]	0.7	<0.01
CP	79.9[b]	88.0[a]	86.2[a]	85.1[a]	1.3	<0.01
OM	90.4[a]	90.2[a]	87.7[b]	86.3[b]	0.7	<0.01
GE	88.2[a]	88.3[a]	86.2[ab]	84.0[b]	0.8	<0.01
Nitrogen balance, g/d						
Intake	19.9[d]	37.0[c]	46.9[b]	50.4[a]	0.6	<0.01
Fecal output	4.0[b]	4.5[b]	6.6[a]	7.5[a]	0.5	<0.01
Urinary output	6.6[b]	10.6[ab]	15.5[a]	15.4[a]	1.4	<0.01
Retention	9.4[b]	21.9[a]	24.8[a]	27.5[a]	1.2	<0.01
Energy utilization, %						
Urinary energy % of DE	1.8[c]	2.7[b]	3.4[ab]	3.9[a]	0.25	<0.01
Methane energy, % DE	0.4	0.7	0.4	0.6	0.08	0.06
ME/DE	97.8[a]	96.6[b]	96.2[bc]	95.5[c]	0.27	<0.01
NE/ME	78.2	76.5	76.3	76.7	0.82	0.16
Energy balance, kJ/(kg BW$^{0.6}$·d)						
ME intake,	2,273[a]	2,160[bc]	2,195[b]	2,120[c]	28	<0.01
THP	1,259	1,308	1,268	1,250	25	0.15
THPc[2]	1,232	1,317	1,265	1,271	22	0.08
RE$_P$[3]	152[c]	337[b]	381[ab]	424[a]	20	<0.01
RE$_L$[4]	862[a]	516[b]	546[b]	446[b]	34	<0.01
RE	1,014[a]	852[b]	928[b]	870[b]	27	<0.01
FHP	769	809	756	766	22	0.18
RQ						
Fed state	1.11[a]	1.05[b]	1.02[b]	1.03[b]	0.01	<0.01
Fasted state	0.77	0.77	0.78	0.78	0.01	0.07
Energy values, MJ/kg DM						
DE	15.98[a]	16.10[a]	16.14[a]	15.42[b]	0.14	<0.01
ME	15.64[a]	15.56[a]	15.52[a]	14.73[b]	0.13	<0.01
NE	12.23[a]	11.90[a]	11.84[ab]	11.30[b]	0.15	<0.01

[1] A total 24 pigs were allotted to 1 of 4 dietary treatments with six pigs per diet and four periods, 16 d per period; SBM, soybean meal; EP-RSM, expelled press rapeseed meal; SE-RSM, solvent extracted rapeseed meal
[2] THPc means that total heat production was adjusted for comparable ME intake by covariance
[3] RE$_P$ = Energy retention as protein [kJ/(kg BW$^{0.6}$·d)] = [N intake (g) – N in feces (g) – N in urine (g)] $\times$ 6.25 $\times$ 23.86 (kJ/g)/BW$^{0.6}$
[4] RE$_L$ = Energy retention as fat [kJ/kg BW$^{0.6}$·d)] = [RE (kJ) – energy retention as protein (kJ)]/BW$^{0.6}$
[abcd] Means in the same row with differing superscripts differ ($P < 0.05$)

markedly greater than in the corn diet (24.7 vs. 9.4 g per day, $P < 0.01$). Methane energy averaged 0.54% of DE and it tended to be greater in the SE-RSM diet than in the EP-RSM diet ($P = 0.07$).

Compared to the pigs fed the rapeseed meal diet, the pigs fed the corn-soybean meal diet had a comparable ME intake while the pigs fed the corn diet had the greatest ME intake. There was little difference in THP (average: 1,270 kJ/(kg BW$^{0.6}$·d)) between diets despite

differences in ME intake. However, it tended to be lower in the corn diet when THP was adjusted for similar ME intake (Table 4). Consequently, the pigs fed the corn diet had the greatest RE compared to other diets (1,014 vs. 883 kJ/(kg BW$^{0.6}$·d), $P < 0.01$). In connection with their lower nitrogen gain and their higher RE, the greatest RE$_L$ was found in the pigs fed the corn diet. The mean FHP was 775 kJ/(kg BW$^{0.6}$·d) and was not affected by diet composition ($P = 0.19$). The RQ was the greatest in the corn diet ($P < 0.01$) in connection with the higher RE_L. There was no difference in NE to ME ratio among the four diets ($P = 0.18$) despite difference in chemical composition. Lower energy values (DE, ME and NE) were observed in the SE-RSM diets compared to the other diets that had comparable energy values.

The ingredients values calculated from the diets values are shown in Table 5. The ATTD of CP, OM and GE of the four ingredients were quite variable with the lowest value for SE-RSM. The ATTD of OM of the corn was the greatest. Compared to other ingredients, the corn had the greatest ME to DE ratio (97.8%) and NE to ME ratio (78.4%), while the SE-RSM had the least ME to DE ratio (90.7%); the soybean meal had the lowest NE to ME ratio (70.2%). The NE of corn, soybean meal, EP-RSM and SE-RSM diets were 12.46, 11.34, 11.71 and 8.83 MJ/kg DM, respectively.

Exp.2: Growth performance experiment
During the 28 days period, the NE intake averaged 21.15 MJ/d and was similar for the five diets. Similarly, final BW, ADG, ADFI and G:F did not change with the inclusion of EP-RSM or SE-RSM in the diets (Table 6). The mean caloric efficiencies of DE, ME and NE for BW gain were 31.5, 30.3 and 23.1 MJ/kg, respectively and were not affected by diet characteristics ($P > 0.10$).

Discussion
In the current experiment, the two rapeseed meals are not from the same batch. Therefore, we can't evaluate the oil extraction processing and its impact on the composition of the non-fat fraction of the meal. However, the lower CP in the high fat RSM is consistent with the dilution effect of residual oil. Overall, the chemical composition and amino acids contents of the four

ingredients are within the range of values reported in our previous work [2, 3, 17, 18]. However, the NDF content (33.9% DM basis) in the two rapeseed meals was greater than the values published by Sauvant et al. (31.9% DM basis) [19] and NRC (25.2% DM basis) [20].

As in other literature studies [21, 22], the digestibility of DM, OM and GE were lower in the rapeseed meal diets in connection with their higher NDF content. But this negative effect of dietary fiber from RSM is attenuated in the EP-RSM diet in connection with its high and digestible oil fraction. This resulted in similar digestibility of GE between the EP-RSM and corn-soybean meal diets. These results are also found in the four ingredients with the lowest digestibility DM, OM and GE in the SE-RSM in connection with its high NDF content.

The ME to DE ratio was lower in the rapeseed diets in connection with their high urinary nitrogen output due to the high CP in the rapeseed diets, which also confirmed the result that CP is negatively correlated with ME to DE ratio [15]. The methane energy measured in our study is consistent with data obtained in similar body weight pigs [9, 10, 15] and indicates that the methane energy loss in growing pigs is little variable with differences in diets composition.

The THP values measured in the current experiment are in agreement with the range of THP observed at comparable ME intakes for 61 diets by Noblet et al. [15]. In the current experiment, in order to eliminate the effect of ME intake, the THP was adjusted for comparable ME intakes by covariance. In this way, the THP for the corn diet became lower in connection with the high starch content of the diet and also the higher proportion of energy retained as fat, which is fully consistent with the rather high efficiency of starch for NE [15] and the higher efficiency for fat gain than for protein gain [23]. In the current experiment, the value of FHP (average: 775 kJ/(kg $BW^{0.6}$·d)) was estimated as the nocturnal heat production after a period of feed deprivation of 31 h. The value is remarkably close to our previous work using the same method [10, 24, 25] and the results of Noblet et al. [15]. In addition, in agreement with other literature data [10, 26, 27], the FHP was not affected by diet composition (dietary CP or EE).

In the current Exp.1, the corn diet was supplemented with essential AA to meet the optimal amino acids balance, however, the daily supply was insufficient and the nitrogen gain and RE_P are lower. In addition, RE was higher than in the other diets and RE_L was then higher in the corn diet. However, we supplemented corn with free AA in order to attenuate this effect of shortage of AA and CP. In this way, the energy value (ME and NE) of corn measured in the Exp.1 could be representative of normal situations where corn is used with optimal levels of essential AA (Exp.2).

Table 5 Energy utilization and energy value of the four ingredients (Exp. 1)[a]

Item	Corn	SBM	EP-RSM	SE-RSM
Digestibility coefficients, %				
CP	82.4	94.8	81.6	78.9
OM	93.1	90.5	76.4	68.0
GE	88.2	89.8	77.6	65.6
Energy utilization, %				
ME/DE	97.8	92.1	94.9	90.7
NE/ME	78.3	70.2	74.7	76.5
NE/DE	76.6	64.6	71.0	69.4
Energy values, MJ/kg DM				
DE	16.27	17.51	16.55	12.82
ME	15.92	16.10	15.71	11.61
NE	12.46	11.34	11.71	8.83

[a] There were six pigs per treatment; SBM, soybean meal; EP-RSM, expelled press rapeseed meal; SE-RSM, solvent extracted rapeseed meal

Table 6 Effect of rapeseed meal on growing pigs performance (Exp.2)[a]

Item	Basal	EP-RSM[b]		SE-RSM[c]		SEM	P-value
		10%	20%	10%	20%		
Initial BW, kg	35.9	36.0	36.1	36.1	36.1	2.0	0.99
Final BW, kg	62.0	62.2	61.3	61.0	61.5	2.4	0.99
ADFI, g/d	2,093	2,059	1,996	2,019	2,086	75	0.86
ADG, g/d	929	937	901	891	910	24	0.64
G:F	0.446	0.458	0.454	0.442	0.436	0.010	0.71
NE intake, MJ/d	21.96	21.58	20.91	20.60	20.69	0.78	0.68
Caloric efficiency, MJ/kg							
DE	31.93	31.24	31.60	31.47	31.16	0.72	0.95
ME	30.80	30.10	30.39	30.21	29.77	0.70	0.88
NE	23.60	23.00	23.16	23.13	22.78	0.53	0.86

[a] A total of 175 pigs were used in an 28-d study. There were five pigs per pen and seven pens per treatment
[b]EP-RSM, expelled press rapeseed meal
[c]SE-RSM, solvent extracted rapeseed meal

The soybean meal had the lowest NE to ME ratio, while the corn had the greatest NE to ME ratio. This indicates that the efficiency of utilization of ME for NE depends on chemical composition [27]. The NE to ME ratio of corn is also similar to the value in Sauvant et al. (78.3% vs. 80.1%) [28]. However, the NE to ME ratio of soybean meal and rapeseed meal obtained in the present trial are greater (70.2% vs. 60.5% for soybean meal; 76.5% vs. 59.7% for rapeseed meal) than the values in Sauvant et al. [28]. The difference in NE to ME ratio may be explained by the slightly greater FHP measured in the current experiment than value used by Sauvant et al. [28] (775 vs. 750 kJ/(kg BW$^{0.6}$·d)).

The DE and ME values of the four ingredients measured in the current experiment are similar to those obtained in our previous studies [2, 3, 17, 18]. In agreement with the values of NE to ME ratio in the present study and in Sauvant et al. [28], the NE value of corn measured in Exp. 1 is similar to the value in Sauvant et al. [28]. But the NE values of soybean meal and rapeseed meal are greater than the table values in Sauvant et al. (11.34 vs. 9.22 MJ/kg DM for SBM, 8.83 vs. 7.14 MJ/kg DM for SE-RSM) [28]. The NE contents in corn, soybean meal were similar to the values reported by Liu et al. (12.46 vs. 13.21 MJ/kg DM for corn, 11.34 vs. 10.62 MJ/kg DM for SBM) [24] using the same methodology. The NE content in SE-RSM measured using indirect calorimetry by Heo et al. (8.83 vs. 8.80 MJ/kg DM) [29] was similar to the value in the present experiment. The greater NE value found in EP-RSM than in SE-RSM is associated with its high EE value, which was in agreement with the results by Woyengo et al. [30].

In the growth trial, diets were supplemented with crystalline amino acids in order to maintain a constant SID Lys/NE ratio (0.98 g/MJ) and balanced for the other essential amino acids. Therefore, the efficiencies of energy for BW gain were calculated without the confounding effect of amino acids deficiency. Several research projects reported that there was no impairment of pig performance when EP-RSM or SE-RSM replaced soybean meal in growing pig diets formulated to equal NE and SID amino acids [31, 32], or that contained equal quantities of digestible CP and digestible amino acids [4]. In the current experiment, we also failed to detect any statistically significant differences in growth performance with increasing inclusion rates of EP-RSM or SE-RSM.

In the current research, we used the same ingredients (corn, soybean meal, EP-RSM and SE-RSM) in the two experiments. Therefore, the NE of the five diets in Exp.2 was calculated according to the values measured in Exp. 1. Increasing inclusion level of EP-RSM or SE-RSM did not affect caloric efficiency of NE, which indicated that the NE value measured in Exp.1 by indirect calorimetry was close to the actual value as indicated by the caloric efficiency results in Exp. 2. In the growth trial (Exp. 2), we also failed to detect any statistically significant differences in caloric efficiency of DE and ME with increasing inclusion rates of EP-RSM or SE-RSM. Therefore, these data do not show any advantage of NE system on DE or ME systems. In fact, the diets in the growth trial were very similar in terms of chemical composition, energy value, CP and AA contents with then little expected differences in ME/DE and NE/ME ratios. Furthermore, our research to measure the caloric efficiency of NE was only according to ADG. This should also be adjusted for differences in BW gain composition (lipid or fat contents) for being more accurate.

Conclusions

The DE, ME and NE values in EP-RSM were greater than values in SE-RSM. The efficiency of utilization of ME for NE depends on chemical composition. The NE values measured in Exp.1 are confirmed by results of Exp. 2 with comparable caloric efficiencies of NE for all diets.

Appendix

Table 7 Design of net energy experiment (Exp.1)

	Chambers	A1	A2	B1	B2	C1	C2
Period 1	Diets	I	II	III	IV	I	II
	Pigs	1	2	3	4	5	6
Period 2	Diets	II	III	IV	I	II	III
	Pigs	7	8	9	10	11	12
Period 3	Diets	III	IV	I	II	III	IV
	Pigs	13	14	15	16	17	18
Period 4	Diets	IV	I	II	III	IV	I
	Pigs	19	20	21	22	23	24

Abbreviations
AA: Amino acid; ADF: Acid detergent fiber; ADFI: Average daily feed intake; ADG: Average daily gain; ATTD: Apparent total tract digestibility; BW: Body weight; CP: Crude protein; DE: Digestible energy; DM: Dry matter; EE: Ether extract; EP-RSM: Expeller-pressed rapeseed meal; F:G: Gain to feed ratio; FHP: Fasting heat production; GE: Gross energy; HP: Heat production; ME: Metabolizable energy; N: Nitrogen; NDF: Neutral detergent fiber; NE: Net energy; OM: Organic matter; RE: Retention energy; RE_L: Retention energy as lipid; RE_P: Retention energy as protein; RQ: Respiratory quotient; SEM: Standard error of the mean; SE-RSM: Solvent-extracted rapeseed meal; SID: Standard ileal digestibility; THP: Total heat production

Acknowledgements
This project was financially supported by the Modern Agricultural Industry Technology System (CARS-36), Developing key equipment for digital management and monitoring environment in animal production (2013AA10230602), National Natural Science Foundation of China (31372317) and the 111 Project (B16044). I would like to express my heartfelt gratitude to Dr. Michael A. Brown for helping me edit the paper.
Submitting author: Zhongchao Li: No.2, Yuanminyuan west road, Haidian district, China Agricultural University, Beijing 100193, China.

Funding
This project was financially supported by the Modern Agricultural Industry Technology System (CARS-36), Developing key equipment for digital management and monitoring environment in animal production (2013AA10230602), National Natural Science Foundation of China (31372317) and the 111 Project (B16044). The funders had no role in study design, data collection and analysis, decision to publish, or preparation of the manuscript.

Authors' contributions
ZCL carried out the animal trial, performed the statistics and drafted the manuscript. YKL, ZQL, HL and JBZ participated in the experiments. JN, CHL, FLW and DFL critically evaluated the manuscript. All authors read and approved the final manuscript.

Competing interests
The authors declare that they have no competing interests.

Author details
[1]State Key Laboratory of Animal Nutrition, Ministry of Agriculture Feed Industry Centre, China Agricultural University, No. 2, Yuanminyuan west road, Haidian district, Beijing 100193, China. [2]INRA, UMR Pegase, 35590 Saint-Gilles, France.

References
1. USDA: Oilseeds: World markets and trade, USDA Foreign Agricultural Service. http://apps.fas.usda.gov/psdonline/circulars/production.pdf. Accessed 18 Sept 2016.
2. Li PL, Wang FL, Wu F, Wang JR, Liu L, et al. Chemical composition, energy and amino acid digestibility in double-low rapeseed meal fed to growing pigs. J Anim Sci Biotechnol. 2015;6:37–47.
3. Li PL, Wu F, Chen YF, Wang JR, Guo PP, Li ZC, et al. Determination of the energy content and amino acid digestibility of double-low rapeseed cakes fed to growing pigs. Anim Feed Sci Technol. 2015;210:243–53.
4. Little KL, Bohrer BM, Maison T, Liu YH, Stein HH, Boler DD. Effects of feeding canola meal from high-protein or conventional varieties of canola seeds on growth performance, carcass characteristics, and cutability of pigs. J Anim Sci. 2015;93:1284–97.
5. Graham AB, Goodband RD, Tokach MD, Dritz SS, DeRouchey JM, Nitikanchana S. The effects of medium-oil dried distillers grains with solubles on growth performance, carcass traits, and nutrient digestibility in growing-finishing pigs. J Anim Sci. 2014;92:604–11.
6. Graham AB, Goodband RD, Tokach MD, Dritz SS, DeRouchey JM, Nitikanchana S, et al. The effects of low-, medium-, and high-oil distillers dried grains with solubles on growth performance, nutrient digestibility, and fat quality in finishing pigs. J Anim Sci. 2014;92:3610–23.
7. De Jong JA, DeRouchey JM, Tokach MD, Dritz SS, Goodband RD. Effects of dietary wheat middlings, corn dried distillers grains with solubles, and net energy formulation on nursery pig performance. J Anim Sci. 2014;92:3471–81.
8. Adeola O, Mahan DC, Azain MJ, Baidoo SK, Cromwell GL, Hill GM, et al. Dietary lipid sources and levels for weanling pigs. J Anim Sci. 2013;91:4216–25.
9. Zhang GF, Liu DW, Wang FL, Li DF. Estimation of the net energy requirements for maintenance in growing and finishing pigs. J Anim Sci. 2014;92:2987–95.
10. Li ZC, Li P, Liu DW, Li DF, Wang FL, Su YB, et al. Determination of the energy value of corn distillers dried grains with solubles containing different oil levels in growing pigs. J Anim Physiol Anim Nutr. 2015. doi:10.1111/jpn.12445.
11. AOAC Official Methods of Analysis. 18th ed. Arlington: Association of Official Chemists; 2007.
12. Thiex NJ, Anderson S, Gildemeister B. Crude fat, diethyl exther extraction in feed, cereal grains, and forge: Collaborative study. J AOAC Int. 2003;86:888–98.
13. Van Soest PJ, Robertson JB, Lewis BA. Methods for dietary fiber, neutral detergent fiber, and nonstarch polysaccharides in relation to animal nutrition. J Dairy Sci. 1991;74:3583–97.
14. Brouwer E. Report of sub-committee on constants and factors. Proceedings of the 3rd EAAP Symposium on Energy Metabolism; Troonn, Publ. 11. London: Academic; 1965. p. 441–3.
15. Noblet J, Fortune H, Shi XS, Dubois S. Prediction of net energy value of feeds for growing pigs. J Anim Sci. 1994;72:344–54.
16. Adeola O. Digestion and balance techniques in pigs. In: Lewis AJ, Southern LL, editors. Swine nutrition. New York: CRC Press; 2001. p. 903–16.
17. Li Q, Zang J, Liu DW, Piao XS, Lai CH, Li DF. Predicting corn digestible and

metabolizable energy content from its chemical composition in growing pigs. J Anim Sci Biotechnol. 2014;5:11–9.

18. Li ZC, Wang XX, Guo PP, Liu L, Piao XS, Stein HH, et al. Prediction of digestible and metabolisable energy in soybean meal produced from soybeans of different origins fed to growing pigs. Arch Anim Sci. 2015;69:473–86.

19. Sauvant D, Perez JM, Tran G. Tables of composition and nutritional value of feed materials: pigs, poultry, cattle, sheep, goats, rabbits, horses and fish. Wageningen Academic Pub; 2004.

20. NRC. Nutrient requirements of swine. 11th ed. Washington: National Academy Press; 2012.

21. Noblet J, Le Goff G. Effect of dietary fibre on the energy value of feeds for pigs. Anim Feed Sci Technol. 2001;90:35–52.

22. Le Goff G, Dubois S, van Milgen J, Noblet J. Influence of dietary fibre level on digestive and metabolic utilization of energy in growing and finishing pigs. Anim Res. 2002;51:245–60.

23. Noblet J, Karege C, Dubois S, van Milgen J. Metabolic utilization of energy and maintenance requirements in growing pigs: effects of sex and genotype. J Anim Sci. 1999;77:1208–16.

24. Liu DW, Jaworski NW, Zhang GF, Li ZC, Li DF, Wang FL. Effect of experimental methodology on fasting heat production and the net energy content of corn and soybean meal fed to growing pigs. Arch Anim Nutr. 2014;68:281–95.

25. Liu DW, Liu L, Li DF, Wang FL. Determination and prediction of the net energy content of seven feed ingredients fed to growing pigs based on chemical composition. Anim Prod Sci. 2015;55:1152–63.

26. Le Bellego L, Van Milgen J, Dubois S, Noblet J. Energy utilization of low-protein diets in growing pigs. J Anim Sci. 2001;79:1259–71.

27. Noblet J, Le Bellego L, Van Milgen J, Dubois S. Effects of reduced dietary protein level and fat addition on heat production and nitrogen and energy balance in growing pigs. Anim Res. 2001;50:227–38.

28. Sauvant D, Perez J M, Tran G. 2004. Tables of composition and nutritional value of feed materials: pigs, poultry, cattle, sheep, goats, rabbits, horses and fish. Wageningen Academic Pub.

29. Heo JM, Adewole D, Nyachoti M. Determination of the net energy content of canola meal from Brassica napus yellow and Brassica juncea yellow fed to growing pigs using indirect calorimetry. Anim Sci J. 2014;85:751–56.

30. Woyengo TA, Sánchez JE, Yáꞌnez J, Beltranena E, Cervantes M, Morales A, et al. Nutrient digestibility of canola co-products for grower pigs. Anim Feed Sci Technol. 2016;222:7–16.

31. Landero JL, Beltranena E, Cervantes M, Morales A, Zijlstra RT. The effect of feeding solvent-extracted canola meal on growth performance and diet nutrient digestibility in weaned pigs. Anim Feed Sci Technol. 2011;170:136–40.

32. Landero JL, Beltranena E, Cervantes M, Araiza AB, Zijlstra RT. The effect of feeding expeller-pressed canola meal on growth performance and diet nutrient digestibility in weaned pigs. Anim Feed Sci Technol. 2012;171:240–5.

8

Analysis of the lifetime and culling reasons for AI boars

Damian Knecht, Anna Jankowska-Mąkosa and Kamil Duziński*

Abstract

Background: The aim of the study was to analyze the lifetime and culling reasons for boars used in insemination centers (AI centers).

Methods: The data collected from 355 culled boars from 1998 to 2013 included: age at start of semen collection, boar herd life, culling reason, daily gain and lean meat content, and number of ejaculates not meeting sales requirements after dilution. Culling reasons were divided into 7 groups: low semen value (LSV), low or lack of libido (LL), leg problems (LP), infectious diseases (ID), old age (OA), reduced demand for semen from the given boar (RD), and others (OT).

Results: The most common culling reasons for boars were LSV (23.7%) and RD (22.5%). It was observed that the lowest daily gains were noted in boars culled due to OA. Boars culled due to OA and RD were maintained in production for the longest time (over 1000 d), for LSV and ID retention was about 700 d, and due to LL below 400 d. The survival probability was over 0.9 until 1.5 yr, and just over 0.2 until 4 yr. The highest relative frequency was observed in the 36th and 42nd mo of life (over 16%). Hazard risk analysis revealed a more than 10 times higher risk of culling in the case of LL, ID or OT, in comparison to OA.

Conclusions: The results can be used as a direct point of reference for the identification of emerging problems in AI boar exploitation and the development of an appropriate culling policy in AI centers.

Keywords: AI centers, Boars, Culling, Exploitation, Management

Background

The reasons for culling of boars demonstrate a close relationship with the efficiency and economic profitability of operations in commercial herds and AI centers. With increasing interesting in AI, the most important role started to play AI centers. It was estimated that already 70% of Polish sows are inseminated and 99% using fresh semen, and almost 85% of this demand come from AI centers. An understanding of the reasons for culling is necessary information for the planning and rational management of such units. It should be noted that there are two main forms of boar culling, i.e. unplanned and planned. Unplanned culling (forced), i.e. culling as a result of diseases, sudden falls, behavioral problems, and lameness, clearly adversely affects productivity. In terms of productivity, Safranski [1] noted that the collection

and production processes of insemination doses is dependent on a number of factors. These factors hinder the long-term accurate estimation of exploitation predispositions of boar, and thus this creates uncertainty in production. On the other hand, planned culling (decision-making) (old age, poor production results), despite the initial costs incurred, over the long term appears to be the most appropriate and cost-effective decision. Unfortunately, decisions about culling boars from a herd are still taken reluctantly by owners and are also often postponed. Only an appropriate culling schedule and herd replacement program can satisfy the essential prerequisite for the maintenance of production stability and repeatability [2].

Boars are very sensitive to abrupt changes which impair correct semen formation [3]. Preparing detailed records about culling within a specified time (e.g. annual balance) allows for the identification of risks to the boar population, especially in terms of possible diseases,

* Correspondence: kamil.duzinski@upwr.edu.pl
Institute of Animal Breeding, Wroclaw University of Environmental and Life Sciences, Chelmoskiego 38C, 51– 630 Wroclaw, Poland

disorders, and behavioral problems [4]. It is important for AI centers to have an active population of boars that produce appropriate ejaculates with high quality and quantity parameters [5], which is also connected with proper preparation of doses [6].

Analysis of culling reasons enables a clear increase in the production efficiency of AI units. Although it might seem that in terms of the profitability of pig production all avenues have already been investigated, research into the control, improvement and acceleration of time required to obtain effective production capacity of boars is still valid. Progress in the production results of modern piggeries may be supported by AI [7]. The development of AI has been made possible due to the constant production of high value semen doses [8]. Smital [9] stated that the economic effectiveness of AI centers is closely dependent on the productivity of boars during exploitation, and is limited in the highest degree by the construction of the testicles, libido, and physical activity (limb, spine defects). However, according to the results of previous studies, the problems of sperm quality are not the only reason for boar culling [3, 4, 10, 11], and this has necessitated a deeper analysis of all possible causes.

Given the above consideration, it can be clearly stated that the choice of boar for AI centers and their accurate monitoring during exploitation are very important issues. Once selected, boars should remain within the active productive population for as long as possible, compensating costs and generating incomes for AI centers [4]. Recuperation of the maintenance and exploitation costs of AI boar stations usually occurs after 2–3 yrs of a boar's life [12]. Only the direct identification of culling reasons can skillfully define the problems faced in herds of boars maintained at AI centers. Unfortunately, this subject is still overlooked in global scientific publications. Most articles focus on an analysis of culling reasons and lifetime of sows and only some concern boars, but these mostly relate to those on breeding farms [3, 4, 10, 11] and not AI boars. Therefore, the aim of our study was to analyze lifetime and culling reasons for boars used in AI center.

Methods
Experimental location and animals
The study was carried out between 1998–2013 at the Boar Exploitation Station in Częstochowa. The study population included 355 culled boars, whose histories of exploitation were followed from birth until death. The presented population was representative in proportion to the most common breed components used for AI in Poland, such as: Polish Landrace, Polish Large White, Duroc, Pietrain, Hampshire, Duroc × Pietrain, Hampshire × Pietrain, Duroc × Hampshire. The collected data from 1998 to 2013 included: age at start of semen collection, boar herd

life, culling reason, daily gain and lean meat content (evaluated before purchase of boar), and number of ejaculates not meeting the requirements of sales after dilution (ejaculates incompatible with requirements). Reasons for culling were divided into 7 groups: low semen value (LSV), low or lack of libido (LL), leg problems (LP), infectious diseases (ID), old age (OA), reduced demand for semen from the given boar (RD), and others (OT). The whole population of boars was divided into experimental groups based on the reason for culling.

Decision process for unplanned and planned culling were different. The structure of unplanned culling were simple and based on the first observation of boars by employees receiving ejaculates. Information were directed to the supervisors and the director, the director usually took the sole decision about unplanned culling. In the case of the planned culling decision-making process was more complex. The long-term replacement plan was developed and approved by the owner and the supervisory board. After all, plan was also constantly being upgraded under the flowing marketing information and the demand for a certain product. On this basis planned culling were made.

Therefore, there were finally 7 groups of culling reasons (LSV, LL, LP, ID, OA, MU, OT). In subsequent years the ratios of culling reasons were similar therefore, the sample period was treated jointly. The overall characteristics of the study population are presented in Table 1.

Daily gain and lean meat content
The assessment of daily gain and lean meat content, before purchase of the boar, was made between d 170 and 210 of life. Daily gains were calculated by dividing the body weight of young boars (during assessment time) by the age on the assessment day. The measurements of the animals for the estimation of daily gains were made using a Mensor WM150P1 electronic scales. Daily gains were standardized to 180 d using a model developed for standardization in accordance with the methodology of Mucha and Różycki [13], in order to reduce age differences during assessment. The percentage of meat content in carcasses was estimated

Table 1 The overall characteristics of the study population of boars ($n = 355$)

Trait	Mean	SD	Min	Max
Age of semen collection entry, d	259.74	33.51	201	458
Boar herd life, d	835.57	503.5	21	2350
Daily gain, g	775.81	94.32	554	1170
Lean meat content, %	61.34	1.94	55	67
Ejaculates incompatible with requirements, n	3.36	5.49	0	37

SD standard deviation, *Min* minimum, *Max* maximum

intravitally on the basis of two backfat thickness (points P2 and P4) and lion eye height (point P4) measurements. The measurements were made using a PIGLOG 105 (SFK) ultrasonic device, positioned behind the last rib (between the thoracic and lumbar vertebrae), 3 cm off the midline (point P2) and 8 cm off the midline (point P4). To increase the accuracy of the assessment, standardization of traits was performed to 110 kg body weight. The measurement values were inserted into a specially developed equation [14], which allowed an estimation of the intravital proportion of meat in the carcass on the assessment day. The results for meat content were also standardized to d 180 [13].

Boars performance

Before the start of semen collection, all boars were held in quarantine, the length of which was approximately 37.24 ± 5.02 d. During quarantine semen was collected once a wk just to observation only and not for insemination. Additionally, all boars were exploited in the same manner, developed and adopted according to the methodology of the AI station. Ejaculates were collected by masturbation via the manual method using a container with a filter. The gelatinous fraction was separated. Immediately after collection, the volume of semen was measured using a scalar cylinder. The concentration of spermatozoa was evaluated using a SpermaCue device, Model 12300/0500 (Minitube International, Verona, USA). Based on the semen volume and spermatozoa concentration, the total number of spermatozoa in the ejaculate was calculated. Semen dilution was effected using the same semen extender. Boars until 10 mon gave ejaculates once a week, at the age of 10–14 mon this was three times in two week and from 15 mon twice a week. The average annual replacement rate was 49.8%. Ejaculates were classified as normal for further dilution, when the following requirements were met: color from gray to milky white, flavor specific, lack of foreign admixtures, more than 70% of progressive motile sperm cells, pH 7.0–7.9, morphological abnormality changes to 15% (5% primary, 10% secondary). Insemination doses of 80 mL contained a constant 2.8×10^9 spermatozoa. Semen was stored at 15 °C for not longer than 48 h.

Housing and feeding

Boars were single-housed and maintained in accordance with the principles of animal welfare [15]. Each individual pen area was 8 m^2/boar. Boars were kept on a solid concrete floor, which was covered with straw. The air temperature in all the boar pens was close to 15 °C (min 12 °C, max 20 °C). Relative humidity was close to 75% (min 65%, max 85%). The air circulation inside the building was equal to 0.15 m/s in Winter and 0.20 m/s in Summer. Preventive care and vaccination was carried

out regularly in accordance with the methodology of the unit. The microclimate of the area and ventilation were controlled by computer. Over the whole study period, boars were fed the same all-mash mixture, dosed according to the recommended nutrition standard for boars, with permanent access to water (Table 2).

Statistical analysis

The data was analyzed using the STATISTICA (2014) statistical program. The values in the tables are

Table 2 The nutritional value of 1 kg of compound feed for boars

Item	Value
Dry matter, g	887
Metabolizable energy, MJ	12.5
Crude protein, g	169.63
Fat, g	41.98
Crude fiber, g	59.58
Lysine, g	9.38
Methionine + cystine, g	6.14
Threonine, g	5.9
Tryptophan, g	1.8
Valine, g	5.9
Isoleucine, g	4.5
Ca, g	7.13
P total, g	5.6
Na, g	1.8
Vitamin A, IU	15,000
Vitamin D$_3$, IU	2000
Vitamin E, mg	100
Vitamin C, mg	1000
Vitamin K, mg	5
Vitamin B$_1$, mg	2
Vitamin B$_2$, mg	7
Vitamin B$_6$, mg	4
Vitamin B$_{12}$, µg	30
Biotin, µg	400
Folic acid, mg	4
Niacin, mg	30
Pantothenic acid, mg	15
Choline, mg	1185
Mn, mg	50
Fe, mg	145
I, mg	1
Zn, mg	100
Cu, mg	20
Co, µg	500
Se, µg	300

arithmetical means ($\bar{x}$) and standard deviations (SD). A Chi-squared test was used to examine the significance of differences for the relative frequency (%) of boars removed from the removal groups. Other permanent collected data were checked for normality with the Kolmogorov-Smirnov (K-S) test with the Lilliefors correction. In addition, the Brown-Forsythe test (B-F) determined whether the distributions of the variables had the same variance. An advanced mixed model using the GLM procedure was used and an analysis of variance was conducted for factorial designs to determine the effect of tested parameters by the culling reason. The significance of differences was calculated on the basis of Tukey's multiple range test. The levels of significance of differences were given conventionally: significant $0.01 < P \leq 0.05$ and highly significant $P \leq 0.01$. Pearson's correlation coefficient (r) was calculated between daily gains, lean meat content and age at start of semen collection entry, and boar herd life. Survival analysis was performed using the Cox proportional hazard model. The time variable was defined as age in months from birth. The year effect in herd was included in the analysis. The Cox proportional hazard model expressed risk during t for the tested independent variable system and was expressed by the equation:

$$h(t) = h_0 e^{\beta_i X_i}$$

where: h(t) was the risk at a given time interval, h_0 the base risk (the risk obtained if there were no risk factors in the model), and β_i the regression coefficient for the i-th removal reasons. The hazard ratio (HR) for the situation when the risk factor X is present versus the situation when the risk factor X is absent was calculated as:

$$HR = \frac{h(t, x = 1)}{h(t, x = 0)} = e^{\beta_i X_i}$$

Results

The results of the assessment of the selected parameters of boar performance, depending on the culling reasons, are presented in Table 3. The most common culling reasons were LSV, followed closely by RD. Differences amounting to only 1.2% ($P > 0.05$) were noted between these reasons. However, of these the two most common culling reasons demonstrated statistically significant differences with other reasons, i.e. LP ($P \leq 0.05$) and ID, OA, LL, OT ($P \leq 0.01$). Boars culled due to LL were characterized by the highest age at the start of semen collection. The difference from the youngest age at the start of semen collection observed for boars culled due to ID was over half mo ($P \leq 0.05$). The longest boar herd life expectancies were noted for boars culled due to OA, and differences with other groups were statistically

proven at the level of $P \leq 0.01$. Boar herd life expectancy in excess of 1000 d were achieved by boars culled due to RD. Boars with LSV and ID were held in production for over 700 d, and the shortest period was noted for boars with LL, i.e. less than 400 d. It was observed that the lowest daily gains were noted in boars culled due to OA, and differences with other reasons were at a level from 53.32 g to 68.14 g ($P \leq 0.05$), with the exception of RD ($P > 0.05$). Similar observations have been made for the lean meat content parameter and differences for OA were statistically confirmed with LSV, ID and OT ($P \leq 0.05$). The highest number of ejaculates incompatible with sales requirements after dilution were recorded in boars culled due to ID and the lowest were for LL, RD and OT boars (all below 3; $P \leq 0.01$).

The estimated survival probability of boars in the AI station and the relative frequency of boar removal in each month are presented in Fig. 1. A survival probability for boars of over 0.9 was noted until 1.5 yr. A drastic drop in the probability was observed to the age of 4 yr, achieving a value a little over 0.2. From this point, survival probability gently fell (expired) to 7 yr, when there was a complete replacement of the herd. The highest relative frequencies were observed at the 36th and 42nd month (over 16%). Indicators of relative frequency over 10% were also noted in the 18th and 30th mo of life.

Figure 2 shows the percentage of boars culled both below and above 4 yr, depending on the reason. It was noted that all boars with LL were culled by 4 yr. Equally large culling rates by 4 yr were reported for boars with OT and LP. More than 70% of boars from the ID, LSV or RD groups were culled by 4 yr. The highest levels of culling over 4 yr were obtained for the OA group.

The hazard analysis for culled boars depending on the reason is presented in Table 4. The highest hazard ratios for culling were achieved for OT, LL and ID. The lowest hazard ratios (i.e. 3.08 to 4.38) were noted for LSV, RD and LP. The highest parameter values were observed for LL and OT, and the lowest negative values occurred for RD.

Table 5 shows correlation coefficients of selected parameters related to culling reasons. Statistically confirmed ($P \leq 0.05$) positive correlations between daily gain and age at the start of semen collection were observed for boars from OA, ID and LSV groups. The remaining statistically significant positive correlations were reported for OA between daily gain and boar herd life expectancy, and also lean meat content and age of semen collection entry. Additionally, for LP boars positive correlations were calculated between lean meat content and age at the start of semen collection or boar herd life expectancy ($P \leq 0.05$). Three negative correlations were found ($P < 0.05$). Two of these were noted for boars culled due to LL between daily gains and boar herd life expectancy, and also lean meat content and

Table 3 Selected parameters of boar performance by culling reasons (mean ± SD)

Item	LSV	LL	LP	ID	OA	RD	OT
Proportionate rate, %	23.7[Aa]	9.3[B]	14.9[b]	9.6[B]	9.3[B]	22.5[Aa]	10.7[B]
Age of semen collection entry, d	259 ±37	267 ±31[b]	261 ±32	251 ±29[a]	258 ±48	261 ±27	259 ±31
Boar herd life, d	738 ±463[B]	399 ±321[A]	695 ±381[AB]	759 ±415[B]	1605 ±433[D]	1105 ±309[C]	460 ±276[AB]
Daily gain, g	786.13 ±86.89[a]	788.55 ±129.72[a]	783.91 ±107.62[a]	797.29 ±83.4[a]	729.15 ±94.03[b]	761.3 ±73.41	782.47 ±93.02[a]
Lean meat content, %	61.62 ±1.82[a]	61.29 ±2.32	61.27 ±1.44	61.65 ±1.77[a]	60.43 ±2.2[b]	61.25 ±1.95	61.54 ±2.18[a]
Ejaculates incompatible with requirements, n	3.42 ±5.72	1.85 ±2.76[A]	3.92 ±5.58	5.79 ±8.64[B]	3.73 ±5.68	2.54 ±4.66[A]	2.97 ±3.74[A]

LSV low semen value, *LL* low or lack of libido, *LP* leg problems, *ID* infectious diseases, *OA* old age, *RD* reduced demand for semen of the given boar, *OT* others
[a,b] – in the same row signifies statistically significant differences between reasons of culling, with $P \le 0.05$
[A,B,C,D] – in the same row signifies statistically significant differences between reasons of culling, with $P \le 0.01$

boar herd life expectancy. The last negative correlations were observed between lean meat content and age at start of semen collection for RD boars.

Discussion

The results presented here provide valuable information on the culling reasons for boars. A proportionate rate of culling reasons is a useful indicator of the strength of reasons and helps prioritize the improvement of a herd [10]. Observed culling of boars due to LSV and LL combined amounted to 33%, which was not as high as in the results of previous studies, which amounted to 19–25% [4, 11]. Such discrepancies could be explained by the higher culling ratio in experimental units and farms using AI, for which the requirements in this regard are much greater [10]. Additionally, current capabilities for the accurate microscopic assessment of ejaculates contribute to changes in the classification of culling reasons especially semen value [11]. The more precise the characterization of culling reasons, the more accurate

the analysis of the herd and the problems appearing therein and also the expected herd life of boars.

The average age at the start of semen collection and boar herd life approached the results obtained in piggery farms [4, 11], and was also much longer than in breeding herds from the early 1990s [10].

In contrast to the study of Koketsu and Sasaki [11], age at the start of semen collection had an effect on the reason for culling. Boars culled due to LL were the latest to begin semen collection. This is probably because the jump reflex was poorly noticeable, and this consequently led to a delay in the introduction of boars to ménage and a later analysis of the reason for culling. Libido is determined by many factors, including the level of circulating testosterone in the male body [16]. AI stations do not typically prefer to maintain boars demonstrating low libido, because it impacts upon operational costs and therefore such boars are eliminated from an active herd as soon as possible. Berger and Conley [17] even stated that boars with low libido and demonstrating a low

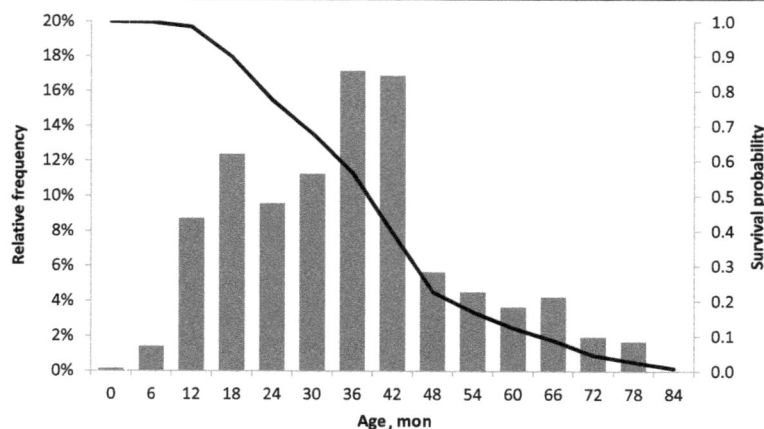

Fig. 1 Survival probability of boars in the AI station and relative frequency (column) of boar removal by age in mon

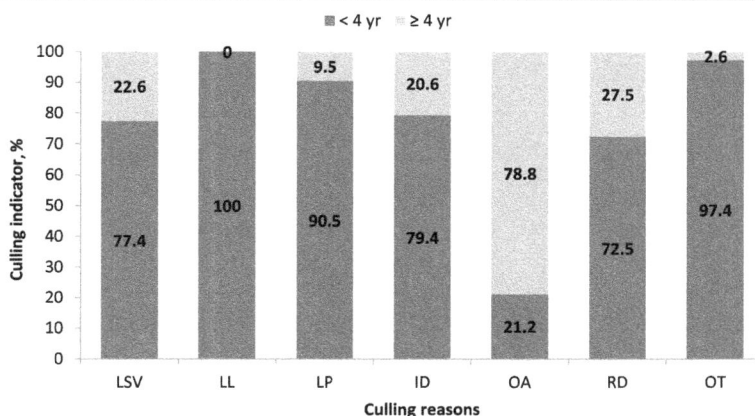

Fig. 2 The percentage of boars culled before and after 4 yr, by culling reasons. LSV- low semen value, LL- low or lack of libido, LP- leg problems, ID- infection diseases, OA- old age, RD- reduced demand for semen of the given boar, OT- others

ability of sperm to fertilize are useless for the AI industry, as are diseased animals.

Boars with diagnosed infectious diseases were characterized by the youngest age at start of semen collection. A longer period of quarantine, acclimatization and isolation before starting semen collection is recommended to reduce the occurrence of new pathogens and deaths of animals as a result of infection [18]. In our study, quarantine time was unified; therefore, in the case of the analyzed place, and the age at the start of semen collection could be extended by a few days without any negative effect on the remaining culling reasons. Such preventive action can lead to a reduction in boars culled due to ID.

Herd management in AI stations is significant, especially health status. Health reasons for boar culling include not only infectious diseases but also limb problems or unexpected falls. The health status of boars is a complex factor and results not only from veterinary care but also nutrition, management, exploitation, maintenance conditions or individual characteristics [19]. During culling analysis it is reasonable to separate the major determinants of health, i.e. infectious diseases and limb problems. The high levels of culling due to LP in our study remained close to results from a study conducted

by D'Allaire and Leman [10], although other authors have noted both higher [3] and lower values [4, 11]. Regardless of the research, herd life expectancy of boars was similar. Important for the diagnosis of locomotor problems in boars is rapid preventive and curative action, because these disorders are classified as being painful for animals [15].

Due to the use of pigs for high quality pork, particular importance during production parameter analysis should be placed on daily gains and lean meat content [20]. It was observed in our study that boars with higher daily gains and lean meat content had a greater predisposition to infectious diseases. On the other hand, boars culled due to OA were characterized by the lowest daily gains and lean meat content. It can be concluded, therefore, that improvement of AI boars in terms of daily gains and lean meat content is justified only to a certain point. Our earlier (unpublished) results have shown that differences between semen parameters for boars with daily gains 700–750 g and 750–800 g are almost the same, are differences between lean meat content. Such precise analyses help us to conclude, that selection of AI boars with lower daily gains allows for longer boar herd life expectancy and culling herds with greater size due to old age. Wolf [21] stated that production traits (daily gains and lean meat content) have a negligible impact on semen characteristics, but play an important role in boar exploitation, which is especially important for AI centers. Confirmation of this thesis may be presented in our study in the form of the statistically confirmed correlations for exploitation parameters between daily gain and lean meat content. Inheritance of important production traits is essential, particularly in AI.

A significant emphasis is placed on the impact of sires on offspring in all currently existing insemination programs, where an extremely large number of offspring are obtained from each sire [22]. The importance of the proper and

Table 4 Hazard analysis for culled boars depending on the reason

Item	Parameters	Standard error	Hazard ratio	Confidence interval
LSV	0.031	0.112	3.08	1.86; 5.01
LL	1.063	0.143	10.47	5.45; 20.1
LP	0.191	0.131	4.38	2.49; 7.72
ID	1.002	0.157	10.42	5.79; 18.75
OA	0	0	1	-----
RD	−0.575	0.113	3.29	2.01; 5.38
OT	1.055	0.156	16.81	9.51; 29.73

LSV low semen value, *LL* low or lack of libido, *LP* leg problems, *ID* infectious diseases, *OA* old age, *RD* reduced demand for semen of the given boar, *OT* others

Table 5 Correlation coefficients for selected parameters by culling reasons

Item	LSV	LL	LP	ID	OA	RD	OT
Daily gain/age of semen collection entry	0.26*	−0.07	0.09	0.21*	0.32*	−0.16	0.14
Daily gain/boar herd life	0.12	−0.44*	0.03	0.05	0.28*	0.08	−0.01
Lean meat content/age of semen collection entry	0.08	0.17	0.34*	−0.14	0.23*	−0.34*	−0.11
Lean meat content/boar herd life	0.16	−0.49*	0.31*	0.18	0.05	0.01	0.13

LSV low semen value, LL low or lack of libido, LP leg problems, ID infectious diseases, OA old age, RD reduced demand for semen of the given boar, OT others
* − correlation statistically significant, with $P \leq 0.05$

efficient functioning of AI stations can be seen in the sale of insemination portions consistent with the expectations of customers [23]. Therefore, great emphasis should be placed on the selection of a boar to achieve profitability of purchase. The identification of the market factor (RD) as one of the main culling reasons clearly shows that artificial insemination has become so popular, and competition in the market so large that in order to maintain the stability of a company adaption to customer requirements and market flexibility is required. As presented in the results, the culling of boars for reasons including market factors (reduced demand) represents an important element in a reasonable culling policy for AI boars and was ranked second. Previous studies on culling boars did not characterize this factor alone and the factor was probably included in the "other reasons" category. Most recent work in this area seems to indicate a higher proportion of culling for "other reasons" (c. 20%) [4, 11] in comparison with our results. Selection of boars is mainly based on the selected parameters of growth and carcass traits, with minimal emphasis on production traits [5]. However, AI stations should not be limited only to selection of boars due to the best production traits for pig producers, but must also take into account factors that affect the individuals' performance, i.e. the construction, temperament, quantity and quality of ejaculate [24]. A company operating on the market and focused on profitability has to meet the needs of customers. The constantly growing popularity of AI contributes to problems with fulfilling needs in terms of the appropriate quality of insemination portions [25]. The best solution is, therefore, to agree upon a compromise choice. If there is excessive pressure to choose traits favoring the AI stations, then market traits may be overlooked. A lack of demand for the offered insemination portions from a specific boar means its maintenance is unjustified. Culling for economic reasons may also take place in the case of new boars of the same genotype which are characterized by much better production parameters from their predecessors. In such a situation a decrease in demand for insemination portions from predecessor may be observed and this raises questions concerning its further use or culling, because the economy affects the company's balance sheet. Production farms are located in a particular economic environment, thus economic culling reasons should be taken into account in

research. Hence, it was fully justified for our study to determine the reduced demand for semen of a boar to be a reason which closely affects the survival analysis of the herd.

Survival analysis is the recommended method for the study of boar stayability in herds [26]. Our results are similar to those in Segura-Correa et al. [4] but were only observed in farm A. Other farms in the study by these authors were characterized by much lower survival probability. On the other hand, in research conducted by Koketsu and Sasaki [11] a survival function with a similar shape was shown, although with increasing age the survival probability was higher in our study. However, boar herd life expectancy in high-performing herds and in our AI centre ended after 78 mon (6.5 yr). It is believed that culling boars from a high production herd should take place no later than the age of 6 yr. Nevertheless, it is suggested that culling of boars with low semen values and andrology-fertility indicators significantly below expectations should occur as soon as possible [5]. Culling in the AI center fits the shape of the survival curve. A higher share of culling due to old age proves the probity of the decisions concerning choosing boar and proper management in the unit. The critical point in the economic viability of boars exploitation is the age of 2–3 yrs [12], which in the case of our own research has been confirmed by relative frequency. Additionally, a detailed distribution of culling reasons both before and above 4 yr has been presented to characterize the problems concerning the direct contribution of boars to AI centre profitability.

Analyzing the above-mentioned distribution, the best solution would be quick culling of diseased boars, and those with reduced libido and low sperm quality. However, sometimes individuals suffer from infectious diseases or limb problems in later years, because the disease can appear in all ages and is a time-independent parameter. In the same regard, low semen quality should be considered, because despite the development of boars producing semen with more favorable traits [24, 27], this problem can also still affect boars with a long history of exploitation and at an advanced productive age. The most unpredictable culling reason for boars to be noted from the graph is the market factor (RD), because reduced customer demand for specific semen may result from the appearance of better individuals or arising trends and fashions in pig production.

The hazard function focuses on the occurrence of an event and reflects the instantaneous potential at the time of the event. An important feature of the hazard function is the reference to a specific time [28]. In our study, the hazard risk ratio described the occurrence of boar culling in comparison with the culling of a boar due to its age. A hazard risk higher than 1 indicates an increased risk of culling boars from the active herd in comparison with culling due to age. A hazard risk ratio below 1 indicates a lower risk of culling. Therefore, one can see that there was a more than 10-fold increased risk of boar culling in the case of LL, ID or OT compared to OA. This indicates the significance of these factors in shaping the overall predictability of possible problems in a herd of boars.

Conclusions

An understanding of culling reasons contributes to the ability to identify problems relating to AI center functioning. The results of the above study may be used as a direct point to develop appropriate strategies for boar culling in AI centers. An effective program of culling affects the economic viability of centers. A too frequent exchange of individuals and a large share of young boars negatively affects the cost of production and the health status of the herd. A large share of culling due to age, and a low rate in the case of diseases, hormonal disorders and quality of sperm provide a good selection of material production and careful management of the unit. An example of good practice is also the separate classification of economic reasons for culling (e.g. reduced demand for semen of the boar), because these reflect the changing preference of customers and the relevancy of products offered on the market. Each decision about boar culling, regardless of the reason, is difficult but necessary. However, if made at the right time and based on relevant observations, such decisions enable the preservation of the liquidity and profitability of AI centers.

Abbreviations
AI: Artificial insemination; GLM: General linear model; HR: Hazard ratio; ID: Infectious diseases; LL: Low or lack of libido; LP: Leg problems; LSV: Low semen value; OA: Old age; OT: Others; RD: Reduced demand; SD: Standard deviation

Acknowledgements
The authors wish to thank Franciszek Wiśniowski, Director of the Boar Exploitation Station in Częstochowa, Małopolska Biotechnology Center for the possibility to conduct research and Karl Bernhardt for revising the English version of the manuscript.

Funding
This experiment and article preparation were financed by Wroclaw University of Environmental and Life Sciences, statutory project no. 2014.

Authors' contributions
DK design of the study and methodology, preparation of manuscript, obtaining results, coordination of research group; AJ-M preparation of manuscript, participating in obtaining results, design of discussion; KD statistical analysis, presentation of results and conclusions, supervision of manuscript submission. All authors read and approved the final manuscript.

Competing interests
The authors declare that they have no competing interests.

References
1. Safranski TJ. Genetic selection of boars. Theriogenology. 2008;70:1310–6.
2. de Jong E, Appeltant R, Cools A, Beek J, Boyen F, Chiers K, et al. Slaughterhouse examination of culled sows in commercial pig herds. Livest Sci. 2014;167:362–9.
3. Acosta MJ, Rueda M. A note on causes of boar removal in Cuban pig farms. Livest Res Rural Dev. 2009;21(52):1–5.
4. Segura-Correa JC, Alzina-López A, Santos Ricalde R. Stayability in the herd and culling causes of boars in four pig farms of Yucatan, Mexico. Trop Subtrop Agroecosystems. 2010;12:411–6.
5. Robinson JAB, Buhr MM. Impact of genetic selection on management of boar replacement. Theriogenology. 2005;63:668–78.
6. Goldberg AMG, Argenti LE, Faccin JE, Linck L, Santi M, Lourdes Bernardi M, et al. Risk factors for bacterial contamination during boar semen collection. Res Vet Sci. 2013;95:362–7.
7. Duziński K, Knecht D, Środoń S. The use of oxytocin in liquid semen doses to reduce seasonal fluctuations in the reproductive performance of sows and improve litter parameters – a 2-year study. Theriogenology. 2014;81:780–86.
8. Knecht D, Jasek S, Procak A, Krzyżewski P. Efficiency of inseminating sows with pure breed and crossbreed boars. Med Wet. 2004;60:1208–11.
9. Smital J. Effects influencing boar semen. Anim Reprod Sci. 2009;110:335–46.
10. D'Allaire S, Leman AD. Boar culling patterns in swine breeding herds in Minnesota. Canadian Vet J. 1990;31:581–3.
11. Koketsu Y, Sasaki Y. Boar culling and mortality in commercial swine breeding herds. Theriogenology. 2009;71:1186–91.
12. Vinent-Duany NJ, Parra C, Sagaró-Zamora F, Garzón-Gómez V. Estudio de desecho de verracos en el centro genetic porcino. . 2007. Consulted March 31, 2009 URL: http://www.ilustrados.com/publicaciones/EEZZVEAAVZIANFBfrv.php.
13. Mucha A, Różycki M. Standaryzacja cech określających mięsność tusz w ocenie przyżyciowej świń. Rocz Nauk Zoot. 2005;32:45–50.
14. Szyndler-Nędza M, Różycki M. Opracowanie równań regresji do przyżyciowego szacowania procentowej zawartości mięsa w tuszy knurów. Rocz Nauk Zoot. 2005;32:51–60.
15. Ordinance of the Minister of Agriculture and Rural Development. Requirements and how to proceed while maintaining livestock species for which protection standards are provisions of the European Union. 2010. http://isap.sejm.gov.pl/DetailsServlet?id=WDU20100560344. Accessed 15 Feb 2010.
16. Okere C, Joseph A, Ezekwe M. Seasonal and genotype variations in libido, semen production and quality in artificial insemination boars. J Anim Vet Adv. 2005;4:885–8.
17. Berger T, Conley AJ. Reducing endogenous estrogen during prepuberal life does not affect boar libido or sperm fertilizing potential. Theriogenology. 2014;82:627–35.
18. Dee S, Joo H, Pijoan C. Controlling the spread of PRRS virus in the breeding herd through management of the gilt pool. J Swine Healt Prod. 1995;3:64–9.
19. Anil SS, Anil L, Deen J. Evaluation of pattern of removal and associations among culling because of lameness and sow productivity traits in swine breeding herds. J Am Vet Med Assoc. 2005;226:956–61.
20. Knecht D, Jankowska-Mąkosa A, Duziński K. Does the activity of producer group organizations improve the production of pigs? Ann Anim Sci. 2015;15:759–74.
21. Wolf J. Genetic correlations between production and semen traits in pig. Animal. 2009;3:1094–9.
22. Knecht D, Środoń S, Szulc K, Duziński K. The effect of photoperiod on selected parameters of boar semen. Livest Sci. 2013;157:364–71.

23. Knecht D, Środoń S, Duziński K. The influence of boar breed and season on semen parameters. S Afr J Anim Sci. 2014;44:1–9.

24. Wolf J, Smital J. Quantification of factors affecting semen traits in artificial insemination boars from animal model analyses. J Anim Sci. 2009;87:1620–7.

25. Knecht D, Środoń S, Duziński K. Does a boar's season of birth determine semen parameters and reproductive performance? Reprod Domest Anim. 2014;49:183–90.

26. Yazdi MH, Rydhmer L, Ringmar-Cederberg E, Lundeheim N, Johansson K. Genetic study of longetivity in Swedish landrace sows. Livest Prod Sci. 2000;63:255–64.

27. Kondracki S, Banaszewska D, Wysokińska A, Radomyska M. Effect of age on semen traits of Duroc breed used in insemination. Anim Sci Pap Rep. 2004;22:281–8.

28. Hoge MD, Bates RO. Developmental factors that influence sow longevity. J Anim Sci. 2011;89:1238–45.

9

Effect of three lactobacilli with strain-specific activities on the growth performance, faecal microbiota and ileum mucosa proteomics of piglets

Yating Su[1], Xingjie Chen[2], Ming Liu[3] and Xiaohua Guo[1*]

Abstract

Background: The beneficial effects of *Lactobacillus* probiotics in animal production are often strain-related. Different strains from the same species may exert different weight-gain effect on hosts in vivo. Most lactobacilli are selected based on their in vitro activities, and their metabolism and regulation on the intestine based on strain-related characters are largely unexplored. The objective of the present study was to study the in vivo effects of the three lactobacilli on growth performance and to compare the differential effects of the strains on the faecal microbiota and ileum mucosa proteomics of piglets.

Methods: Three hundred and sixty piglets were assigned to one of four treatments, which included an antibiotics-treated control and three experimental groups supplemented with the three lactobacilli, *L. salivarius* G1-1, *L. reuteri* G8-5 and *L. reuteri* G22-2, respectively. Piglets were weighed and the feed intake was recorded to compare the growth performance. The faecal lactobacilli and coliform was quantified using quantitative PCR and the faecal microbiota was profiled by denaturing gradient gel electrophoresis (DGGE). The proteomic approach was applied to compare the differential expression of proteins in the ileum mucosa.

Results: No statistical difference was found among the three *Lactobacillus*-treated groups in animal growth performance compared with the antibiotics-treated group ($P > 0.05$). Supplementation of lactobacilli in diets significantly increased the relative 16S rRNA gene copies of *Lactobacillus* genus on both d 14 and d 28 ($P < 0.05$)., and the bacterial community profiles based on DGGE from the lactobacilli-treated groups were distinctly different from the antibiotics-treated group ($P < 0.05$). The ileum mucosa of piglets responded to all *Lactobacillus* supplementation by producing more newly expressed proteins and the identified proteins were all associated with the functions beneficial for stabilization of cell structure. Besides, some other up-regulated and down-regulated proteins in different *Lactobacillus*-treated groups showed the expression of proteins were partly strain-related.

Conclusions: All the three lactobacilli in this study show comparable effects to antibiotics on piglets growth performance. The three lactobacilli were found able to modify intestinal microbiota and mucosa proteomics. The regulation of protein expression in the intestinal mucosa are partly associated with the strains administrated in feed.

Keywords: Faecal microbiota, Growth, *Lactobacillus*, Mucosa proteomics, Probiotics

* Correspondence: guo_xh@hotmail.com
[1]Provincial Key Laboratory for Protection and Application of Special Plants in Wuling Area of China, College of Life Science, South-Central University for Nationalities, No. 182, Minyuan Road, Hongshan District, Wuhan, Hubei Province 430074, China
Full list of author information is available at the end of the article

Background

As living microorganisms, probiotics act in the intestine to modulate the host microbiota [1]. Among the strains of probiotics, lactic acid bacteria (LAB), especially from *Lactobacillus* and *Bifidobacteria* species, are recognized as one of the main sources and are widely used in food, drugs and feed additives as intestinal flora improvers [2]. In animal production, probiotics are expected to improve performance and to produce high-qualified meat without drug residues as an alternative to antibiotics [3].

Generally, *Lactobacillus* species selected for probiotics are highly diverse in the phenotypic and genetic characteristics [4]. Different strains may exert different weight-gain effect on hosts in vivo even if in the same species. Million et al. assessed the effect of lactobacilli-containing probiotics on weight based on 51 studies on farm animals and suggested that the weight-gain effect was greatly associated with strains of the genus [3]. Simon et al. showed similar results after summarizing above 20 published papers on lactobacilli used in feed additives [5]. The phenomena suggest that *Lactobacillus* strains may benefit their hosts through different mechanisms and more work should be done to explore the relationship between the choice of strains and their in vivo behaviours [6].

Nowadays the selection of *Lactobacillus* is often based on the strains' activities in vitro, which is expected to show corresponding effectiveness in vivo. The strains with bacteriocin-producing activity showed specific anti-infective effect in the gut [7]. The strains with enzyme activities including amylase, protease and α-galactosidase had the potential to stimulate feed digestion [8–10]. However, the gut ecosystem was so complicated and the in vivo activities often depended on the strains' survival and metabolism in the gastrointestinal tract (GIT) [10]. In our previous studies, three *Lactobacillus* strains (*Lactobacillus salivarius* G1-1, *Lactobacillus reuteri* G8-5 and *Lactobacillus reuteri* G22-2) were selected from swine faeces for probiotic use. They shared strain-related in vitro functional properties, including antimicrobial activity, amylolytic activity and bile-salt-hydrolase activity, respectively [11]. Meanwhile, from the in vivo studies in rats, the three *Lactobacillus* species showed some similar beneficial effects and some of the functionalities to rats were strain-specific [12]. When used in swine nutrition, the lactobacilli were hypothesized to interact with the intestinal flora and with the host mucosa, which might be associated with the mechanism of lactobacilli as probiotics. The objective of the present study was to study the in vivo effects of the three lactobacilli on growth performance and to compare the differential effects of the strains on the faecal microbiota and ileum mucosa proteomics of piglets.

Methods

Lactobacillus strains and freeze-dried powder preparation

Three strains, *L. salivarius* G1-1, *L. reuteri* G8-5 and *L. reuteri* G22-2 were isolated for probiotics based on the strain-specific functional properties in vitro [11]. All strains were incubated in DeMan Rogosa Sharp broth under anaerobic conditions at 37 °C for about 24 h. The microbial cells were collected by centrifugation at 11,000×g for 10 min, washed twice and mixed with protective additives for freeze-drying. The freeze-dried sample was smashed and diluted with dextrin as a carrier. The concentration of viable cells from each strain was determined by agar-plate assay and adjusted to 0.5×10^9 colony forming unit per gram (CFU/g) by the carrier before animal trial.

Animals, diets, experimental design and sampling

Three hundred and sixty castrated male, crossbred (Landrace × Large White) piglets, 35–40 days old, were randomly assigned to one of four treatments, which included an antibiotics control (Group A) and three experimental groups supplemented with *L. salivarius* G1-1 (Group B), *L. reuteri* G8-5 (Group C) and *L. reuteri* G22-2 (Group D), respectively. The piglets were housed with 15 piglets per pen and six pens of piglets received each treatment ($n = 6$). The pigs had free access to feed and water throughout the feeding trial with the environmental temperature 25–28 °C. The diet composition was listed in Table 1. The diet in the antibiotics control was supplemented with 200 mg/kg flavomycin. The three experimental diets consisted of the basal diet supplemented with 200 mg/kg *Lactobacillus* powder (10^9 CFU/kg of feed) from each strain.

Piglets were weighed and the feed intake was recorded during the trial term to calculate the average daily weight gain (ADG), average daily feed intake (ADFI) and feed conversion ratio (F:G). Fresh faecal samples (4–5 g) from 3–4 individual piglets were collected and pooled from three randomly chosen pens were collected for each treatment on d 14 and 28. The samples after collection were immediately stored at −20 °C until the molecular analysis for microbiota. At the end of the trial, three randomly chosen piglets were selected from each treatment and slaughtered for ileum sampling. About 20 cm ileum at the same place of each pig were rapidly cut and the chyme was washed out using sterile water. The mucosa was carefully scrapped by coverslips and kept in 1.5 mL Eppendorf tube. The samples were frozen immediately by liquid nitrogen and stored at −80 °C for proteomics analyses. All surgical and animal care procedures in the study followed the protocols approved by Experimental Animal Care and Use Guidelines (Chinese Science and Technology Committee, 1988).

Table 1 Basal diet formula and nutrient levels

Ingredients	Percentage, %	Nutrient levels	
Extruded corn, soybean and sorghum with the proportion (3:1:1)	30.37	DE, Mcal/kg	3.30
High protein flour	17.40	Crude protein, %	18.2
Extruded soybean	5.00	Crude fat, %	5.05
Concentrated soybean meal	8.00	Crude ash, %	5.69
Limestone	1.00	Crude fiber, %	1.79
Calcium phosphate	0.69	Ca, %	0.75
Diamond V XP Yeast Culture	0.50	Total phosphorus, %	0.54
Mineral premix	1.50	Salt, %	0.63
Vitamins premix	1.00		
Lysine, 98%	0.55		
N-carbamoylglutamate	0.06		
Threonine	0.32		
Methionine	0.26		
Skim milk powder	7.50		
Whey powder	12.50		
Fatty powder	3.33		
Proprietary milk substitute	5.00		
Dextrose	5.00		
Antibiotics or lactobacilli powder	0.02		
Total	100.00		

[a] Vitamins provided per kilogram diets: vitamin A, 8000 IU; vitamin D3, 1800 IU; vitamin E, 30 IU; vitamin K3, 3.56 mg; vitamin B1, 1.8; vitamin B2, 6 mg; vitamin B6, 1.26 mg; vitamin B12, 0.02 mg; folic acid, 0.3 mg; biotin, 0.44 mg; niacin, 32 mg; pantothenic acid, 15 mg
[b] Minerals provided per kilogram diets: Cu, 250 mg; Fe, 130 mg; Zn, 130 mg; Mn, 60 mg; Se, 0.3 mg; I, 0.4 mg

DNA extraction, real-time quantitative PCR and PCR-DGGE analyses

The total genomic DNA was extracted from faeces (about 1.0 g) based on the method of bead-beating and following phenol-chloroform extraction [13, 14]. Total lactobacilli and coliform were detected by real-time quantitative PCR, respectively. The lactobacilli were quantified using primer Lac1 (5′-AGCAGTAGGGAATCTTCCA-3′), and Lab0677 (5′- CACCGCTACACATGGAG –3′) [15]. Two primers, EcoliFimH2F (5′-AGCAGTAGGGAATCTTCCA-3′) and EcoliFimH2R (5′- TCATCCCTGTTATAGTTGYYGGTCT-3′) were used to amplify 16S rRNA gene of coliform [16]. The reverse transcription PCR (RT-PCR) system was quantified using the ABI 7500 system (Applied Biosystems, US). The optimum thermal cycles were performed as follows: pre-denaturation at 95 °C for 10 min, 40 cycles of 95 °C for 15 s and 60 °C for 1 min, and followed by the stage of melting curve. The relative 16S rRNA gene copies were calculated through the $2^{-\Delta\Delta CT}$ method according to the report of Livak [17]. The

results were compared based on the three paralleled values of faeces from each treatment.

A set of universal primers, U968-GC (5′-CGCCCG GGGCGCGCCCCGGGCGGGGCGGGGGCACGGGG GGAACGCGAAGAACCTTAC-3′), L1401 (5′-CGGTGT GTACAAGACCC-3′) [18], Bact 1369 F (5′-CGGTGAA TACGTTCYCGG-3′), and 1492R (5′-GGWTACCTT GTTACGACTT-3′) [19] were employed to amplify the total bacteria. The amplicons were separated by DGGE according to the specification as described previously [20]. Briefly, DGGE was performed in 8% polyacrylamide gels (acrylamide-bis, 37.5:1). The gels with a 38–51% denaturing gradient was used for the separation of PCR products based on the primers U968-GC and L1401, while gradients of 30–45% were applied for the separation of the Bact 1369 F and 1492R generated amplicons. The electrophoresis procedures were performed at 70 V for 16 h at 60 °C and the gel was finally stained with SYBR Green I for 30 min after electrophoresis. The DGGE gels were scanned using an image scanner and analysed with Bio-rad gel imaging system through Quantity One software (Version 4.6.2).

The similarities among DGGE profiles were determined by Dice coefficient based on the unweighted pair group method with arithmetic average (UPGMA) clustering algorithm [21]. The faecal samples from the antibiotics group were evenly mixed and conducted for DGGE profiles used as the control band. The bands from three paralled faecal samples of each *Lactobacillus* group were profiled and compared with the control band ($n = 3$).

2-dimensional electrophoresis (2-DE), image analysis and protein identification

Isoelectric focusing (IEF) was performed using immobilized pH gradient (IPG) Strips (pH 4–7; 7 cm long; Pharmacia Biotech.). Samples were diluted with IEF buffer containing 7 mol/L urea, 2 mol/L thiourea, 4% CHAPS, 20 mmol/L Tris–HCl, pH 8.5, 20 mmol/L DTT, 0.5% carrier ampholyte (pH 4–7) and a trace of bromophenol blue. The desired protein amount in buffer was 50 μg. After equilibration, the immobilized pH gradient strips were loaded onto 12.5% (w/v) homogeneous acrylamide gels and sealed with 1% (w/v) agarose. The electrophoretic separation of proteins was conducted as described previously [22, 23]. Upon completion of 2-dimensional SDS-PAGE, the gels were stained by silver or Coomassie Brilliant Blue G-250. The high-resolution gel images (200 dpi) from silver-stained gels were obtained using an image scanner (Powerlook1100, UMAX) for image analysis. The gels stained by silver were run in triplicate, and spots that appeared consistently in all three runs were selected for analysis. Spot detection and analysis were performed using the PDQuest version 6.1 software (Bio-Rad) according to the protocols provided by the manufacturer. Some differentially expressed protein spots with 3.0-fold differences in volume detected by the software were selected for protein identification. The protein spots of interest were confirmed in the Coomassie Brilliant Blue stained gels and manually excised for the treatment of digestion by trypsin. The matrix-assisted laser desorption/ionization time of flight mass spectrometry (MALDI-TOF MS) was used for protein identification as described by early reports [24, 25]. The peptide fragments produced from each protein spot were employed to produce peptide-mass mapping (PMM) data. The protein identification was carried out by peptide mass fingerprinting (PMF) analysis through the MASCOT server (www.matrixscience.com; Matrix Science, UK). The search parameters were as follows, database: Swiss-Prot Sus (34361 sequences); species: sus; enzyme: trypsin; fixed modifications: carbamidomethylation; variable modifications: oxidation (M). The gene name, accession code and function of each protein were determined using the Mascot V2.1 software protein database search engine and the Swiss-Prot Sus protein database.

Statistical analyses

All quantitative data were expressed as the mean and standard deviation of replicates. The differences among antibiotics-treated and lactobacilli-treated groups were considered statistically significant at $P < 0.05$ using one-way analysis of variance (One-way ANOVA) through JMP software (JMP; SAS Institute Inc., Cary, NC). $0.5 < P < 0.1$ was considered a trend towards significance.

Results

Growth performance

Over the 4-week feeding trial, there was no statistical difference in ADG, ADFI and F:G between piglets supplemented with lactobacilli and the antibiotics group (Table 2). Among the three *Lactobacillus* groups, the diet containing *L. reuteri* G8-5 tended to show lower ADG and ADFI than that of the other two *Lactobacillus* groups $(0.5 < P < 0.1)$.

Relative 16S rRNA gene copies by RT-PCR

A comparison of the relative 16S rRNA gene copies of *Lactobacillus* and coliform in faeces on d 14 and d 28 was shown in Fig. 1. Supplementation of lactobacilli in diets significantly increased the counts of *Lactobacillus* genus on both d 14 and d 28 compared with the antibiotics group $(P < 0.05)$. However, no significant difference in the relative 16S rRNA gene copies of coliform was observed in all groups $(P > 0.05)$.

PCR-DGGE profiles

The representative DGGE profiles were presented in Fig. 2. The DGGE patterns were transformed into graphs by the Bio-Rad Quantity One™ software, which calculated the Dice similarity among lanes (Fig. 2). The similarities among four treatments on d 14 and d 28 were listed in Table 3. On d 14, the dendrogram based on the banding patterns showed low similarities and the bacterial community profiles form the lactobacilli were distinctly different from the antibiotics group $(P < 0.05)$. Meanwhile, the similarities in *L. reuteri* G8-5 group were significantly lower than those in *L. salivarius* G1-1 group $(P < 0.05)$. On d 28, the percentage of similarity in all *Lactobacillus* groups increased but was still significantly lower than that of antibiotics group. There was no marked difference in similarities in all *Lactobacillus*-treated groups on d 28 $(P > 0.05)$.

2-DE profiles of differentially expressed proteins

By comparing the 2-DE profiles of differentially expressed proteins in the ileum of piglets between the antibiotics-treated and *Lactobacillus*-treated groups, supplementation of lactobacilli significantly increased the counts of newly expressed proteins. 4, 6 and 8 new proteins were expressed only in the antibiotics group compared with the three *Lactobacillus* groups, respectively. Nevertheless, 32, 40 and 27 new proteins only existed in the three *Lactobacillus* groups compared with the antibiotics group, respectively (Fig. 3a). Among the differentially expressed proteins, 4 protein spots which were up-regulated in all the *Lactobacillus*-treated groups were selected for the identification by MALDI-TOF.

Table 2 The effect of three lactobacilli on the growth performance of weaned piglets during a 4-week feeding trial

Treatments	Antibiotics (A)	*L. salivarius* G1-1 (B)	*L. reuteri* G8-5 (C)	*L. reuteri* G22-2 (D)	*P*-value					
					A vs. B	A vs. C	A vs. D	B vs. C	B vs. D	C vs. D
Initial body weight, kg	7.44 ± 1.22	7.51 ± 1.00	7.56 ± 0.92	7.51 ± 0.92	0.913	0.871	0.922	0.957	0.992	0.949
Final body weight, kg	14.48 ± 1.19	14.81 ± 0.56	13.88 ± 1.13	14.71 ± 1.87	0.727	0.516	0.807	0.325	0.916	0.376
ADG, g/d	270.9 ± 17.1	280.3 ± 20.5	243.3 ± 15.8	276.9 ± 38.8	0.603	0.141	0.743	0.056	0.846	0.080
ADFI, g/d	408.6 ± 33.4	422.8 ± 15.4	338.5 ± 36.9	418.1 ± 20.5	0.486	0.331	0.639	0.109	0.817	0.161
F:G	1.51 ± 0.13	1.51 ± 0.08	1.60 ± 0.09	1.54 ± 0.11	0.976	0.338	0.859	0.337	0.872	0.420

Values are means ± S.D, $n = 6$

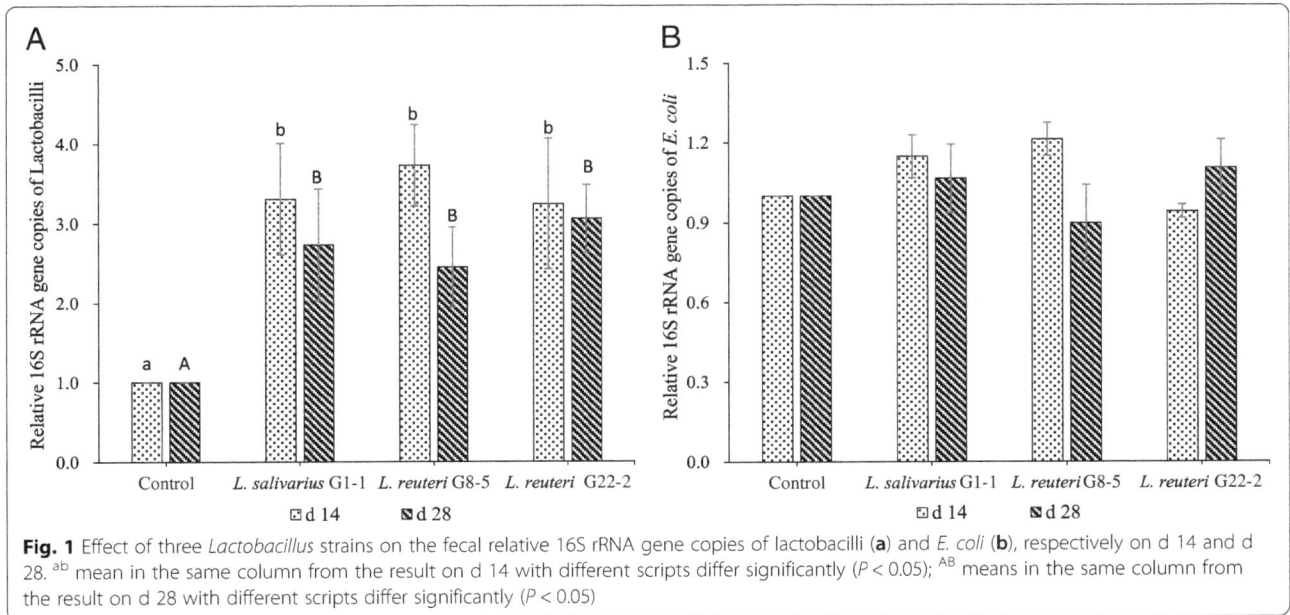

Fig. 1 Effect of three *Lactobacillus* strains on the fecal relative 16S rRNA gene copies of lactobacilli (**a**) and *E. coli* (**b**), respectively on d 14 and d 28. [ab] mean in the same column from the result on d 14 with different scripts differ significantly ($P < 0.05$); [AB] means in the same column from the result on d 28 with different scripts differ significantly ($P < 0.05$)

These proteins included tropomyosin beta chain (TPM2, Spot R1), vimentin (VIM, Spot R2), keratin type I cytoskeletal 19 (KRT19, Spot R3), tropomyosin alpha-1 chain (TPM1, Spot R4) (Table 4; Fig. 3b). Other six protein spots were chosen because they were specifically affected by different *Lactobacillus* strains (Table 4; Fig. 3c). The proteins in *L. salivarius* G1-1 group included the up-regulation of phosphatidylinositol 4,5-bisphosphate 3-kinase catalytic subunit gamma isoform (PIK3CG, Spot UB1) and

cofilin-1 (CFL1, Spot UB2), which were only detectable in *L. salivarius* G1-1-treated group. The proteins expressed in *L. reuteri* G8-5 group included the up-regulation of Rho GDP-dissociation inhibitor 2 (ARHGDIB, Spot UC1; only detectable in lactobacilli group) and the down-regulation of nucleophosmin (NPM1, Spot UC2). The proteins in *L. reuteri* G22-2 group included the up-regulation of Rho GDP-dissociation inhibitor 2 (ARHGDIB, Spot UD1; only detectable in

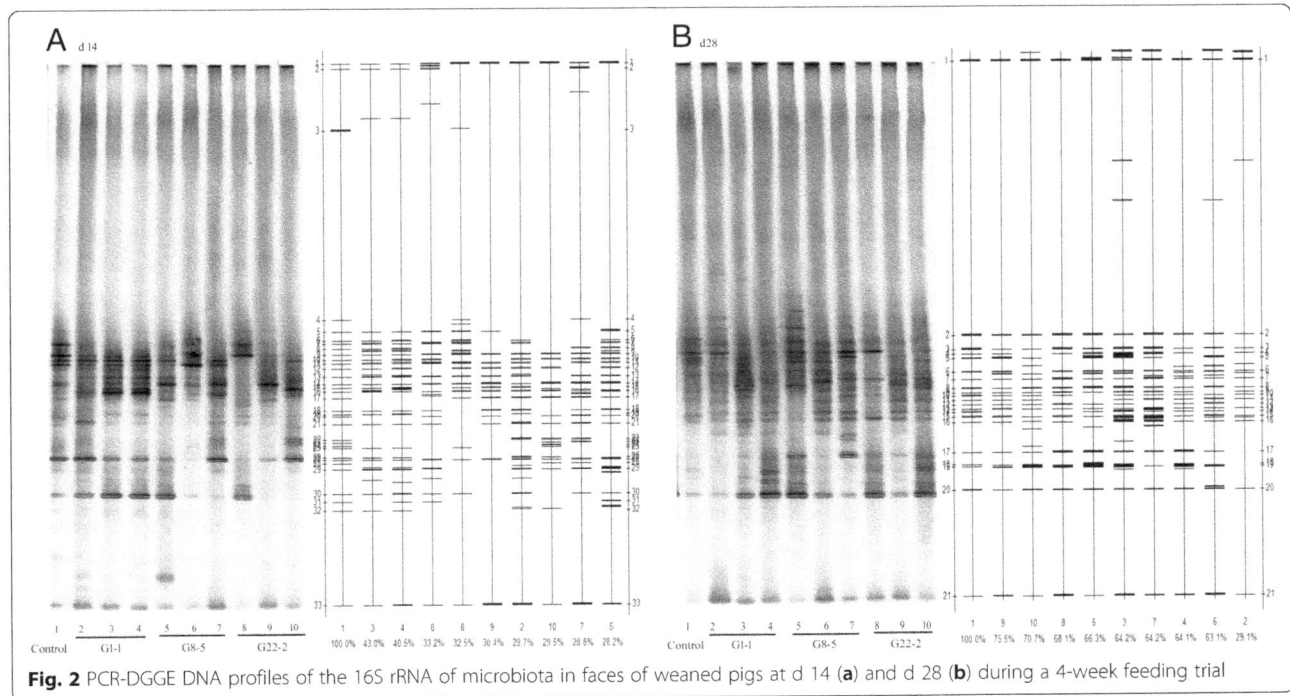

Fig. 2 PCR-DGGE DNA profiles of the 16S rRNA of microbiota in faces of weaned pigs at d 14 (**a**) and d 28 (**b**) during a 4-week feeding trial

Table 3 Effect of three *Lactobacillus* strains on the similarities among digitalized DGGE profiles of PCR-amplified 16S rRNA from fecal DNA after Bio-Rad Quantity One software comparison

Treatments	Similarity, %	
	d 14	d 28
Antibiotics	100.00 ± 0.00[a]	100.00 ± 0.00[a]
L. salivarius G1-1	37.73 ± 7.07[b]	52.47 ± 20.24[b]
L. reuteri G8-5	29.83 ± 2.33[c]	64.53 ± 1.63[b]
L. reuteri G22-2	31.03 ± 1.93[bc]	71.43 ± 3.75[b]

Values are means ± S.D, n = 3. [a, b, c] Mean in a same column with different superscripts differ significantly ($P < 0.05$)

lactobacilli-treated group) and the down-regulation of actin cytoplasmic 1 (ACTB, UD2; only detectable in antibiotics group).

Discussion

The supplementation of lactobacilli in animal diets affects gastrointestinal tract health and growth performance of piglets [1, 5]. However, different *Lactobacillus* strains used as probiotics may achieve the beneficial effects on hosts through different mechanisms [12, 26]. The present study was conducted to compare the different efficacies among three lactobacilli with strain-specific activities in growth performance, faecal microbiota and ileum mucosa proteomics of piglets.

No significant differences in growth performance among *Lactobacillus*-treated groups were observed

compared with the antibiotics-treated group. The result showed that all the three lactobacilli had the same potential as alternative to antibiotics in feed. However, among the three lactobacilli, the supplementation of *L. reuteri* G8-5 was the least effective in enhancing the growth performance of piglets, which was in line with the previous study in the rat experiment [12]. The reason is probably associated with the strain's lower antimicrobial activity compared with the other two strains, which was reported in the previous study [11].

Increased lactobacilli in faeces from lactobacilli-treated piglets on both d 14 and d 28 in this study verified the ability of the three lactobacilli to maintain the balance of microbiota, which was one of the possible mechanisms of lactobacilli as probiotics in vivo [27]. Meanwhile, the modulation of intestinal microbiota by lactobacilli might be strain-insensitive since all the three lactobacilli used in the study showed the same ability as intestinal flora improvers. No difference in coliform counts was observed in whole feeding period compared with the antibiotics group. The result suggested the antibiotics used in the study and lactobacilli had the similar resistance to pathogens and kept them in low level in the gastrointestinal tract. It is assumed that the increasing intestinal microbial abundance caused by antibiotics or lactobacilli has more power to resist the disruption of microbial balance [28, 29]. Further analysis on the microbiota in the gastrointestinal tract treated by lactobacilli and antibiotics by PCR-DGGE was investigated for the comparison

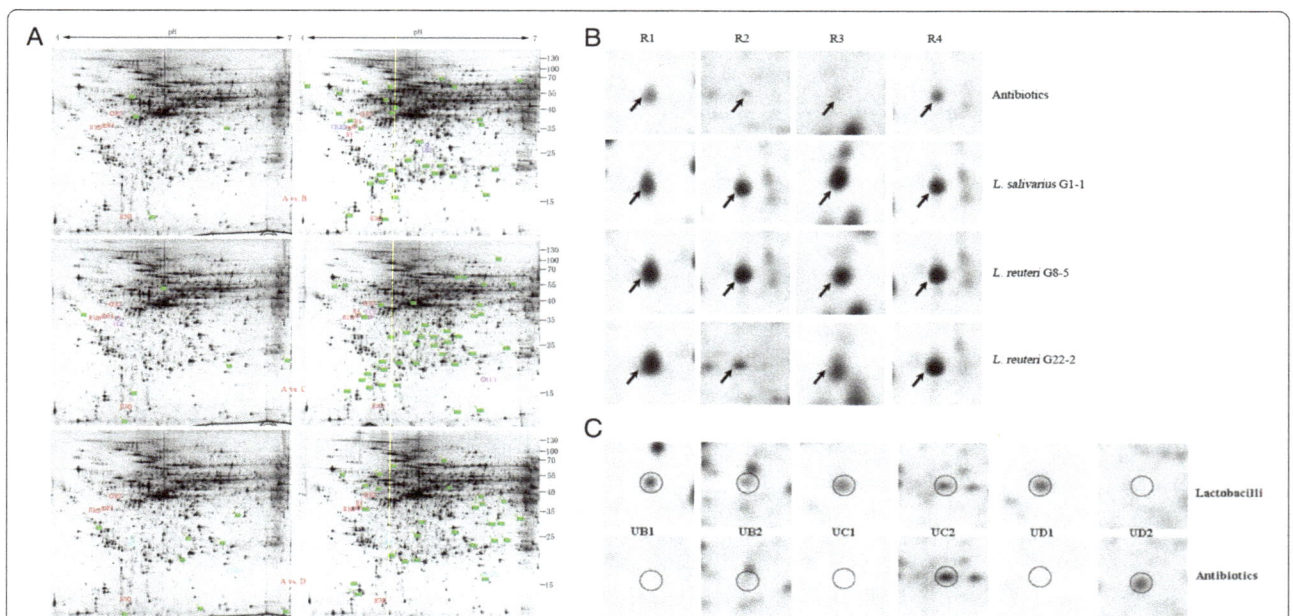

Fig. 3 Representative 2-DE profiles of differentially expressed proteins in the small intestinal mucosa of piglets administrated by lactobacilli or antibiotics. (**a**): Distribution of differentially expressed proteins in antibiotics group (**a**) and each *Lactobacillus* group (**b, c, d**; G1-1, G8-5, G22-2, respectively); (**b**): Up-regulated protein spots in all *Lactobacillus*-treated piglets compared with antibiotics-treated piglets; (**c**): Differentially expressed proteins spots varying from *Lactobacillus* and antibiotics-treated piglets

Table 4 Differentially expressed proteins in the ileum mucosa of piglets supplemented by three lactobacilli in diets compared with antibiotics

Category	Spot No.[a]	Gene	Accession code	Protein	Expression change (Lactobacilli VS. Antibiotics)	Score[b]	Putative function
Lactobacilli-insensitive spots compared with antibiotics	R1	TPM2	F1SG00	Tropomyosin beta chain	Up-regulation	261	Stabilizing cytoskeleton actin filaments
	R2	VIM	P02543	Vimentin	Up-regulation	138	Maintaining cell shape, integrity of the cytoplasm, and stabilizing cytoskeletal interactions
	R3	KRT19	F1S0J8	Keratin type I cytoskeletal 19	Up-regulation	185	Maintaining structural integrity of epithelial cells
	R4	TPM1	F2Z5B6	Tropomyosin alpha-1 chain	Up-regulation	277	Stabilizing cytoskeleton actin filaments
Lactobacilli-related spots compared with antibiotics	UB1(G1-1)	PIK3CG	O02697	Phosphatidylinositol 4,5-bisphosphate 3-kinase catalytic subunit	Only detectable in lactobacilli	137	Maintaining structural and functional integrity of epithelia
	UB2(G1-1)	CFL1	P10668	Cofilin-1	Only detectable in lactobacilli	121	Regulation of cell morphology and cytoskeletal organization
	UC1(G8-5)	ARHGDIB	F1SQW8	Rho GDP-dissociation inhibitor 2	Only detectable in lactobacilli	109	Small GTPase regulator activity receptor binding;
	UC2(G8-5)	NPM1	F1RRY2	Nucleophosmin	Down-regulation	99	Ribosome biogenesis and transport
	UD1(G22-2)	ARHGDIB	F1SQW8	Rho GDP-dissociation inhibitor 2	Only detectable in lactobacilli	116	Small GTPase regulator activity receptor binding;
	UD2(G22-2)	ACTB	Q6QAQ1	Actin cytoplasmic 1	Only detectable in antibiotics	139	Involved in cell motility, structure, and integrity

[a] Spot No. refers to protein spot numbers that were labeled in Fig. 3
[b] Protein score generated by MS identification platform MASCOT; a score > 65 is considered significant

of microbial diversity. On d 14 and d 28, the similarities in all *Lactobacillus*-treated groups were significantly different from the antibiotics-treated group. The results suggested that the mechanisms of antibiotics and lactobacilli on regulating intestinal microbiota were through different ways and lactobacilli contributed to comparatively complex bacterial community. Some similar results were also shown in other reports [30, 31]. The results in Table 4 and Fig. 2 showed the discrepancy in similarities between lactobacilli and antibiotics treatments tended to decrease from d 14 to d 28. This indicates the bacterial diversity tended to be stable and not sensitive to extraneous drugs or introduced bacteria during animals' growth. The significantly lower Dice similarity in *L. reuteri* G8-5 compared with *L. salivarius* G1-1 was observed in this study, and the result was in line with that in the growth performance.

Proteomics play an important role in the assessment of specific health-promoting activities exerted by *Lactobacillus* species [32, 33]. The ileum mucosa samples were collected to compare the differentially expressed proteins through 2-DE profiles. From the result in Table 4, the supplementation of lactobacilli all greatly increased the number of expressed protein spots compared with the antibiotics group. Similar result was also observed in the study of Wang et al. [32]. Up-regulation of four proteins including TPM2, VIM, KRT19 and TPM1 in all three *Lactobacillus* groups are all associated

with the functions of maintaining and stabilizing cell structure and stabilization. The four proteins were inferred to be *Lactobacillus*-insensitive, and the mutual mechanisms for *Lactobacillus* as probiotics were to enhance the expression of proteins beneficial for stabilization of cell structure. Both TPM1 and TPM2 bind to actin filaments and up-regulation of the two proteins benefit to stabilizing cytoskeleton actin filaments [34]. Meanwhile, increased level of VIM is responsible for maintaining cell shape, integrity of the cytoplasm, and stabilizing cytoskeletal interactions [35]. The up-regulation of KRT19 is responsible for the structural integrity of epithelial cells [36]. The increased expression of KRT19 in lactobacilli groups can contribute to more opportunities for living cells to adhere to the epithelial and exclusively inhibit pathogen infection [37, 38]. Similar result was also observed in the study of Wang et al. [32], in which KRT10 was higher in the intestinal mucosa of piglets supplemented with *L. fermentum* I5007 compared with that in antibiotics piglets [32]. Both KRT10 and KRT19 belong to the keratin family which are intermediate filament proteins responsible for the structural integrity of epithelial cells [36].

There were six extra proteins differently expressed in different *Lactobacillus* groups, which were inferred to be *Lactobacillus*-related. The different expression of protein might be caused by the characters of specific strains. In the groups of *L. salivarius* G1-1, two proteins, PIK3CG and CFL1, detected only in *Lactobacillus* group were also associated with cell structure and stability.

ARHGDIB was only detectable in the ileum mucosa of piglets in response to the supplementation of both *L. reuteri* G8-5 and *L. reuteri* G22-2. The high expression of the protein enhances the recycling and distribution of activated Rho GTPases in the cell and play a role in regulating cell motility through the modulation of Rho proteins [39]. NPM1 help cells survive environmental stresses, such as drug attack [40]. Up-regulation of NPM1 in antibiotics might be associated with the intake of flavomycin. The increase in ACTB found in vivo would indicate drastic oxidative modification leading to functional impairments [41], which might be the side-effect of antibiotics supplemented in feed. More experiments are needed in order to document the potential beneficial effects of the lactobacilli strains for the piglets, notably in terms of mucosal health.

Conclusions

In conclusion, this study provides a comprehensive comparison of three lactobacilli with strain-specific activities through the supplementation in piglet diets. All the three lactobacilli show the potential as alternatives to antibiotics and no statistical difference in animal growth performance compared with the antibiotics group. Supplementation of lactobacilli in diets could significantly increase the relative 16S rRNA gene copies of lactobacilli genus on both d 14 and d 28, and the bacterial community profile based on PCR-DGGE from the lactobacilli are distinctly different from the antibiotics group. The ileum mucosa piglets respond to all lactobacilli supplementation by more newly expressed proteins and the identified proteins are all associated with the functions beneficial for stabilization of cell structure. Besides, some other up-regulated and down-regulated proteins in different *Lactobacillus* groups showed the expression of proteins were partly strain-related.

This comparative study helps to explore the mutual mechanisms for *Lactobacillus* as probiotics on altering intestinal abundance of microbiota and expression of mucosa proteins in piglets and provides information for strain-specific screening in application.

Abbreviations

2-DE: 2-dimensional electrophoresis; ACTB: Actin cytoplasmic 1; ADFI: Average daily feed intake; ADG: Average daily weight gain; ARHGDIB: Rho GDP-dissociation inhibitor 2; CFU: Colony forming unit; DGGE: Denaturing gradient gel electrophoresis; DNA: Deoxyribose nucleic acid; F:G: Feed conversion Ratio; GIT: Gastrointestinal tract; IEF: Isoelectric focusing; IPG: Immobilized pH gradient; KRT19: Keratin type I cytoskeletal 19; LAB: Lactic acid bacteria; MALDI-TOF MS: Matrix-assisted laser desorption/ionization time of flight mass spectrometry; NPM1: Nucleophosmin; PCR: Polymerase chain reaction; PIK3CG: Phosphatidylinositol 4,5-bisphosphate 3-kinase catalytic subunit gamma isoform; PMF: Peptide mass fingerprinting; PMM: Peptide-mass mapping; rRNA: Ribosomal ribonucleic acid; RT-PCR: Reverse transcription-polymerase chain reaction; SDS-PAGE: Sodium dodecyl sulfate-polyacrylamide gel electrophoresis; TPM1: Tropomyosin alpha-1 chain; TPM2: Tropomyosin beta chain;

UPGMA: Unweighted pair group method with arithmetic average; VIM: Vimentin

Acknowledgements
We would like to thank Jingjing Yang, Huang Ren and Xiaoping Fu for their lab analysis.

Funding
This work was financially supported by the National Natural Science Foundation of China (No. 31372348; No. 31672455) and the Fundamental Research Funds for the Central Universities (CZY15026).

Authors' contributions
XHG was involved in the study design, data analysis, interpretation and drafting the manuscript; YTS participated in all laboratory analyses; XJC and ML performed the animal management. All authors read and approved the final manuscript.

Competing interests
The authors declare that they have no competing interests.

Author details
[1]Provincial Key Laboratory for Protection and Application of Special Plants in Wuling Area of China, College of Life Science, South-Central University for Nationalities, No. 182, Minyuan Road, Hongshan District, Wuhan, Hubei Province 430074, China. [2]Guangxi Yang-Xiang Animal Husbandry Co. Ltd., Guigang, Guangxi Province 537100, China. [3]Beijing China-agri Hong-Ke Biotechnology Co., Ltd., Beijing 102206, China.

References
1. Fuller R, Gibson GR. Modification of the intestinal microflora using probiotics and prebiotics. Scand J Gastroenter. 1997;32:28–31.
2. Naidu A, Bidlack W, Clemens R. Probiotic spectra of lactic acid bacteria (LAB). Crit Rev Food Sci. 1999;39:13–126.
3. Million M, Angelakis E, Paul M, Armougom F, Leibovici L, Raoult D. Comparative meta-analysis of the effect of *Lactobacillus* species on weight gain in humans and animals. Microb Pathog. 2012;53:100–8.
4. Klaenhammer TR. Functional activities of *Lactobacillus* probiotics: genetic mandate. Int Dairy J. 1998;8:497–505.
5. Simon O, Jadamus A, Vahjen W. Probiotic feed additives-effectiveness and expected modes of action. J Anim Feed Sci. 2001;10:51–67.
6. Abbott A. Microbiology: gut reaction. Nature. 2004;427:284–6.
7. Corr SC, Li Y, Riedel CU, O'Toole PW, Hill C, Gahan CG. Bacteriocin production as a mechanism for the antiinfective activity of *Lactobacillus salivarius* UCC118. P Natl Acad Sci USA. 2007;104:7617–21.
8. Lee H, Gilliland S, Carter S. Amylolytic cultures of *Lactobacillus acidophilus*: potential probiotics to improve dietary starch utilization. J Food Sci. 2001;66:338–44.
9. LeBlanc JG, Piard JC, Sesma F, de Giori GS. *Lactobacillus fermentum* CRL 722 is able to deliver active α-galactosidase activity in the small intestine of rats. FEMS Microbiol Lett. 2005;248:177–82.
10. Oozeer R, Goupil-Feuillerat N, Alpert CA, Van de Guchte M, Anba J, Mengaud J, et al. *Lactobacillus casei* is able to survive and initiate protein synthesis during its transit in the digestive tract of human flora-associated mice. Appl Environ Microbiol. 2002;68:3570–4.
11. Guo XH, Kim JM, Nam HM, Park SY, Kim JM. Screening lactic acid bacteria from swine origins for multistrain probiotics based on *in vitro* functional properties. Anaerobe. 2010;16:321–6.
12. Guo XH, Zhao ZD, Nam HM, Kim JM. Comparative evaluation of three Lactobacilli with strain-specific activities for rats when supplied in drinking water. Anton Leeuw Int J G. 2012;102:561–8.
13. Zoetendal EG, Akkermans AD, De Vos WM. Temperature gradient gel electrophoresis analysis of 16S rRNA from human fecal samples reveals stable and host-specific communities of active bacteria. Appl Environ Microbiol. 1998;64:3854–9.
14. Su Y, Yao W, Perez-Gutierrez ON, Smidt H, Zhu WY. 16S ribosomal RNA-based methods to monitor changes in the hindgut bacterial community of piglets after oral administration of *Lactobacillus sobrius* S1. Anaerobe. 2008;14:78–86.

15. Su Y, Yao W, Perez-Gutierrez ON, Smidt H, Zhu WY. Changes in abundance of *Lactobacillus* spp. and *Streptococcus suis* in the stomach, jejunum and ileum of piglets after weaning. FEMS Microbiol Ecol. 2008;66:546–55.

16. Li J, Wang J, Wang F, Wang A, Yan P. Evaluation of gaseous concentrations, bacterial diversity and microbial quantity in different layers of deep litter system. Asian Australas J Anim Sci. 2016;30:275–83.

17. Livak KJ, Schmittgen TD. Analysis of relative gene expression data using real-time quantitative PCR and the $2^{-\Delta\Delta CT}$ method. Methods. 2001;25:402–8.

18. Nübel U, Engelen B, Felske A, Snaidr J, Wieshuber A, Amann RI, et al. Sequence heterogeneities of genes encoding 16S rRNAs in *Paenibacillus polymyxa* detected by temperature gradient gel electrophoresis. J Bacteriol. 1996;178:5636–43.

19. Suzuki MT, Taylor LT, DeLong EF. Quantitative analysis of small-subunit rRNA genes in mixed microbial populations via 5'-nuclease assays. Appl Environ Microbiol. 2000;66:4605–14.

20. Muyzer G, de Waal EC. Determination of the genetic diversity of microbial communities using DGGE analysis of PCR-amplified 16S rDNA. Microbial Mats. 1994;35:207–14.

21. Wu X, Ma C, Han L, Nawaz M, Gao F, Zhang X, et al. Molecular characterisation of the faecal microbiota in patients with type II diabetes. Curr Microbiol. 2010;61:69–78.

22. Echan LA, Tang HY, Ali-Khan N, Lee K, Speicher DW. Depletion of multiple high-abundance proteins improves protein profiling capacities of human serum and plasma. Proteomics. 2005;5:3292–303.

23. Izquierdo E, Horvatovich P, Marchioni E, Aoude-Werner D, Sanz Y, Ennahar S. 2-DE and MS analysis of key proteins in the adhesion of *Lactobacillus plantarum*, a first step toward early selection of probiotics based on bacterial biomarkers. Electrophoresis. 2009;30:949–56.

24. Cordero H, Morcillo P, Cuesta A, Brinchmann MF, Esteban MA. Differential proteome profile of skin mucus of gilthead seabream (*Sparus aurata*) after probiotic intake and/or overcrowding stress. J Proteomics. 2016;132:41–50.

25. Yang F, Wang J, Li X, Ying T, Qiao S, Li D, et al. 2-DE and MS analysis of interactions between *Lactobacillus fermentum* I5007 and intestinal epithelial cells. Electrophoresis. 2007;28:4330–9.

26. Du Toit M, Franz C, Dicks L, Schillinger U, Haberer P, Warlies B, et al. Characterisation and selection of probiotic lactobacilli for a preliminary minipig feeding trial and their effect on serum cholesterol levels, faeces pH and faeces moisture content. Int J Food Microbiol. 1998;40:93–104.

27. Brashears M, Jaroni D, Trimble J. Isolation, selection, and characterization of lactic acid bacteria for a competitive exclusion product to reduce shedding of *Escherichia coli* O157: H7 in cattle. J Food Prot. 2003;66:355–63.

28. Wohlgemuth S, Haller D, Blaut M, Loh G. Reduced microbial diversity and high numbers of one single *Escherichia coli* strain in the intestine of colitic mice. Environ Microbiol. 2009;11:1562–71.

29. Kim HB, Borewicz K, White BA, Singer RS, Sreevatsan S, Tu ZJ, et al. Microbial shifts in the swine distal gut in response to the treatment with antimicrobial growth promoter, tylosin. P Natl Acad Sci USA. 2012;109:15485–90.

30. Sung V, Hiscock H, Tang ML, Mensah FK, Nation ML, Satzke C, et al. Treating infant colic with the probiotic *Lactobacillus reuteri*: double blind, placebo controlled randomised trial. BMJ. 2014;348:g2107.

31. Pieper R, Janczyk P, Urubschurov V, Hou Z, Korn U, Pieper B, et al. Effect of *Lactobacillus plantarum* on intestinal microbial community composition and response to enterotoxigenic *Escherichia coli* challenge in weaning piglets. Livest Sci. 2010;133:98–100.

32. Wang X, Yang F, Liu C, Zhou H, Wu G, Qiao S, et al. Dietary supplementation with the probiotic *Lactobacillus fermentum* I5007 and the antibiotic aureomycin differentially affects the small intestinal proteomes of weanling piglets. J Nutr. 2012;142:7–13.

33. Ruiz L, Hidalgo C, Blanco-Míguez A, Lourenço A, Sánchez B, Margolles A. Tackling probiotic and gut microbiota functionality through proteomics. J Proteomics. 2016;147:28–39.

34. Perry SV. Vertebrate tropomyosin: distribution, properties and function. J Muscle Res Cell M. 2001;22:5–49.

35. Wang N, Stamenovic D. Mechanics of vimentin intermediate filaments. J Muscle Res Cell M. 2002;23:535–40.

36. Bragulla HH, Homberger DG. Structure and functions of keratin proteins in simple, stratified, keratinized and cornified epithelia. J Anat. 2009;214:516–59.

37. Dhanani AS, Bagchi T. *Lactobacillus plantarum* CS24. 2 prevents *Escherichia coli* adhesion to HT-29 cells and also down-regulates enteropathogen-induced tumor necrosis factor-α and interleukin-8 expression. Microbiol Immunol. 2013;57:309–15.

38. Li X, Yue L, Guan X, Qiao S. The adhesion of putative probiotic lactobacilli to cultured epithelial cells and porcine intestinal mucus. J Appl Microbiol. 2008;104:1082–91.

39. Garcia-Mata R, Boulter E, Burridge K. The 'invisible hand': regulation of RHO GTPases by RHOGDIs. Nat Rev Mol Cell Bio. 2011;12:493–504.

40. Yang YX, Hu HD, Zhang DZ, Ren H. Identification of proteins responsible for the development of adriamycin resistance in human gastric cancer cells using comparative proteomics analysis. BMB Rep. 2007;40:853–60.

41. Li G, Chang M, Jiang H, Xie H, Dong Z, Hu L. Proteomics analysis of methylglyoxal-induced neurotoxic effects in SH-SY5Y cells. Cell Biochem Funct. 2011;29:30–5.

10

Solid-state fermentation of corn-soybean meal mixed feed with *Bacillus subtilis* and *Enterococcus faecium* for degrading antinutritional factors and enhancing nutritional value

Changyou Shi, Yu Zhang, Zeqing Lu and Yizhen Wang[*]

Abstract

Background: Corn and soybean meal (SBM) are two of the most common feed ingredients used in pig feeds. However, a variety of antinutritional factors (ANFs) present in corn and SBM can interfere with the bioavailability of nutrients and have negative health effects on the pigs. In the present study, two-stage fermentation using *Bacillus subtilis* followed by *Enterococcus faecium* was carried out to degrade ANFs and improve the nutritional quality of corn and SBM mixed feed. Furthermore, the microbial composition and in vitro nutrient digestibility of inoculated mixed feed were determined and compared those of the uninoculated controls.

Results: During the fermentation process, *B. subtilis* and lactic acid bacteria (LAB) were the main dominant bacteria in the solid-state fermented inoculated feed, and fermentation produced a large amount of lactic acid (170 mmoL/kg), which resulted in a lower pH (5.0 vs. 6.4) than the fermented uninoculated feed. The amounts of soybean antigenic proteins (β-conglycinin and glycinin) in mixed feed were significantly decreased after first-stage fermentation with *B. subtilis*. Inoculated mixed feed following two-stage fermentation contained greater concentratioin of crude protein (CP), ash and total phosphorus (P) compared to uninoculated feed, whereas the concentrations of neutral detergent fiber (NDF), hemicellulose and phytate P in fermendted inoculated feed declined (*P* < 0.05) by 38%, 53%, and 46%, respectively. Notably, the content of trichloroacetic acid soluble protein (TCA-SP), particularly that of small peptides and free amino acids (AA), increased 6.5 fold following two-stage fermentation. There was no difference in the total AA content between fermented inoculated and uninoculated feed. However, aromatic AAs (Phe and Tyr) and Lys in inoculated feed increased, and some polar AAs, including Arg, Asp, and Glu, decreased compared with the uninoculated feed. In vitro dry matter and CP digestibility of inoculated feed improved (*P* < 0.05) compared with the uninoculated feed.

Conclusions: Our results suggest that two-stage fermentation using *B. subtilis* followed by *E. faecium* is an effective approach to improve the quality of corn-soybean meal mixed feed.

Keywords: Antinutritional factors (ANFs), In vitro digestibility, Mixed feed, Nutritional value, Two-stage fermentation

* Correspondence: yzwang321@zju.edu.cn
Institute of Feed Science, College of Animal Science, Zhejiang University,
Yuhangtang Road 866#, Hangzhou, Zhejiang Province 310058, People's
Republic of China

Background

Corn and soybean meal (SBM) are the most common feedstuffs used in pig production in China. However, conventional corn-SBM diets contain a variety of antinutritional factors (ANFs), such as soybean antigenic proteins, phytic acid, oligosaccharides, and other factors that can interfere with the bioavailability of nutrients and have negative health effects in pigs. Soybean antigenic proteins in the diets of weaned pigs provoked a transient hypersensitivity associated with the abnormal morphology of the small intestine [1]. These morphological changes can cause malabsorption syndrome, growth depression, and diarrhea [2, 3]. Phytate in diets may reduce mineral bioavailability and protein digestibility when it is fed to pigs [4]. Furthermore, high levels of soy oligosaccharides, in particular, stachyose and raffinose, can cause intestinal disorder in weaning piglets [5].

Previous research has indicated that fermentation can improve the nutritional quality of animal feed by increasing nutrient bioavailability and reducing ANFs [6]. In China, fermented feed is mainly produced through solid state fermentation (SSF), in which the focus is to decrease the ANFs in single feed ingredients, such as soybean meal [7], cottonseed meal [8], and rapeseed meal [9]. There have been few studies on the use of fermented mixed feed manufactured with SSF. However, the ability of SSF to effectively enhance the nutritional value of mixed feed should be further evaluated. Feeding pigs with fermented liquid feed (FLF) is a useful feeding strategy. Although the growth performance of piglets fed FLF compared with those fed dry feed or non-FLF has been shown to be variable, high lactic acid concentration and low pH in FLF can improve the gastrointestinal health of piglets [10]. In the present study, corn and SBM mixed feed was inoculated with *Bacillus subtilis* in the first stage of fermentation. The aim of the first stage was to decrease ANFs in mixed feed. Subsequently, *Enterococcus faecium* was used in the second stage fermentation to produce lactic acid and reduce the pH of the mixed feed.

Methods

Microorganisms and basal substrate

B. subtilis ZJ12-1 was isolated from a traditional fermented food (pickled vegetables). *B subtilis* ZJ12-1 was selected with a specific screening plate in which the soybean antigenic protein was the sole nitrogen source extracted from SBM. This strain was identified with Gram's dye and conventional biochemical tests including sugar fermentation, Voges-Proskauer, starch hydrolysis, gelatin liquefaction, salt tolerance etc., then confirmed with 16S rDNA sequencing (Additional file 1: Figures S1, S2, S3 and Table S1). *E. faecium* NCIMB 10415 was obtained from Baolai-leelai Bio-tech Co. Ltd. (Taian, China). *E. faecium*, which is an authorized feed additive

for piglets in the EU and China, was isolated from healthy piglet intestines. Dried corn and SBM sieved through 40 mesh sieves were used in SSF.

Preparation of inoculated mixed feed

A schematic outline showing the manufacturing process of the two-stage fermented feed is provided in Fig. 1. Before fermentation, *B. subtilis* was cultured in Luria broth (LB) liquid medium at 37 °C for 12 h. *E. faecium* was cultured in de Man, Rogosa and Sharp (MRS) liquid medium at 37 °C for 16 h. The basal substrate (150 g) included 45% corn, 45% SBM and 10% wheat bran, which was mixed and placed in a 500 mL Erlenmeyer flask covered with cotton plugs and supplemented with sterile water to achieve a 40% moisture content. The wet mixed substrate was inoculated with *B. subtilis* (8.0 log cfu/g) and fermented at 37 °C for 24 h. After the first-stage of fermentation, the fermented mixture was transferred to a plastic bag equipped with a one-way valve (Rou Duoduo Biotechnology Co., Beijing, China), inoculated with *E. faecium* (8.0 log cfu/g), and incubated under anaerobic conditions at 37 °C (the second-stage of fermentation). Uninoculated flasks served as controls. In uninoculated samples, all the experimental procedures were the same as those for inoculated feed, except for the addition of sterile medium (LB and MRS) instead of inoculated bacteria. Inoculated and uninoculated samples (control) were set up in triplicate. After 48 h of anaerobic fermentation, wet samples (approximately 100 g) were collected and treated at 105 °C for 30 min to prevent continuous fermentation. Then, the inoculated and uninoculated samples were dried at 65 °C for 24 h, cooled and ground. Treated samples were subjected to sodium dodecyl sulfate-polyacrylamide gel electrophoresis (SDS-PAGE), in vitro digestibility, and chemical analysis. Moreover, the remaining inoculated feed continued to ferment under anaerobic conditions for 48 h (total 96 h) at 37 °C. Moisture samples were collected at different inoculation times for microbial, pH, and lactic acid analysis.

Microbial determinations

Inoculated feed at different incubation times (0, 24, 48, 72, 96, and 120 h) were taken, 2 g of samples was diluted 1:9 (w/v) with sterile water. The suspension was homogenized in a stomacher blender (Interscience, St. Nom, France) for 2 min. Ten-fold dilutions were prepared in sterile water and 0.1 mL samples were plated on selective media. Lactic acid bacteria (LAB) were measured on MRS agar following anaerobic incubation at 37 °C for 2 d. The number of *Enterobacteriaceae* was determined on eosin methylene blue agar following aerobic incubation at 37 °C for 1 d. Yeasts and molds were counted on yeast extract peptone dextrose agar with 50 mg/L chloramphenicol (SR0177E,

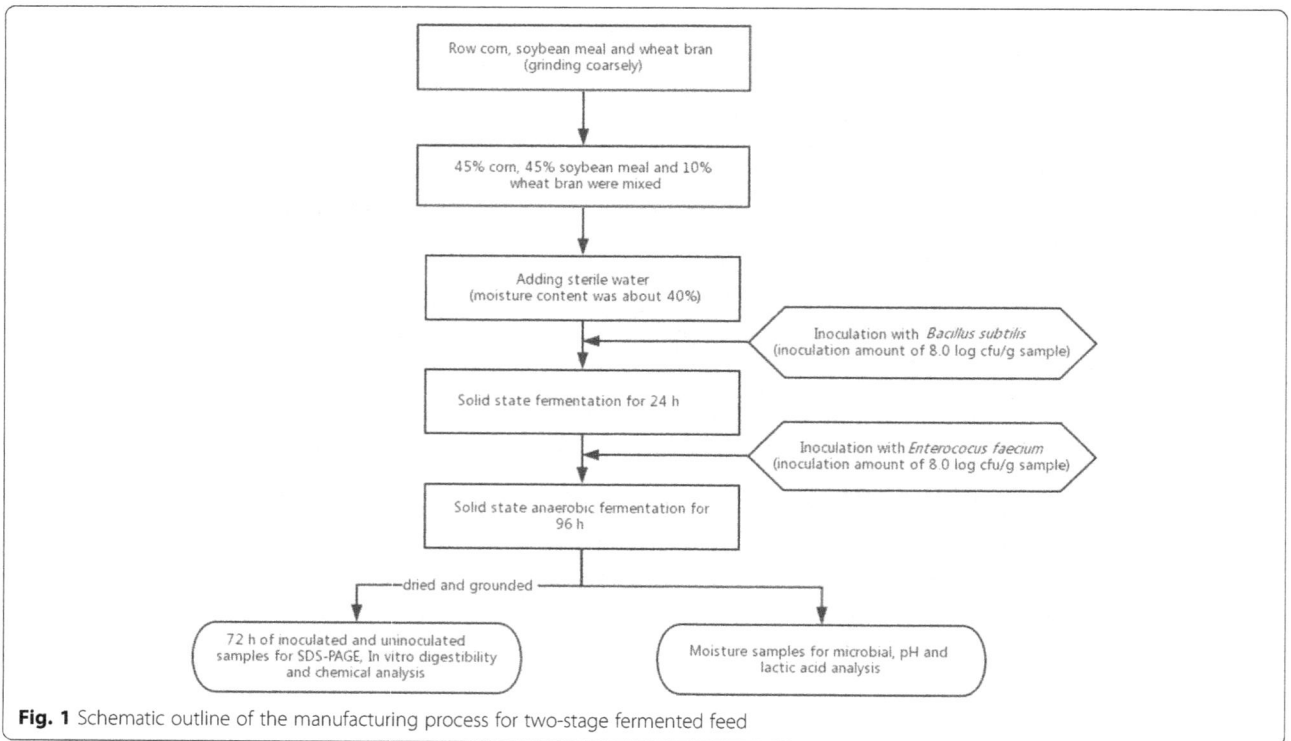

Fig. 1 Schematic outline of the manufacturing process for two-stage fermented feed

Oxoid LTD, Basingstoke, Hampshire, England) following aerobic incubation at 30 °C for 2 d. Nutrient Broth agar was used to count *Bacillus* spp. by morphological and biochemical identification after aerobic incubation at 37 °C for 1 d. The biochemical tests were the same as strain identification.

Chemical analysis

Uninoculated and inoculated (24 h with *Bacillus subtilis* and 48 h with *Enterococcus faecium*) feeds were analyzed for dry matter (DM), crude protein (CP), ether extract, neutral detergent fiber (NDF), acid detergent fiber (ADF), ash, calcium (Ca) and total phosphorus (P) using the AOAC International guidelines (2005). The trichloroacetic acid soluble protein (TCA-SP) of the sample was determined using the methods described by Ovissipour et al. [11]. The phytic acid content was measured according to the procedures described by Nair and Duvnjak [12]. Phytate-bound P was calculated as 28.2% of phytate. The amino acid profile was analyzed using an automatic amino acid analyzer (L-8900; Hitachi, Tokyo, Japan). Before analysis, the samples were hydrolyzed with 6 mol/L HCl at 110 °C for 24 h. Methionine and cysteine were analyzed as Met sulfone and cysteic acid after cold performic acid oxidation overnight before hydrolysis. The pH values (at different incubation times) were measured using a HI 99163 pH meter (Hanna instruments, Woonsocket, RI, USA) using 2 g of sample mixed with 18 mL of distilled water. The lactic acid content was determined using a

lactic acid enzymology assay kit (Nanjing - Jiancheng Bio Co., Nanjing, China) according to the manufacturer's protocol. The contents of glycinin and β-conglycinin in uninoculated and inoculated feed were analyzed using an indirect competitive enzyme-linked immunosorbent assay (ELISA) kit (Longzhoufangke Bio Co., Beijing, China) according to the manufacturer's protocol.

Sodium dodecyl sulfate-polyacrylamide gel electrophoresis (SDS-PAGE)

Soluble proteins in fermented uninoculated and inoculated feed were extracted according to the protocol described by Faurobert [13] with some minor modifications. The samples were ground finely to pass through a 60-mesh sieve, and 1.5 mL of Tris–HCl buffer (20 mmol/L, pH 8.3), including 0.1% SDS, 5 mmol/L dithiothreitol and 5 μg/mL protease inhibitor was added to each 0.1 g sample, and then homogenized on ice for 30 min. The homogenized samples were centrifuged at 14,000 × g for 10 min at 4 °C (5804R, Eppendorf, Germany), and the supernatants were transferred to Eppendorf tubes. The protein concentration in each sample was determined using the Bio-Rad Protein Assay Kit (Bio-Rad, USA). Soluble protein was fractionated using an SDS-PAGE system as previously described [14]. The electrophoresis system was based on 4 - 12% polyacrylamide gradient separating gels containing 0.1% SDS in Tris-glycine buffer. Approximately 20 μg of extracted protein sample was loaded into each well and separated at 65 mV for 120 min. A Thermo

26616 page ruler pre-stained protein ladder (10–170 kDa) was used as a size marker. After electrophoresis, the gel was stained using Coomassie Brilliant Blue R-250 (Bio-Rad, USA) for 45 min and de-stained with 7% acetic acid.

In vitro digestibility

An in vitro two-stage enzyme hydrolysis procedure was performed as described by Sakamoto et al. [15], with some modifications. In brief, fermented inoculated feed or uninoculated feed (3 g) was placed in 150 mL Erlenmeyer flasks. Thirty milliliters of 10,000 U/mL pepsin (activity: 3,000 U/mg, Sigma) solution (0.05 mol/L KCl-HCl buffer, pH 2.0) was mixed and incubated at 39 °C at 150 revolutions per min (rpm) for 4 h. The pH was adjusted to 7.0 with 1 mol/L NaOH, and 150 mg trypsin (activity: 250 U/mg, Sigma) was added to each sample, which were then mixed again, and incubated at 39 °C at 150 rpm for 4 h. After the digestion was complete, 5 mL of 20% sulfosalicylic acid was added and the samples were settled for 30 min. The digesta slurry samples were centrifuged at 3,000×g for 15 min, and the supernatants were discarded. The resulting pellets were dried at 105 °C for 4 h and analyzed in subsequent CP and AA assays. In vitro nutrient digestibility (%) = (original nutrient amount − residual nutrient amount) / original nutrient amount × 100%.

Statistical analysis

The data were analyzed by a one-way analysis of variance using the General Linear Models in SAS software (SAS, 1999). A value of 0.05 was used to indicate of a significant difference. The results are expressed as the means and standard deviations.

Results

Microbial composition, pH and lactic acid concentration during SSF

Figure 2a and b shows the microbial composition, pH and lactic acid dynamics during SSF. When the fermentation was prolonged, B. bacillus and LAB were the dominant bacteria present in the solid-state fermented inoculated feed (Fig. 2a). The initial density of B. subtilis was 8.0 log cfu/g. After 24 h of incubation, the density increased to 9.6 log cfu/g, and this level was maintained throughout the fermentation experiment. LAB naturally occurring in the mixed feed was low (<3.0 log cfu/g), and the density of LAB at 24 h was 8.1 log cfu/g after inoculation with E. faecium; this increased to 9.6 log cfu/g at 48 h in inoculated feed. During the subsequent fermentation period, the number of LAB was similar to that of B. subtilis. Notably, there was a proliferation of Enterobacteriaceae (potentially pathogens) during the first-stage of fermentation, which reached a maximum level (8.3 log cfu/g) after 24 h of incubation. However,

the number of Enterobacteriaceae gradually decreased as the fermentation time increased, and the final count was below the level of detection (<3.0 log cfu/g). The two-stage process of B. subtilis and E. faecium fermentation had a significant effect on both the pH and lactic acid concentrations of the fermented substrate (Fig. 2b). There was a small increase in pH after incubation with B. subtilis (6.8 in the inoculated feed with 24 h fermentation vs. 6.4 in the raw mixed feed with 0 h fermentation). After inoculation with E. faecium, the pH gradually decreased from 6.8 to 5.0. Almost no change in the lactic acid content was observed during the first-stage of fermentation, and there was a gradual increase in the lactic acid content from 31 to 170 mmol/kg during the second-stage of anaerobic fermentation.

Biodegradation of soybean antigenic proteins of mixed feed after fermentation

Our results showed that the protein profile corresponded to multiple bands in the range of 23–80 kDa in fermented uninoculated feed (Fig. 3). Subunits of soybean antigenic proteins, including β-conglycinin of α, α′ and β, and acidic and basic glycinin in soybean protein were separated. First-stage fermentation with B. subtilis significantly affected the characteristics of proteins in mixed feed. The α and α′ subunits of β-conglycinin and the acidic subunits of glycinin in the mixed feed were almost completely degraded during SSF. In contrast, fermentation increased the number of small peptides (<25 kDa) compared with the fermented uninoculated substrate. However, there was no effect on the protein profile during the second-stage of fermentation with E. faecium compared with the first-stage fermented feed. The contents of soybean antigenic protein in fermented uninoculated and inoculated feed are presented in Table 1. Both β-conglycinin and glycinin contents were significantly decreased after fermentation, and degradation of the antigenic protein had already occurred in the first-stage of fermentation.

Chemical composition

The analyzed nutrient contents of the fermented uninoculated and inoculated feed after 72 h of incubation are presented in Table 2. Compared with uninoculated feed, the inoculated feed contained more CP, ash, and total P, whereas the concentrations of NDF, hemicellulose and phytate P declined ($P < 0.05$) by 39%, 53%, and 46%, respectively. Notably, the content of TCA-SP (<10 kDa) in uninoculated feed was 1.18%, which was increased 6.5-fold in inoculated feed. Inoculating with B. subtilis and E. faecium affected the AA composition patterns of mixed feed; content of some polar amino acids (Arg, Asp, and Glu) decreased, and aromatic amino acids (Phe and Tyr) and Lys increased. Compared to uninoculated

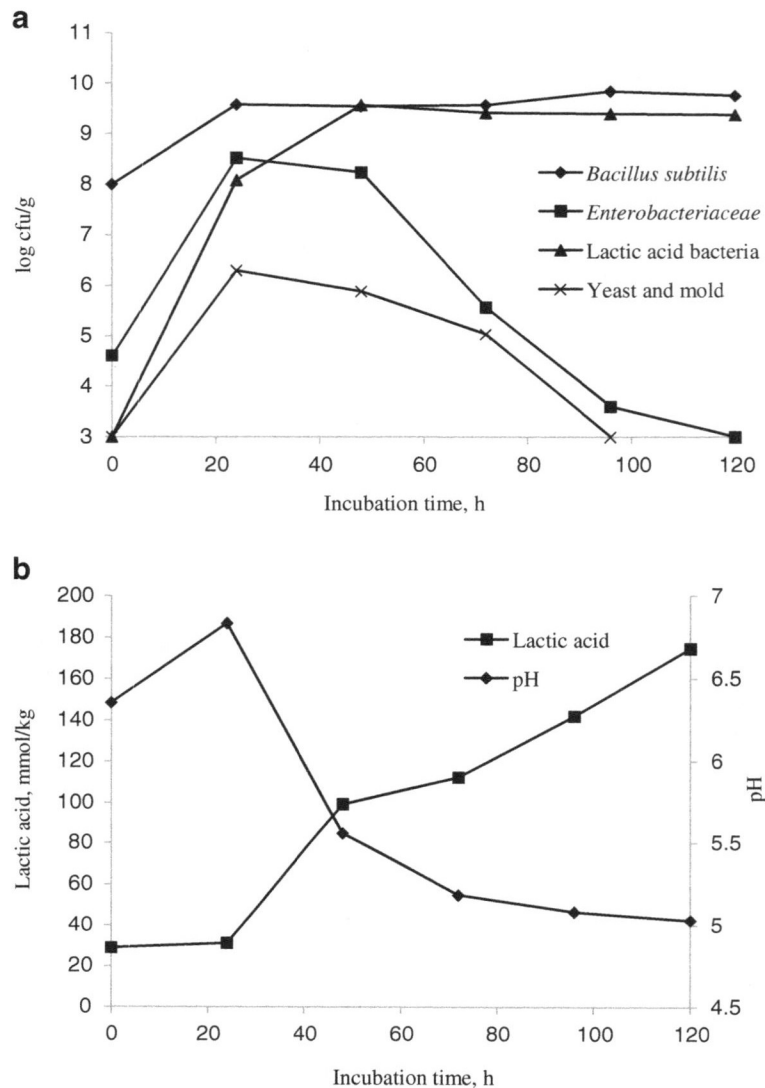

Fig. 2 Microbial composition (log cfu/g) **a**, pH, and lactic acid concentration (mmol/kg) (**b**) in inoculated feed during solid-state fermentation, on a DM basis

feed, the total AA of inoculated feed increased by 7%; however, the difference was not significant.

In vitro amino acid digestibility of the fermented samples

The results of in vitro AA digestibility of fermented inoculated feed with two-stage fermentation are presented in Table 3. In vitro CP and DM digestibility of inoculated feed were improved ($P < 0.05$) by 8% and 11%, respectively, compared with uninoculated substrate. In addition, the in vitro digestibility of 11 amino acids, including six essential amino acids (His, Ile, Leu, Met, Phe and Val), improved greatly ($P < 0.05$). Notably, the in vitro digestibility of three amino acids (His, Phe and Cys) increased by more than 10%.

Discussion

Interest in the fermentation of feed for improving the health of pigs increased dramatically after the European Union banned the use of antibiotics as antimicrobial growth promoters for swine [16, 17]. FLF usually contains >9 log cfu/g of LAB and a high concentration of lactic acid (>150 mmol/L), which can prevent the proliferation of spoilage organisms in the gastrointestinal tracts (GIT) of pigs, such as coliforms and *Salmonella* [18]. Additional advantages of feeding FLF include an increase in nutrient digestibility [19], improved intestinal morphology [20], and a reduction in dust levels in swine barns [21]. Feeding FLF has been shown to improve the performance of piglets and growing-finishing pigs [22],

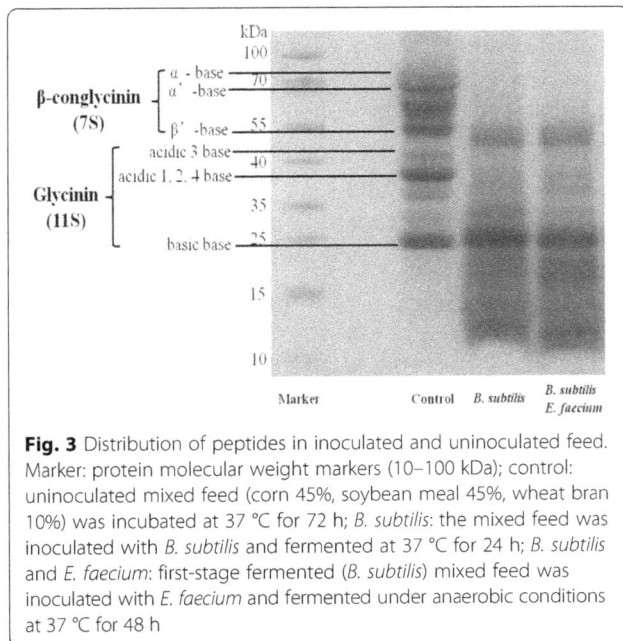

Fig. 3 Distribution of peptides in inoculated and uninoculated feed. Marker: protein molecular weight markers (10–100 kDa); control: uninoculated mixed feed (corn 45%, soybean meal 45%, wheat bran 10%) was incubated at 37 °C for 72 h; B. subtilis: the mixed feed was inoculated with B. subtilis and fermented at 37 °C for 24 h; B. subtilis and E. faecium: first-stage fermented (B. subtilis) mixed feed was inoculated with E. faecium and fermented under anaerobic conditions at 37 °C for 48 h

although the results showed high variation. In the present study, changes in the microbial composition with incubation time were determined. During the fermentation process, B. subtilis and LAB were the main dominant bacteria in the solid-state fermented feed. The final count of LAB (9.4 log cfu/g) in the present study was similar to that achieved with FLF. The B. subtilis count was >9.0 log cfu/g after the first stage of fermentation. However, the source of determined bacteria (exogenous addition or naturally occurring in the feed) is not clear. Notably, a proliferation of Enterobacteriaceae, mainly coliforms (potentially pathogens), also occurred at this stage. This result was consistent with that during the initial fermentation of FLF reported by Canibe and Jensen [10]. An increase in feed pH from 6.4 to 6.8 was observed during first-stage fermentation. This increase may be the result of fermentation with B. subtilis and other microbes, which introduce some new metabolites or changes in the chemical composition of the substrate. However,

additional research is needed to determine the specific reasons for this increase. During the second stage of fermentation, a decrease in pH from 6.8 to 5.0 was most likely the result of increased lactic acid production. Several previously published studies showed that FLF had a pH of 3.8–4.5 [23–25], which was lower than that obtained in the present study. One explanation for the lower pH with FLF may be the difference in the composition of raw materials, since different ingredients have different buffer capacities. The pH may have decreased more rapidly when only the cereals were fermented because cereals have a lower buffering capacity than compound feed [10]. In the present study, the number of Enterobacteriaceae gradually decreased as the anaerobic incubation time increased, and the final count was reduced to levels below detection limits (<3.0 log cfu/g). Coliform was reduced mainly due to the low pH and increased lactic acid in fermented inoculated feed. Feeding fermented feed with low pH and high concentration of lactic acid can prevent the proliferation of pathogens (e.g. Enterobacteriaceae) along the GIT of piglets [18, 23].

Corn-SBM diet is the most commonly used feed for animal production in China. Corn, as the main energy feed, usually accounts for approximately 60% of the animals' diet. Soybean meal is the most important plant protein feed for monogastric animals. Soybean antigenic proteins in the diet of weaned pigs, particularly glycinin and β-conglycinin, promote transient hypersensitivity, which may lead to morphological changes in the small intestine, including villi atrophy and crypt hyperplasia [1]. The use of solid-state fermentation to enhance the nutritional characteristics of raw plant materials has been proposed to improve the use of these materials in animal feeds. Several recent studies have shown that soy antigenic proteins could be degraded during fermentation [26, 27]; consequently, immunoreactivity and allergic reactions caused by soy products were reduced in human and animals [28, 29]. In the present study, the first stage of fermentation with B. subtilis significantly affected the characteristics of proteins in corn-SBM mixed feed. The α and α' subunits of β-conglycinin and

Table 1 Effect of fermentation on the concentration of soybean antigenic protein, as air-dry basis

Item	Glycinin		β-conglycinin	
	Content, mg/g	Degradation[a], %	Content, mg/g	Degradation, %
Raw mixed feed	63.74	-	31.76	-
Uninoculated feed[b]	61.02	-	32.15	-
B. subtilis[c]	7.97	86.94	6.98	78.28
B. subtilis and E. faecium[d]	8.47	86.12	7.12	77.53

[a]Degradation rate = soybean antigenic protein content in uninoculated feed – soybean antigenic protein content in inoculated feed) / soybean antigenic protein content in uninoculated feed × 100%
[b]Uninoculated feed: sterile medium was added to mixed feed (45% corn, 45% soybean meal and 10% wheat bran) instead of inoculated bacteria, other experimental procedures were the same as those of inoculated mixed feed
[c]B. subtilis: mixed feed was inoculated with B. subtilis and fermented at 37 °C for 24 h
[d]B. subtilis and E. faecium: first-stage fermented (B. subtilis) mixed feed was inoculated with E. faecium and incubated under anaerobic conditions at 37 °C for 48 h

Table 2 Analyzed nutrient composition of fermented inoculated and uninoculated feed, as air-dry basis[1]

Item	Inoculated feed	Uninoculated feed[2]	AA compositon	Inoculated feed	Uninoculated feed
DM,%	88.06 ± 1.02	89.09 ± 1.67	Indispensable AA, %		
CP,%	27.61 ± 2.73[a]	24.03 ± 1.93[b]	Arg	1.01 ± 0.15[b]	1.17 ± 0.19[a]
TCA-SP,%	8.85 ± 1.19[a]	1.18 ± 0.12[b]	His	0.58 ± 0.08	0.52 ± 0.11
Fat,%	3.37 ± 0.65	3.67 ± 0.73	Ile	0.78 ± 0.13	0.78 ± 0.16
NDF,%	8.33 ± 0.95[b]	13.64 ± 0.99[a]	Leu	1.50 ± 0.18	1.46 ± 0.23
ADF,%	3.58 ± 0.40	3.49 ± 0.76	Lys	1.17 ± 0.08[a]	0.99 ± 0.10[b]
Hemicellulose[3],%	4.75 ± 0.87[b]	10.15 ± 0.56[a]	Met	0.26 ± 0.05	0.23 ± 0.07
Ash,%	4.71 ± 0.51[a]	3.77 ± 0.38[b]	Phe	1.78 ± 0.26[a]	0.86 ± 0.13[b]
Ca,%	0.18 ± 0.03	0.17 ± 0.02	Thr	0.79 ± 0.12	0.75 ± 0.09
Total P,%	0.55 ± 0.05	0.49 ± 0.07	Val	1.06 ± 0.12	1.04 ± 0.17
Phytate P,%	0.21 ± 0.04[b]	0.39 ± 0.04[a]	Dispensable AA,%		
			Asp	1.68 ± 0.10[b]	1.92 ± 0.17[a]
			Ser	0.79 ± 0.14	0.75 ± 0.16
			Glu	3.23 ± 0.58	3.49 ± 0.44
			Gly	0.85 ± 0.18	0.80 ± 0.13
			Ala	0.98 ± 0.16	0.94 ± 0.10
			Cys	0.48 ± 0.05[a]	0.38 ± 0.06[b]
			Tyr	1.40 ± 0.21[a]	0.67 ± 0.09[b]
			Pro	1.09 ± 0.17	1.17 ± 0.21
			Total AA	19.56 ± 2.33	18.12 ± 2.47

[1]Values are means of three replicates per treatment. Means in a row without common superscript differ significantly ($P < 0.05$)
[2]Uninoculated feed: sterile medium was added to mixed feed (45% corn, 45% soybean meal and 10% wheat bran) instead of inoculated bacteria, other experimental procedures were the same as those of inoculated feed
[3]Hemicellulose = NDF-ADF

acidic subunits of glycinin of compound feed were almost completely degraded. This result was consistent with previous reports [14, 27]. ELISA analysis also showed that the contents of β-conglycinin and glycinin in mixed feed were degraded by 78 and 88%, respectively, after the first stage of fermentation. However, no degradation of soybean antigenic protein occurred during the second stage of anaerobic fermentation. This indicated that the decreased amount of antigenic protein in mixed feed may have been due to the hydrolysis of the proteolytic enzyme secreted by *B. subtilis* during the first stage of fermentation. *B. subtilis* secretes many proteolytic enzymes during fermentation, including aminopeptidases, serine endopeptidases, and metalloproteinases [30]. Recently, we showed that the enzyme activity of neutral protease was significantly increased during *B. subtilis* fermentation (unpublished data).

Inoculated mixed feed with two-stage fermentation contained greater concentrations of CP, ash, and total P than uninoculated feed, which was consistent with previous results for fermented compound feed [31], soybean meal [14, 32], rapeseed meal [33, 34] and cottonseed meal [35]. The loss of dry matter (mainly carbohydrates) in the fermented substrate contributing to a relative increase in the concentration of these nutrients was probably the reason for these results. Rozan et al. showed that the loss of dry matter during fermentation may explain the increase in total protein [36]. Because the CP increased after fermentation, the total AA content of inoculated feed would also be increased. However, the composition of amino acids differed between uninoculated and inoculated feed, and some polar amino acids in inoculated feed, such as Asp and Glu were decreased, whereas Tyr, Phe, and Lys increased compared with uninoculated feed. These results were similar to those of a previous study of fermented soybeans [14]. The total protein content in *B. bacillus* is 62.93%. Among them, the most abundant essential AAs are Leu, Lys, Phe, and Val [37], which are more than two times as much as the uninoculated feed. We hypothesized that part of the vegetable protein in mixed feed was used to synthesize microbial protein during SSF. However, the specific mechanism accounting for differences in amino acid composition between fermented inoculated and uninoculated feeds requires further study. Furthermore, inoculated feed exhibited an increase in TCA-SP (8.8%) compared to uninoculated feed (1.2%). An increase in TCA-SP was probably due to the hydrolysis of

Table 3 In vitro CP and AA digestibility (%) of fermented inoculated feed and uninoculated[1]

Item	Inoculated feed	Uninoculated feed[2]
DM,%	70.60 ± 2.87^a	59.33 ± 2.32^b
CP,%	86.28 ± 2.23^a	78.36 ± 2.04^b
Indispensable AA,%		
Arg	82.50 ± 4.65	82.72 ± 3.87
His	84.91 ± 3.70^a	74.85 ± 3.46^b
Ile	80.49 ± 3.42^a	75.62 ± 2.44^b
Leu	77.30 ± 3.04^a	69.71 ± 2.81^b
Lys	84.59 ± 3.91	81.44 ± 3.60
Met	85.30 ± 3.96^a	70.31 ± 2.74^b
Phe	81.99 ± 4.25^a	65.64 ± 3.63^b
Thr	78.73 ± 4.12	75.03 ± 3.83
Val	80.74 ± 3.77^a	74.49 ± 3.48^b
Mean	81.29 ± 4.09^a	74.80 ± 3.21^c
Dispensable AA,%		
Asp	83.14 ± 5.32	78.54 ± 4.97
Ser	77.86 ± 3.13	74.29 ± 3.74
Glu	85.13 ± 2.47^a	80.25 ± 3.02^b
Gly	80.78 ± 4.21	76.70 ± 4.08
Ala	84.53 ± 3.38^a	75.51 ± 3.66^b
Cys	79.74 ± 3.64^a	67.90 ± 3.87^b
Tyr	81.86 ± 3.43^b	72.28 ± 3.10^c
Pro	79.31 ± 4.28	75.41 ± 3.94
Mean	82.72 ± 3.11^a	77.16 ± 3.04^b
Total AA,%	82.15 ± 3.43^a	76.07 ± 3.35^c

[1]Values are means of three replicates per treatment. Means in a row without common superscript differ significantly ($P < 0.05$)
[2]Uninoculated feed: sterile medium was added to mixed feed (45% corn, 45% soybean meal and 10% wheat bran) instead of inoculated bacteria, other experimental procedures were the same as those of inoculated feed

macromolecular proteins (especially antigenic proteins). TCA-SP was assumed to consist of small molecular peptides (2–20 amino acid residues) and free AAs [38]. Di- and tripeptides in TCA-SP can be directly absorbed in the animal gut system, and transport of AA in the form of small peptides was faster than their constituent AAs in the free form [39]. Furthermore, a decrease in NDF, hemicellulose, and phytic acid in FMF was observed after fermentation. This is might be due to the production of relevant enzymes by micro-organism, such as non-starch polysaccharide (NSP)-degrading enzymes and phytase, which caused the breakdown of these antinutritional substrates. Fermentation of rapeseed meal with *Aspergillus niger* was studied by Shi et al. [33], who showed that the levels of NDF, glucosinolates, and phytic acid in rapeseed meal declined by 14.45%, 43.72%, and 86.08%, respectively, after SSF. Pig feeding showed that fermented rapeseed meal had a greater P and energy

digestibility than rapeseed meal [34]. Therefore, lower NDF and phytic acid indicate that inoculated mixed feed with two-stage fermentation may have higher nutrient digestibility compared with non-fermented mixed feed. The results of in vitro digestibility showed that two-stage fermentation with *B. subtilis* and *E. faecium* may improve the nutritional value of corn-SBM mixed feed.

Conclusions

Two-stage fermentation with *B. subtilis* followed by *E. faecium* effectively reduced ANFs (soy antigenic protein, NDF, and phytic acid) in corn-SBM mixed feed and increased the TCA-SP and CP content. Furthermore, the high lactic acid concentration and low pH in fermented inoculated feed inhibited the proliferation of *Enterobacteriaceae*. The results of in vitro digestion indicated that inoculated feed subjected to two-stage fermentation had higher DM and CP digestibility than fermented uninoculated feed. Therefore, two-stage fermented feed may be used as a novel feed ingredient in animal diets, especially for piglets. Our results suggest that the two-stage SSF method provides an effective approach for improving the quality of corn-soybean mixed feed.

Abbreviations
AA: Amino acid; ANFs: Antinutritional factors; Ca: Calcium; CP: Crude protein; FLF: Fermented liquid feed; LAB: Lactic acid bacteria; NDF: Neutral detergent fiber; NSP: Non-starch polysaccharide; P: Phosphorus; SBM: Soybean meal; SDS-PAGE: Sodium dodecyl sulfate – polyacrylamide gel electrophoresis; SSF: Solid state fermentation; TCA-SP: Trichloroacetic acid soluble protein

Acknowledgments
The authors thank the Specialized Research Fund for the China Pig Modern Industrial Technology System Grant (CARS-36), the China and Zhejiang province Postdoctoral Science Foundation (518000-X91604, 518000-X81601) for supporting this study.

Funding
The design of the study and collection, analysis, and interpretation of data were supported by a China Pig Modern Industrial Technology System Grant (CARS-36), the China and Zhejiang province Postdoctoral Science Foundation (518000-X91604, 518000-X81601).

Author's contributions
YZW and CYS conceived and designed the experiment. CYS and YZ carried out the experiment, including the solid-state fermentation, chemical analysis, and determination of in vitro digestibility. CYS analyzed the data and wrote the manuscript. ZQL verified the validity and checked the results. All authors read and approved the final version of this manuscript.

Competing interests
The authors declare that they have no competing interests.

References

1. Li DF, Nelssen JL, Reddy PG, Blecha F, Hancock JD, Allee GL, et al. Transient hypersensitivity to soybean meal in the early-weaned pig. J Anim Sci. 1990;68:1790–9.

2. Sun P, Li DF, Dong B, Qiao SY, Ma X. Effects of soybean glycinin on performance and immune function in early weaned pigs. Arch Anim Nutr. 2008;62:313–21.

3. Sun P, Li DF, Li ZJ, Dong B, Wang FL. Effects of glycinin on IgE-mediated increase of mast cell numbers and histamine release in the small intestine. J Nutr Biochem. 2008;19:627–33.

4. Guggenbuhl P, Simões-Nunes C. Effects of two phytases on the ileal apparent digestibility of minerals and amino acids in ileo-rectal anastomosed pigs fed on a maize rapeseed meal diet. Livest Sci. 2007;109:261–63.

5. Zhang LY, Li DF, Qiao SY, Wang JT, Bai L, Wang ZY, et al. The effect of soybean galactooligosaccharides on nutrient and energy digestibility and digesta transit time in weaning piglets. Asian-Aust J Anim Sci. 2001;14:1598–604.

6. Song YS, Pérez VG, Pettigrew JE, Martinez-Villaluenga C, de Mejia EG. Fermentation of soybean meal and its inclusion in diets for newly weaned pigs reduced diarrhea and measures of immunoreactivity in the plasm. Anim Feed Sci Technol. 2010;159:41–9.

7. Wang Y, Liu XT, Wang HL, Li DF, Piao XS, Lu WQ. Optimization of processing conditions for solid-state fermented soybean meal and its effects on growth performance and nutrient digestibility of weanling pigs. Livest Sci. 2014;170:91–9.

8. Zhang WJ, Xu ZR, Zhao SH, Sun JY, Yang X. Development of a microbial fermentation process for detoxification of gossypol in cottonseed meal. Anim Feed Sci Technol. 2007;135:176–86.

9. Shi CY, He J, Yu J, Yu B, Huang Z, Mao XB, et al. Solid state fermentation of rapeseed cake with Aspergillus niger for degrading glucosinolates and upgrading nutritional value. J Anim Sci Biotechnol. 2015;6:13–20.

10. Canibe N, Jensen BB. Fermented liquid feed – microbial and nutritional aspects and impact on enteric diseases in pigs. Anim Feed Sci Technol. 2012;173:17–40.

11. Ovissipour M, Abedian A, Motamedzadegan A, Rasco B, Safari R, Shahiri H. The effect of enzymatic hydrolysis time and temperature on the properties of protein hydrolysates from Persian sturgeon (Acipenser persicus) viscera. Food Chem. 2009;115:238–42.12. Nair VC, Duvnjak Z. Reduction of phytic acid content in canola meal by Aspergillus ficuum in solid state fermentation process. Appl Microbiol Biotechnol. 1990;34:183–88.

12. Faurobert M. Application of two-dimensional gel electrophoresis to Prunus armeniaca leaf and bark tissues. Electrophoresis. 1997;17:170–73.

13. Hong KJ, Lee CH, Kim SW. Aspergillus oryzae GB-107 fermentation improves nutritional quality of food soybeans and feed soybean meals. J Med Food. 2004;7:430–35.

14. Sakamoto K, Asano T, Furuya A, Takahashi S. Estimation of in vivo digestibility with the laying hen by an in vitro method using the intestinal fluid of the pig. Brit J Nutr. 1980;43:389–91.

15. Plumed-Ferrer C, Von Wright A. Fermented pig liquid feed: nutritional, safety and regulatory aspects. J Appl Microbiol. 2009;106:351–68.

16. Missotten JAM, Michiels J, Ovyn A, De Smet S, Dierick NA. Fermented liquid feed for pigs. Arch Anim Nutr. 2010;64:437–66.

17. van Winsen RL, Urlings BAP, Lipman LJA, Snijders JMA, Keuzenkamp D, Verheijden JHM. Effect of fermented feed on the microbial population of the gastrointestinal tracts of pigs. Appl Environ Microb. 2001;67:3071–76.

18. Lyberg K, Lundh T, Pedersen C, Lindberg JE. Influence of soaking, fermentation and phytase supplementation on nutrient digestibility in pigs offered a grower diet based on wheat and barley. Anim Sci. 2006;82:853–58.

19. Hong TTT, Thuy TT, Passoth V, Lindberg JE. Gut ecology, feed digestion and performance in weaned piglets fed liquid diets. Livest Sci. 2009;125:232–37.

20. Missotten JAM, Michiels J, Degroote J, Smet SD. Fermented liquid feed for pigs: an ancient technique for the future. J Anim Sci Biotechnol. 2015;6:4.

21. Jensen BB, Mikkelsen LL. Feeding liquid diets to pigs. In: Garnsworthy PC, Wiseman J, editors. Recent Advances in Animal Nutrition. Nottingham, UK: Nottingham University Press; 1998. p. 107–26.

22. Canibe N, Jensen BB. Fermented and non-fermented liquid feed to growing pigs: Effect on aspects of gastrointestinal ecology and growth performance. J Anim Sci. 2003;81:2019–31.

23. Canibe N, Jensen BB. Fermented liquid feed and fermented grain to piglets- effect on gastrointestinal ecology and growth performance. Livest Sci. 2007;108:232–35.

24. Canibe N, Miettinen H, Jensen BB. Effect of adding Lactobacillus plantarum or a formic acid containing product to fermented liquid feed on gastrointestinal ecology and growth performance of piglets. Livest Sci. 2007;114:251–62.

25. Aguirre L, Hebert EM, Garro MS, Giori GSD. Proteolytic activity of Lactobacillus strains on soybean proteins. LWT-Food Sci Technol. 2014;59:780–85.

26. Chi CH, Cho SJ. Improvement of bioactivity of soybean meal by solid state fermentation with Bacillus amyloliquefaciens versus Lactobacillus spp. and Saccharomyces cerevisiae. LWT-Food Sci Technol. 2016;68:619–25.

27. Frias J, Song YS, Martínez-Villaluenga C, González DME, Vidal-Valverde C. Immunoreactivity and amino acid content of fermented soybean products. J Agric Food Chem. 2008;56:99–105.

28. Feng J, Liu X, Xu ZR, Lu YP, Liu YY. The effect of Aspergillus oryzae fermented soybean meal on growth performance, digestibility of dietary components and activities of intestinal enzymes in weaned piglets. Anim Feed Sci Technol. 2007;134:295–303.

29. Simonen M, Palva I. Protein secretion in Bacillus species. Microbiol Mol Biol Rev. 1993;57:109–37.

30. Hu JK, Lu WQ, Wang CL, Zhu RH, Qiao JY. Characteristics of solid-state fermented feed and its effects on performance and nutrient digestibility in growing-finishing pigs. Asian-Aust J Anim Sci. 2008;21:1635–41.

31. Chen CC, Shih YC, Chiou PWS, Yu B. Evaluating nutritional quality of single stage- and two stage-fermented soybean meal. Asian-Aust J Anim Sci. 2010;23:598–606.

32. Chiang G, Lu WQ, Piao XS, Hu JK, Gong LM, Thacker PA. Effects of feeding solid-state fermented rapeseed meal on performance, nutrient digestibility, intestinal ecology and intestinal morphology of broiler chickens. Asian-Aust J Anim Sci. 2010;23:263–71.

33. Shi C, He J, Yu J, Yu B, Mao XB, Zheng P, et al. Amino acid, phosphorus, and energy digestibility of Aspergillus niger fermented rapeseed meal fed to growing pigs. J Anim Sci. 2015;93:2916–25.

34. Sun H, Tang JW, Yao XH, Wu XF, Wang X, Feng J. Improvement of the Nutritional Quality of Cottonseed Meal by Bacillus subtilis and the Addition of Papain. Int J Agric Biol. 2012;14:563–68.

35. Rozan P, Villaume C, Bau HM, Schwertz A, Nicolas JP, Mejean L. Detoxication of rapeseed meal by Rhizopus oligosporus sp-T3: A first step towards rapeseed protein concentrate. Int J Food Sci Technol. 1996;31:85–90.

36. Stokes JL, Gunness M. The amino acid composition of microorganisms. J Bacteriol. 1946;52:195–207.

37. Kuchroo CN, Fox PF. Soluble nitrogen in cheddar cheese: Comparison of extraction procedures. Milchwissenschaft. 1982;37:331–35.

38. Gilbert ER, Wong EA, Webb KE. Peptide absorption and utilization: Implications for animal nutrition and health. J Anim Sci. 2008;86:2135–55.

11

Regularized quantile regression for SNP marker estimation of pig growth curves

L. M. A. Barroso[1], M. Nascimento[1*], A. C. C. Nascimento[1], F. F. Silva[2], N. V. L. Serão[3], C. D. Cruz[4], M. D. V. Resende[1,5], F. L. Silva[6], C. F. Azevedo[1], P. S. Lopes[2] and S. E. F. Guimarães[2]

Abstract

Background: Genomic growth curves are generally defined only in terms of population mean; an alternative approach that has not yet been exploited in genomic analyses of growth curves is the Quantile Regression (QR). This methodology allows for the estimation of marker effects at different levels of the variable of interest. We aimed to propose and evaluate a regularized quantile regression for SNP marker effect estimation of pig growth curves, as well as to identify the chromosome regions of the most relevant markers and to estimate the genetic individual weight trajectory over time (genomic growth curve) under different quantiles (levels).

Results: The regularized quantile regression (RQR) enabled the discovery, at different levels of interest (quantiles), of the most relevant markers allowing for the identification of QTL regions. We found the same relevant markers simultaneously affecting different growth curve parameters (mature weight and maturity rate): two (ALGA0096701 and ALGA0029483) for RQR(0.2), one (ALGA0096701) for RQR(0.5), and one (ALGA0003761) for RQR(0.8). Three average genomic growth curves were obtained and the behavior was explained by the curve in quantile 0.2, which differed from the others.

Conclusions: RQR allowed for the construction of genomic growth curves, which is the key to identifying and selecting the most desirable animals for breeding purposes. Furthermore, the proposed model enabled us to find, at different levels of interest (quantiles), the most relevant markers for each trait (growth curve parameter estimates) and their respective chromosomal positions (identification of new QTL regions for growth curves in pigs). These markers can be exploited under the context of marker assisted selection while aiming to change the shape of pig growth curves.

Keywords: Genome association, Growth curve, Pig, QTL, Regularized quantile regression

Background

In general, the study of growth curves is carried out by fitting nonlinear models to weight (dependent variable) and age (independent variable) data. These models are used because they are flexible and have parameters with biological interpretations, such as maturity rate and adult weight.

With the goal of estimating SNP marker effects on parameter estimates of growth curves, Pong-Wong and Hadjipavlou [1] proposed a two-step approach. In the first step, nonlinear models were fitted to the weight-age data of each animal. In the second step, genomic regression models were fitted while considering the parameter estimates from the previous step as the dependent variable. Such an approach allows for the estimation of marker effects based only on the conditional mean of the dependent variable. Specifically, genomic growth curves are defined only in terms of population mean, i.e., the identification of genetically superior individuals in relation to the growth efficiency is based on population mean distribution (quantile 0.5 of a normal distribution of the sampled data).

An alternative approach for the second step that has not yet been exploited in genomic analyses of growth curves is the Quantile Regression (QR) [2]. This methodology allows for the estimation of marker effects at different levels (quantiles) of the variable of interest. Obtaining these effects in specific quantiles allows for a more informative study on the chromosomal regions affecting the growth curve trajectory.

* Correspondence: moysesnascim@ufv.br
[1]Department of Statistics, Federal University of Viçosa, Av. P H Rolfs, s/n, University Campus, Viçosa, MG 36570-000, Brazil
Full list of author information is available at the end of the article

In general, the larger number of markers and the dependence between them due to linkage disequilibrium leads to multicolinearity estimation problems. Thus, methods such as shrinkage estimation, which highlight the high dimensionality and multicollinearity issues, are required. Under a QR framework, this method is named regularized quantile regression (RQR), since the shrinkage (or penalty) parameter regularizes the variance of the markers' effects, thus performing a direct variable selection framework.

We aimed to propose and evaluate a regularized quantile regression for SNP marker effect estimation of pig growth curves, as well as to identify the chromosome regions of the most relevant markers and to estimate the genetic individual weight trajectory over time (genomic growth curve) under different quantiles (levels).

Methods

Animals and genotyping data

Phenotypic data was obtained from the Pig Breeding Farm of the Department of Animal Science of the Federal University of Viçosa, Minas Gerais, and refer to the weights at birth, 21, 42, 63, 77, 105 and 150 days of age. These weights were measured in 345 animals from a F2 outbred population (Brazilian Piau X commercial). More details about this population are found by Azevedo et al. [3] and Band et al. [4].

DNA was extracted at the Animal Biotechnology Lab from Animal Science Department of Federal University of Viçosa. The low-density customized SNPChip with 384 markers was based on the Illumina Porcine SNP60 BeadChip (San Diego, CA, USA, [5]). The number of SNP markers was distributed as follows in the pig chromosomes: (*Sus scrofa*; SSC): SSC1 (n = 56), SSC4 (n = 54), SSC7 (n = 59), SSC8 (n = 31), SSC17 (n = 25), and SSCX (n = 12), totaling 237 SNPs. These markers were selected according to QTL positions that were previously identified in this population by using meta-analyses [6] and fine mapping [7, 8]. Thus, although a small number of markers have been used, the customized SNPchip based on previously identified QTL positions ensures appropriate coverage of the relevant genome regions in this population.

Statistical analysis

Initially, the logistic nonlinear regression model [9] was fitted to the individual weight-age data:

$$w_{ij} = \frac{\alpha_{1i}}{1 + \exp\left[(\alpha_{2i} - t_j)/\alpha_{3i}\right]} + e_{ij}, \tag{1}$$

where w_{ij} is the weight of the animal i at age $t_j(0, 21, 42, 63, 77, 105$ and $150)$; α_{1i}, α_{2i} and α_{3i} are the parameters. If $\alpha_{3i} > 0$ then α_{1i} is the horizontal asymptote as $t_j \to \infty$

(mature weight) and 0 is the horizontal asymptote as $t_j \to -\infty$. If $\alpha_{3i} < 0$ these roles are reversed. The parameter α_{2i} is the t_j value at which the response is $\alpha_{1i}/2$. It is the inflection point of the curve. The scale parameter α_{3i} (growth scale) represents the distance on the t-axis between this inflection point and the point where the response is $\alpha_{1i}/(1 + e^{-1}) \approx 0.73\alpha_{1i}$; e_{ij} is the independent and normally distributed residual term, $e_{ij} \sim N(0, \sigma_e^2)$. In this parameterization, the growth scale parameter is the reciprocal of growth rate on the model presented by Ratkowsky [10].

After obtaining parameter estimates of the logistic model, they were used as dependent variables in a linear model to carry out fixed effect corrections (sex, lot, and halothane gene). The corrected variables were identified based on the residual of the fitted linear model plus the overall mean. Subsequently, the corrected variables ($\hat{\alpha}_{1i}^*$, $\hat{\alpha}_{2i}^*$ and $\hat{\alpha}_{3i}^*$) were used as dependent variables in a multiple regression model while using SNP markers as the independent variables. This procedure is known in the literature as a two-step approach: in the first step, a growth curve is fitted to the data of each animal, and in the second step, the parameter estimates from the previous step are used as phenotypic values [1, 11].

In the second step, the following genomic model proposed by Meuwissen et al. [12] was fitted separately for each trait (parameter estimates from previous step):

$$y_i = \left[\mu + \sum_{k=1}^{237} x_{ik}\beta_k\right] + \varepsilon_i, \tag{2}$$

in which y_i is the corrected phenotype $\hat{\alpha}_{1i}^*, \hat{\alpha}_{2i}^*$ and $\hat{\alpha}_{3i}^*$ from the first step; μ is the general mean; x_{ik} is the SNP marker, encoded as 2 (AA), 1 (Aa), or 0 (aa); β_k is the effect of the marker k; and ε_i corresponds to the residual term, $\varepsilon_i \sim N(0, \sigma_e^2)$.

To obtain the markers' effects at different levels of the variables (traits defined by $\hat{\alpha}_1^*, \hat{\alpha}_2^*$ and $\hat{\alpha}_3^*$), the regularized quantile regression [13] was used. This method consists of obtaining the marker effects (β_k) that solve the following optimization problem:

$$\hat{\beta}_s = \text{argmin}_\beta \left\{ \sum_{i=1}^{345} \rho_{\tau s}\left[\hat{\alpha}_{si}^* - \left(\mu + \sum_{k=1}^{237} x_{ik}\beta_{sk}\right)\right] + \lambda_s \sum_{k=1}^{237} |\beta_{sk}| \right\},$$

where s = 1, 2, and 3 (respectively for each assumed trait, $\hat{\alpha}_1^*, \hat{\alpha}_2^*$ and $\hat{\alpha}_3^*$); $\sum_{k=1}^{237} |\beta_{sk0}|$ is the sum of the absolute values of the regression coefficients; λ_s is the regularization parameter for each trait; and $\tau \in (0, 1)$ indicates the quantile of interest. This parameter (λ_s) is required to avoid multicollinearity problems that are a result of the larger number of highly dependent markers associated with linkage disequilibrium. It leads to the formulation of the RQR.

The parameter $\rho_{\tau s}(.)$ is denoted as a check function [2] and is defined by:

$$\rho_{\tau s}\left[\hat{\alpha}_{si}^* - \left(\mu + \sum_{k=1}^{237} x_{ik}\beta_{sk}\right)\right]$$

$$= \begin{cases} \tau \cdot \left[\hat{\alpha}_{si}^* - \left(\mu + \sum_{k=1}^{237} x_{ik}\beta_{sk}\right)\right], & \text{if } \hat{\alpha}_{si}^* - \mu + \sum_{k=1}^{237} x_{ik}\beta_{sk} > 0, \\ -(1-\tau) \cdot \left[\hat{\alpha}_{si}^* - \left(\mu + \sum_{k=1}^{237} x_{ik}\beta_{sk}\right)\right], & \text{otherwise.} \end{cases}$$

in which $\tau \in (0, 1)$ indicates the quantile of interest. Thus, the values of $\beta_{sk}(\tau)$ represent the markers' effects in the τth quantile of interest for s^{th} trait.

In this study, for each trait ($\hat{\alpha}_1^*$, $\hat{\alpha}_2^*$ and $\hat{\alpha}_3^*$), the quantiles $\tau = 0.2$, 0.5 and 0.8 were used to generate results at three distinct levels that may characterize the low, average, and high distribution of the phenotypic values under study ($\hat{\alpha}_{1i}^*$, $\hat{\alpha}_{2i}^*$ and $\hat{\alpha}_{3i}^*$). Furthermore, these quantiles were chosen to minimize the residual term in previous studies (pilot analysis) by using the same datasets.

In order to verify whether marker effects differ between the quantile levels of the traits ($\hat{\alpha}_1^*$, $\hat{\alpha}_2^*$ and $\hat{\alpha}_3^*$), the 2.5% most relevant SNPs (highest absolute values) and their p values, based on bootstrapped standard error values, were presented. In addition, these SNPs were used to identify possible QTL regions affecting growth traits in pigs.

The Genomic Estimated Breeding Values (GEBV) from RQR were obtained through $GEBV(\tau) = \hat{u} = \sum_k x_{ik}\hat{\beta}_k$ (τ), in which τ represents the quantile of interest. Subsequently, the genomic growth curves were obtained for each animal based on GEBV ($\hat{u}$) according to the following expression:

$$\hat{y}_{ij} = \frac{\hat{\mu}_{\hat{\alpha}_1^*} + \hat{u}_{\hat{\alpha}_{1i}^*}}{\left\{1 + \exp\left[\left(\hat{\mu}_{\hat{\alpha}_2^*} + \hat{u}_{\hat{\alpha}_{2i}^*}\right) - \left(\hat{\mu}_{\hat{\alpha}_3^*} + \hat{u}_{\hat{\alpha}_{3i}^*}\right)t_{ij}\right]\right\}}, \quad (3)$$

in which $\hat{y}_{ij}$ is the predicted breeding value for each animal i for the weight at each age (t_{ij}) (j = 0 to 150 d); $\hat{\mu}_{\hat{\alpha}_1^*}$, $\hat{\mu}_{\hat{\alpha}_2^*}$ and $\hat{\mu}_{\hat{\alpha}_3^*}$ are the means of each trait (parameter estimates for the logistic model); and $\hat{u}_{\hat{\alpha}_{1i}^*}$, $\hat{u}_{\hat{\alpha}_{2i}^*}$ and $\hat{u}_{\hat{\alpha}_{3i}^*}$ are the GEBV of these traits.

Finally, the genetic parameters for the interpretable traits derived from the logistic model (α_1 and α_3) as well as the original traits associated with slaughter weight (SW) and average daily gain (ADG) were estimated by using the following multi-trait model:

$$\begin{bmatrix} \mathbf{y}_1 \\ \mathbf{y}_2 \end{bmatrix} = \begin{bmatrix} \mathbf{X}_1 & \mathbf{0} \\ \mathbf{0} & \mathbf{X}_2 \end{bmatrix} \begin{bmatrix} \boldsymbol{\beta}_1 \\ \boldsymbol{\beta}_2 \end{bmatrix} + \begin{bmatrix} \mathbf{Z}_1 & \mathbf{0} \\ \mathbf{0} & \mathbf{Z}_2 \end{bmatrix} \begin{bmatrix} \mathbf{g}_1 \\ \mathbf{g}_2 \end{bmatrix} + \begin{bmatrix} \mathbf{e}_1 \\ \mathbf{e}_2 \end{bmatrix},$$

$$(4)$$

where $\begin{bmatrix} \mathbf{y}_1 \\ \mathbf{y}_2 \end{bmatrix}$ is the vector of response variables of traits I and II (α_1 and α_3 with SW and ADG), $\mathbf{X}_1$ and $\mathbf{X}_2$ are the fixed-effects design matrix (Sex, Batch, and Halothane presence), $\mathbf{Z}_1$ and $\mathbf{Z}_2$ are the random-effects design matrix, and $\begin{bmatrix} \mathbf{e}_1 \\ \mathbf{e}_2 \end{bmatrix}$ is the vector of random residuals of the two traits. It is assumed that $\begin{bmatrix} \mathbf{g}_1 \\ \mathbf{g}_2 \end{bmatrix} \sim N(\mathbf{0}, \mathbf{G} \otimes \mathbf{H})$, where $\mathbf{H} = \begin{bmatrix} \sigma_{g_1}^2 & \sigma_{g_{12}} \\ \sigma_{g_{21}} & \sigma_{g_2}^2 \end{bmatrix}$ is the additive genetic variance and covariance matrix of the two traits, and $\begin{bmatrix} \mathbf{e}_1 \\ \mathbf{e}_2 \end{bmatrix} \sim N$ $(\mathbf{0}, \mathbf{I} \otimes \mathbf{R})$, where $\mathbf{R} = \begin{bmatrix} \sigma_{e_1}^2 & \sigma_{e_{12}} \\ \sigma_{e_{21}} & \sigma_{e_2}^2 \end{bmatrix}$ is the residual variance and covariance matrix of the two traits. Finally, $\mathbf{G}$ is the additive relationship matrix constructed by using 501 pigs and $\mathbf{I}$ is the identity matrix.

Computational features

Fitting of the models was carried out by using the *nls* (to fit the logistic nonlinear model in the first step) and *rq* (to fit the regularized quartile regression in the second step) functions of the *stats* and *quantreg* packages [14] of R software [15], respectively. The Mixed Model Analyses were performed in ASReml 3.0 [16].

To obtain the shrinkage parameter values (λ), a grid of λ values between 0 and 50 was utilized, varying in 0.5 increments. The predictive capacity, defined as the correlation between the estimated and observed values (curve parameters that were obtained from fitting the Logistic model to the weight-age data), was used as a criterion to define the optimal value λ.

The computational codes that were implemented in the R software are found on the website of the Statistics Department of the Federal University of Viçosa (2017): https://licaeufv.wordpress.com/scriptrqr_jasb/.

Results

The summary containing the descriptive statistics of the adjusted phenotypic data is presented in Table 1.

The summary containing the correlation and descriptive statistics of the adjusted phenotypic data ($\hat{\alpha}_{1i}^*$, $\hat{\alpha}_{2i}^*$ and $\hat{\alpha}_{3i}^*$) is presented in Table 2.

Considering the aforementioned grid (0 to 50, by 0.5), the shrinkage parameter value that showed the best results in terms of predictive capacity was $\lambda = 0.5$.

Table 1 Means, standard deviations and ranges for weights at seven different ages of F2 outbred population

Age, d	n	Mean weight ± SD, kg	Min, kg	Max, kg
0	345	1.20 ± 0.27	0.53	2.13
21	345	4.90 ± 1.00	2.56	8.00
42	345	8.36 ± 1.81	2.66	12.90
63	345	16.29 ± 3.38	7.43	26.53
77	345	21.44 ± 4.39	9.30	34.50
105	345	36.25 ± 6.64	12.79	55.00
150	345	64.97 ± 5.72	39.09	85.20

Specifically, the predictive capacity ranged between 0.6219 and 0.8252 (Table 3).

The mean and standard error for marker effects ($\hat{\beta}_k$ s) and R^1 goodness of fit measure for each quantile adjusted model are present in Table 4. The goodness of fit ranged between 0.67 and 0.75 (Table 4).

In order to verify whether the most relevant SNPs for the three approaches (RQR (0.2), RQR (0.5), and RQR (0.8)) were the same, the 2.5% most relevant SNPs for each phenotype ($\hat{\alpha}_{1i}^*$, $\hat{\alpha}_{2i}^*$ and $\hat{\alpha}_{3i}^*$) were reported (Table 5).

Table 5 describes the most relevant markers considering the fitting through RQR (0.2). For the mature weight (α_1), the markers are located on chromosomes SSC1, SSC4, SSC7, SSC8, and SSC17 (Table 5). The position of the marker ALGA0096701 on chromosome 17 (55.81 cM) is in accordance with the results of Pierzchala et al. [17], in which the authors found QTL for the slaughter weight at the position 51.1 cM with the cross between Meishan, Pietrain, and European Wild Boar. For birth weight (α_2), the marker ALGA0044519 stands out, which is found in the SSC7 at the position 115.23 cM, next to the QTL for the birth weight found by Guo et al. [18] at the position 120.9 cM for crosses of Large white and Meishan. In terms of growth rate (α_3), the marker that presented with the highest effect is found on chromosome 8. The position of the marker ALGA0049546 at SSC8 (60.04 cM) is close to the position 62.2 cM, as reported by Casas-Carrillo et al. [19] for average daily gain when using families from outbred lines that were selected for high (fast) and low (slow) growth rates.

Table 2 Correlation and descriptive statistics among the adjusted phenotypic data ($\hat{a}_{1i}^*$, $\hat{a}_{2i}^*$ and $\hat{a}_{3i}^*$)

	Correlation			Descriptive statistics		
	$\hat{a}_{1i}^*$	$\hat{a}_{2i}^*$	$\hat{a}_{3i}^*$	Mean ± SD	Min	Max
$\hat{a}_{1i}^*$	1.00	0.82	0.63	89.43 ± 22.32	35.70	149.85
$\hat{a}_{2i}^*$	0.82	1.00	0.83	113.18 ± 17.97	72.83	166.43
$\hat{a}_{3i}^*$	0.63	0.83	1.00	32.03 ± 4.24	22.76	47.29

Table 3 Predictive capacity obtained by means of RQR, considering estimates of the nonlinear regression parameters

Quantile	Trait		
	$a_1(\lambda = 0.5)$	$a_2(\lambda = 0.5)$	$a_3(\lambda = 0.5)$
0.2	0.7143	0.6938	0.6219
0.5	0.8252	0.7889	0.7904
0.8	0.7678	0.7663	0.7636

Considering the RQR (0.5) in Table 5, the most important markers for α_1, α_2 and α_3 are located on chromosomes SSC4 and SSC8 (Table 5; RQR (0.5)). For α_1, the marker ALGA0047992 stands out, which is found on SSC8 at the position 30.17 cM, which is close to the QTL for slaughter weight found by Beeckmann et al. [20], and at the position 33.9 cM on chromosome 8 in pigs obtained from crosses between Meishan, Pietrain, and European Wild Boar. For the birth weight trait (α_2), the marker with the greatest estimated effect was ALGA0026100. The position of this marker at SSC4 (75.53 cM) is close to the position at 74.4 cM reported by Walling et al. [21] for body weight at birth. For α_3, the position of marker ALGA0048131 on SSC8 (35.02 cM) was close to the position 33.1 cM reported by Beeckmann et al. [20] who used data from an experimental cross between Meishan, Pietrain, and European Wild Boar for average daily gain (Table 5; RQR (0.5)).

Considering the RQR (0.8) in Table 5, the most significant SNPs for α_1, α_2 and α_3 are located on chromosomes SSC1 and SSC8 (Table 5; RQR (0.8)). Regarding the mature weight trait (α_1), the marker with the highest absolute value pertaining to the estimates of the parameter effect is ALGA0007216. This marker is located on chromosome 1 (160.61 cM). Chen et al. [22] used a pig population comprised of Yorkshires and Meishans to find significant QTLs for slaughter weight at the position 122.4 cM of SSC1, i.e., close to the position 160.61 cM of the ALGA0007216 marker (Table 5; RQR (0.8)).

Table 4 Mean, standard error for marker effects and Pseudo R^2 for each quantile adjusted model

Model	Trait	Mean (Standard error)	Pseudo R^{2a}
RQR (0.2)	$\hat{a}_1$	0.43(0.37)	0.71
	$\hat{a}_2$	0.44(0.45)	0.69
	$\hat{a}_3$	0.11(0.14)	0.70
RQR (0.5)	$\hat{a}_1$	0.28(0.44)	0.68
	$\hat{a}_2$	0.42(0.40)	0.67
	$\hat{a}_3$	0.10(0.12)	0.68
RQR (0.8)	$\hat{a}_1$	0.48(0.44)	0.75
	$\hat{a}_2$	0.52(0.49)	0.74
	$\hat{a}_3$	0.13(0.09)	0.75

[a] Pseudo R^2 [28]

Table 5 Absolute values of the estimated effects of the 2.5% most relevant SNP by RQR

Phenotype	Quantile	SNP marker	Estimated effect (abs)	P-value[*]	Chromossome (SSC)	Position, cM
	0.20	ALGA0096701	18.93	0.099	17	55.81
	0.20	ALGA0026109	15.29	0.019	4	75.57
	0.20	ALGA0024036	14.98	0.007	4	20.55
	0.20	ALGA0038840	14.50	0.041	7	15.18
	0.20	ALGA0029474	14.15	0.060	4	122.99
	0.20	ALGA0029483	14.07	0.042	4	123.28
	0.50	ALGA0047992	30.89	0.008	8	30.17
Mature	0.50	ALGA0047995	29.47	0.006	8	30.31
Weight,	0.50	ALGA0096701	21.81	0.058	17	55.81
a_1	0.50	ALGA0003761	17.22	0.098	1	50.37
	0.50	ALGA0044299	15.65	0.153	7	110.66
	0.50	ALGA0096707	15.57	0.144	17	55.84
	0.80	ALGA0007216	22.14	0.001	1	160.61
	0.80	ALGA0003761	19.86	0.018	1	50.37
	0.80	ALGA0096701	19.71	0.005	17	55.81
	0.80	ALGA0042986	15.88	0.014	7	90.01
	0.80	ALGA0029474	15.57	0.042	4	122.99
	0.80	ALGA0042863	15.57	0.009	7	86.24
	0.20	ALGA0048131	13.55	0.027	8	35.02
	0.20	ALGA0044519	13.12	0.020	7	115.23
	0.20	ALGA0096701	12.98	0.011	17	55.81
	0.20	ALGA0029483	12.50	0.029	4	123.28
	0.20	ALGA0026109	11.23	0.033	4	75.57
	0.20	ALGA0003761	10.85	0.095	1	50.37
	0.50	ALGA0026100	19.87	0.009	4	75.53
Birth	0.50	ALGA0047995	18.71	0.027	8	30.31
Weight,	0.50	ALGA0048131	18.47	0.029	8	35.02
a_2	0.50	ALGA0047992	16.36	0.062	8	30.17
	0.50	ALGA0039880	14.78	0.047	7	30.13
	0.50	ALGA0021973	14.36	0.015	4	0.28
	0.80	ALGA0048131	17.66	0.007	8	35.02
	0.80	ALGA0005071	17.64	0.002	1	80.44
	0.80	ALGA0042986	16.32	0.005	7	90.01
	0.80	ALGA0029483	15.21	0.010	4	123.28
	0.80	ALGA0003761	15.08	0.025	1	50.37
	0.80	ALGA0026769	14.49	0.073	4	90.18
	0.20	ALGA0049546	3.91	0.015	8	60.04
	0.20	ALGA0029483	3.77	0.005	4	123.28
	0.20	ALGA0096701	3.42	0.011	17	55.81
	0.20	ALGA0021973	3.31	0.014	4	0.28
	0.20	ALGA0048854	3.29	0.031	8	50.17
	0.20	ALGA0048131	3.27	0.035	8	35.02
	0.50	ALGA0048131	5.89	0.004	8	35.02
Growth	0.50	ALGA0021973	4.37	0.023	4	0.28

Table 5 Absolute values of the estimated effects of the 2.5% most relevant SNP by RQR (Continued)

Rate,	0.50	ALGA0048854	4.13	0.058	8	50.17
α_3	0.50	ALGA0096701	3.66	0.075	17	55.81
	0.50	ALGA0027642	3.62	0.054	4	102.39
	0.50	ALGA0027644	3.36	0.087	4	102.41
	0.80	ALGA0003761	4.43	0.008	1	50.37
	0.80	ALGA0048131	3.74	0.018	8	35.02
	0.80	ALGA0024881	3.61	0.005	4	40.50
	0.80	ALGA0044299	3.34	0.052	7	110.66
	0.80	ALGA0026769	3.07	0.105	4	90.18
	0.80	ALGA0048133	3.01	0.034	8	35.04

*P-value calculated using the bootstrap standard error

Another interesting result that was observed through RQR is the simultaneous existence of important markers for different traits (Table 5). This fact is important for breeding, since pleiotropy is the main factor in genetic correlation. Specifically, for RQR (0.5) (Table 5), two markers (ALGA0047992 and ALGA0047995) were simultaneously important for the mature weight (α_1) and birth weight (α_2) traits. In addition, three SNPs for RQR (0.2) (ALGA0096701, ALGA0026109, and ALGA0029483) and one for RQR (0.8) (ALGA0042986) were simultaneously relevant for α_1 and α_2.

Considering the traits α_1 (mature weight) and α_3 (growth rate), two (ALGA0096701 and ALGA0029483), one (ALGA0096701), and one (ALGA0003761) markers were simultaneously important for the methodologies RQR (0.2), RQR (0.5), and RQR (0.8), respectively. For the traits α_2 (birth weight) and α_3 (growth rate), three markers in the RQR (0.2) methodology (ALGA0048131, ALGA0096701, and ALGA0029483), two in the RQR (0.5) methodology (ALGA0048131 and ALGA0021973), and three in the RQR (0.8) methodology (ALGA0003761, ALGA0026769, and ALGA0048131) were simultaneously relevant for these two traits (Table 5).

The three genomic growth curves ($\tau = 0.2, 0.5, 0.8$) that were obtained based on all of the data are shown in Fig. 1b. The estimated curve based on the three quantiles showed a similar pattern until 100 d. After that, differences in the estimated growth curves increased with time (Fig. 1b). This result was expected given the increase in the heterogeneity of variances that were presented at the final evaluated times, 100 and 150 d (Fig. 1a).

The genomic growth curves for each RQR, for quantiles 0.2, 0.5, and 0.8 and their confidence intervals showed significant differences (based on non-overlapping confidence intervals) only in terms of mature weight (Fig. 2a). These differences are highlighted in Fig. 2b.

Estimates of genetic parameters (heritability, and genetic and phenotypic correlations) are presented in Table 6. Estimates of heritability for growth curve parameters were

moderate, with 0.447 ± 0.200 and 0.4991 ± 0.164, for parameters α_1 and α_3, respectively. The original traits (SW and ADG) had low heritability estimates, with 0.214 ± 0.127 and 0.094 ± 0.087, for SW and ADG, respectively.

Estimates of genetic and phenotypic correlations are presented in the off-diagonals (Table 6). Between the interpretable growth curve parameters (α_1 and α_3) with the original correspondent traits (SW and ADG), correlations were, respectively, highly positive and negative, with a positive genetic correlation estimated for parameters α_1 and SW (0.404 ± 0.113) and a negative genetic correlation estimated for α_3 with ADG (-0.681 ± 0.229). Phenotypic correlations between interpretable growth curve parameters with slaughter weight (SW) and average daily gain (ADG) traits were also moderately positive and negative, with 0.662 ± 0.051 for α_1 with SW and -0.451 ± 0.06 for α_3 with ADG.

Discussion

In this study, we aimed to propose and evaluate a regularized quantile regression (RQR) for SNP marker effect estimation on pig growth curves and to estimate the genetic weight trajectory over time (genomic growth curve) under different quantiles (levels). In order to do so, a real data set consisting of 345 animals from an F2 outbred population with information on 237 SNP markers, randomly distributed over six chromosomes, was used. The phenotypic data refers to the weight at birth, 21, 42, 63, 77, 105, and 150 days of age. To estimate SNP marker effects for growth curves, we used a two-step approach [1]. In the first step, we fitted logistic nonlinear models to the data of each animal, and in the second step, genomic regression models were fitted while considering the estimated parameters from the previous step as the phenotypic values. We obtained the three genomic growth curves for the three evaluated quantiles ($\tau = 0.2, 0.5, 0.8$). Finally, the genetic parameters for the interpretable traits of the logistic model

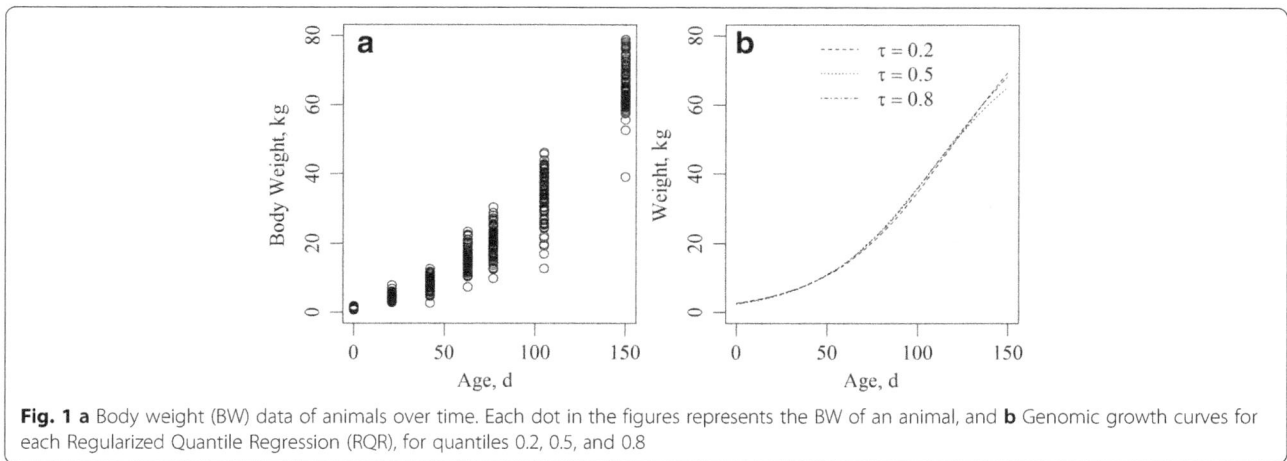

Fig. 1 a Body weight (BW) data of animals over time. Each dot in the figures represents the BW of an animal, and **b** Genomic growth curves for each Regularized Quantile Regression (RQR), for quantiles 0.2, 0.5, and 0.8

(α_1 and α_3) and the original traits, slaughter weight and average daily gain, were estimated.

Quantile regression (QR) can be used to provide a more complete statistical analysis of the stochastic relationships among random variables. In general, the chosen quantiles depend entirely on the purpose of the study, i.e., we can study all distributions or only some parts by defining specific quantiles. In this study, with the aim of representing three distinct levels that characterize low, average, and high distributions of the phenotypic values (estimated parameters while considering a logistic nonlinear model), we choose, $\tau = 0.2$, 0.5, 0.8.

The use of RQR to estimate SNP marker effects and obtain the estimated genomic growth curve was efficient since it was possible to construct genomic growth curves and find the most relevant markers, which thus allows for the identification of QTL regions at different levels of interest. Besides that, R^1 goodness of fit measures ranging from 0.67 to 0.75 indicating that the model fits well for the observations.

Unlike traditional methods that are based on conditional expectations, $E(Y|X)$, RQR allows us to fit regression models on different parts of the distribution of the variable response, therefore enabling a more complete understanding of the phenomenon under study [2, 23]. Besides, the heterogeneous variance over time (Fig. 1a) indicates that there is not a single rate of change that characterizes changes in the probability distribution, therefore indicating that RQR is a good tool to deal with those situations. Also, the predictive capacity that was obtained by means of RQR (Table 3) was better than that obtained by Silva et al. [24].

The advantages of RQR, such as studying different parts of the distribution of the variable response, can be combined with those from the two-step approach. Specifically, the two-step approach enables us to obtain the genomic values for each observed time (t_j), as well as to estimate the weight for any other time of interest within the measured range before this weight is attained [24].

Based on the results, it is possible to note that RQR allows for the identification of markers close to QTLs at

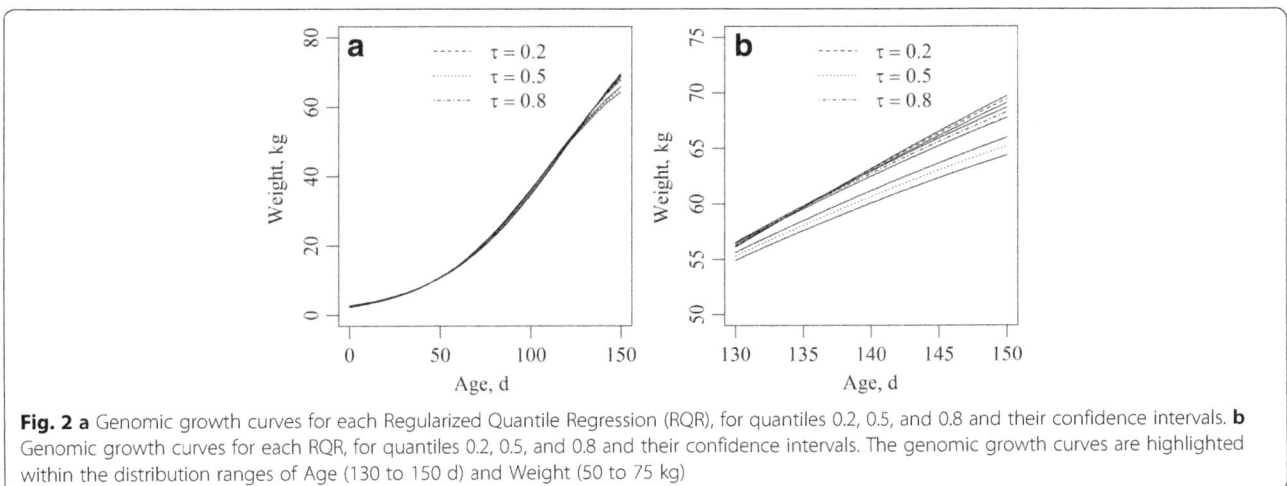

Fig. 2 a Genomic growth curves for each Regularized Quantile Regression (RQR), for quantiles 0.2, 0.5, and 0.8 and their confidence intervals. **b** Genomic growth curves for each RQR, for quantiles 0.2, 0.5, and 0.8 and their confidence intervals. The genomic growth curves are highlighted within the distribution ranges of Age (130 to 150 d) and Weight (50 to 75 kg)

Table 6 Genetic parameters[a] (standard error) for growth curve parameters, ADG, and SW

Traits[b]	a_1	a_3	ADG	SW
a_1	0.447 (0.200)	0.809 (0.191)	−0.613 (0.390)	0.404 (0.113)
a_3	0.759 (0.030)	0.491 (0.164)	−0.681 (0.229)	-
ADG	0.047 (0.090)	−0.451 (0.06)	0.214 (0.127)	0.892 (0.677)
SW	0.662 (0.051)	−0.191 (0.080)	0.687 (0.039)	0.094 (0.087)

[a]Heritability, and genetic and phenotypic correlations presented on the diagonal, lower off-diagonal, and upper off-diagonal, respectively
[b]a_1 asymptotic weight (mature body weight), a_3 inflection point, ADG average daily gain, SW slaughter weight

different distribution levels of the phenotypic values of interest. The regions indicated by RQR coincide with the results of several studies in which the authors found QTL for the traits that were evaluated in this study.

The use of quantile regression to estimate genomic curves based on three contrasting quantiles in our population was efficient when it came to producing distinct growth curves. Specifically, we can see in Fig. 2b that the final BW of the genomic growth curves was statistically different; in other words, the growth behavior over time changed in terms of mature weight. In fact, this result shows that RQR is a statistical method that could be effectively used to estimate more than a single mean behavior, thereby providing a more complete picture of the relationships between variables.

The genetic correlations between α_1 and α_3 with BW and ADG had, respectively, a high positive and negative genetic correlation, which indicates that α_1 and α_3 have the potential to be used as selection tools to improve SW and ADG. Additionally, the high genetic correlation between α_1 with α_3 and SW with ADG enable us to understand causes of SNPs' pleiotropic effects. These results are in agreement with Silva et al. [24], who found significant genetic correlation between the interpretable traits of logistic model ($r_{\alpha_1,\alpha_3} = -0.69$) in the same populations that were used in this study. The difference between the signals of genetic correlation estimates observed in the present study is due to the different Logistic model parameterizations. Specifically, our approach uses the parameterization presented in Pinheiro and Bates [9], where the growth scale parameter (α_3) is the reciprocal of growth rate [10, 24].

The study of different distribution levels of the variable of interest using QR has been successfully performed in medicine by Beyerlein et al. [25], who used QR in GWAS (Genome-Wide Association Study) analysis in human genetics where they emphasized statistical and biological advantages when estimating marker effects in different quantiles of the phenotypic distribution. Sun et al. [26] proposed to use QR to identify hypermethylated CpG islands (CGIs) that can be associated with breast and ovarian cancer. They concluded that the quantile

level between 80 and 90% is the best strategy to identify methylated and unmethylated CGIs. Moreover, regularized quantile regression has already been successfully evaluated for analyzing ultra-high dimension data [27]. These authors demonstrated that QR greatly enhances existing tools for large dimensional data analysis, since it revealed a substantial reduction in model complexity when compared with alternative methods.

However, even though the use of RQR is promising and efficient, more studies are needed to address the choice of the shrinkage parameter value, which is always critical to find as it can be defined by using a grid of values, cross-validation, or by using a Bayesian approach. Another issue about the use of RQR is the choice of the quantile. There are a lot of quantiles that can be used; therefore, finding the best one to explain the functional relationship is a challenge.

Conclusions

The proposed model enabled the discovery, at different levels of interest (quantiles), of the most relevant markers for each trait (growth curve parameter estimates) and their respective chromosomal positions (identification of new QTL regions for growth curves in pigs). Furthermore, RQR enabled the construction of genomic growth curves, which identified genetically superior individuals in relation to growth efficiency.

Abbreviations
ADG: Average daily gain; BW: Body weight; GEBV: Genomic estimated breeding value; GWAS: Genome-Wide Association Study; QR: Quantile Regression; QTL: Quantitative trait loci; RQR: Regularized Quantile Regression; SNP: Single-nucleotide polymorphism; SSC: *Sus scrofa*; SW: Slaughter weight

Acknowledgements
To Coordination for the Improvement of Higher Education Personnel (Capes) and Foundation Arthur Bernardes (Funarbe) for Laís Barroso and Moysés Nascimento scholarships.

Funding
This research was supported by Coordination for the Improvement of Higher Education Personnel (Capes), Foundation Arthur Bernardes (Funarbe), and Foundation of research Support of the state of Minas Gerais (FAPEMIG).

Authors' contributions
LMAB, MN, ACCN, FFS and NVLS conceived the study, participated in the statistical analysis and drafted the manuscript. CDC, MDVR, FLS and CFA checked the results, participated in the study design and helped to draft the manuscript. PSL and SEFG participated in the animal studies and helped to draft the manuscript. All authors read and approved the final manuscript.

Competing interests
The authors declare that they have no competing interests.

Author details
[1]Department of Statistics, Federal University of Viçosa, Av. P H Rolfs, s/n, University Campus, Viçosa, MG 36570-000, Brazil. [2]Department of Animal Science, Federal University of Viçosa, Av. P H Rolfs, s/n, University Campus, Viçosa, MG 36570-000, Brazil. [3]Department of Animal Science, Iowa State

University, Kildee Hall 50011 Ames, Iowa, USA. [4]Department of General Biology, Federal University of Viçosa, Av. P H Rolfs, s/n, University Campus, Viçosa, MG 36570-000, Brazil. [5]Embrapa Forestry, Estrada da Ribeira, km 111, Colombo, PR, Brazil. [6]Department of Plant Science, Federal University of Viçosa, Av. P H Rolfs, s/n, University Campus, Viçosa, MG 36570-000, Brazil.

References

1. Pong-Wong R, Hadjipavlou GA. A two-step approach combining the Gompertz growth with genomic selection for longitudinal data. BMC Proc. 2010;4:S4.
2. Koenker R, Basset G. Regression Quantiles. Econometrica. 1978;46:33–50.
3. Azevedo CF, Nascimento M, Silva FF, Resende MDV, Lopes PS, Guimarães SEF. Comparison of dimensionality reduction methods to predict genomic breeding values for carcass traits in pigs. Genet Mol Res. 2015;14:12217–27.
4. Band GO, Guimarães SEF, Lopes PS, Peixoto JO, Faria DA, Pires AV, et al. Relationship between the porcine stress syndrome gene and carcass and performance traits in F2 pigs resulting from divergent crosses. Genet Mol Biol. 2005;28:92–6.
5. Ramos AM, Crooijmans RPMA, Affara NA, Amaral AJ, Archibald AL, Beever JE, et al. Design of a high density SNP genotyping assay in the pig using SNPs identified and characterized by next generation sequencing technology. PLoS One. 2009;4:e6524.
6. Silva KM, Knol EF, Merks JWM, Guimarães SEF, Bastiaansen JWM, Van Arendonk JAM, et al. Meta-analysis of results from quantitative trait loci mapping studies on pig chromosome 4. Anim Genet. 2011;42:280–92.
7. Hidalgo AM, Lopes PS, Paixão DM, Silva FF, Bastiaansen JWM, Paiva SR, et al. Fine mapping and single nucleotide polymorphism effects estimation on pig chromosomes 1, 4, 7, 8, 17 and X. Genet Mol Biol. 2013;36:511–9.
8. Verardo L, Silva FF, Varona L, Resende MDV, Bastiaansen JWM, Lopes PS, et al. Bayesian GWAS and network analysis revealed new candidate genes for number of teats in pigs. J Appl Genet. 2015;56:123–32.
9. Pinheiro JC, Bates DM. Mixed-effects models in S and S-PLUS. New York: Springer; 2000.
10. Ratkowsky DA. Nonlinear regression modeling. New York: Marcel Dekker; 1983.
11. Varona L, Moreno C, Garcia-Cortés LA, Yague G, Altarriba J. Two-step vs. joint analysis of von Bertalanffy function. J Anim Breed Genet. 1999;116:331–8.
12. Meuwissen THE, Hayes BJ, Goddard ME. Prediction of total genetic value using genome wide dense marker maps. Genetics. 2001;157:1819–29.
13. Li Y, Zhu J. L1-Norm Quantile Regression. J Comput Graph Stat. 2008;17:1–23.
14. Koenker R. quantreg: Quantile Regression. R package version 5.29. 2016. https://cran.r-project.org/web/packages/quantreg/index.html. Accessed 19 Oct 2016.
15. R Core Team. R: a language and environment for statistical computing. R Foundation for Statistical Computing, Vienna, Austria. 2014. https://www.r-project.org. Accessed 19 Oct 2016.
16. Gilmour AR, Gogel BJ, Cullis BR, Thompson R. ASReml User Guide Release 3. 0 VSN International Ltd, Hemel Hempstead, HP1 1ES, UK. 2009. https://www.vsni.co.uk/downloads/asreml/release3/UserGuide.pdf. Accessed 14 Mar 2017.
17. Pierzchala M, Cieslak D, Reiner G, Bartenschlager H, Moser G, Geldermann H. Linkage and QTL mapping for Sus scrofa chromosome 17. J Anim Breed Genet. 2003;120:132–7.
18. Guo YM, Lee GJ, Archibald AL, Haley CS. Quantitative trait loci for production traits in pigs: a combined analysis of two Meishan x large white populations. Anim Genet. 2008;39:486–95.
19. Casas-Carrillo E, Prill-Adams A, Price SG, Clutter AC, Kirkpatrick BW. Mapping genomic regions associated with growth rate in pigs. J Anim Sci. 1997;75:2047–53.
20. Beeckmann P, Mose G, Bartenschlager H, Reiner G, Geldermann H. Linkage and QTL mapping for Sus scrofa chromosome 8. J Anim Breed Genet. 2003;120:66–73.
21. Walling GA, Visscher PM, Andersson L, Rothschild MF, Wang L, Moser G, et al. Combined analyses of data from quantitative trait loci mapping studies. Chromosome 4 effects on porcine growth and fatness. Genetics. 2000;155:1369–78.
22. Chen K, Hawken R, Flickinger GH, Rodriguez-Zas SL, Rund LA, Wheeler MB, et al. Association of the Porcine Transforming Growth Factor Beta Type I Receptor (TGFBR1) Gene with growth and carcass traits. Anim Biotechnol. 2012;23:43–63.
23. Cade BS, Noon BR. A gentle introduction to quantile regression for ecologists. Front Ecol Environ. 2003;1:412–20.
24. Silva FF, Resende MDV, Rocha GS, Duarte DAS, Lopes PS, Brustolini OJB, et al. Genomic growth curves of an outbred pig population. Genet Mol Biol. 2013;36:520–7.
25. Beyerlein A, Von Kries R, Ness AR, Ong KK. Genetic markers of obesity risk: stronger associations with body composition in overweight compared to normal-weight children. PLoS One. 2011;6:e19057.
26. Sun S, Chen Z, Yan PS, Huang Y-W, Huang THM, Lin S. Identifying hypermethylated cpg islands using a quantile regression model. BMC Bioinformatics. 2011;12:54.
27. Wang L, Wu Y, Li R. Quantile regression for analyzing heterogeneity in ultra-high dimension. J Am Stat Assoc. 2012;107:214–22.
28. Koenker R, Machado JAF. Goodness-of-fit and related inference processes for Quantile regression. J Am Stat Assoc. 1999;94:1296–310.

12

Impact of dietary L-arginine supply during early gestation on myofiber development in newborn pigs exposed to intra-uterine crowding

Johannes Gulmann Madsen[1,2], Camilo Pardo[1,2], Michael Kreuzer[2] and Giuseppe Bee[1]* (ID)

Abstract

Background: Intra-uterine crowding (IUC) observed in hyperprolific sows impairs myofiber hyperplasia and overall fetal growth. Arginine supplementation (ARG) in gestation diets has been shown to positively affect litter and muscle development. The study objective was to assess whether the effect of ARG on offspring characteristics, with special emphasis on myofiber hyperplasia, differs under IUC conditions from these responses, because in that situation growth retardation is particularly prevalent due to reduced fetal nutrient supply. Unilateral oviduct ligation (OL) was used as a model for an uncrowded and hyperprolificacy (IN) as a model for a crowded intra-uterine environment.

Methods: Five OL and five IN sows were fed a diet supplemented daily with either 43 g L-alanine (Ctrl) or 25 g L-arginine from d 14 to 28 of gestation in a cross-over design (two periods: 5th and 6th parity). At farrowing, two male and two female offspring, respectively, with a low and intermediate birth weight (BtW) were selected. After euthanization, the *Semitendinosus* muscle (STM) was removed and weighed, and the light and dark portions (STM$_d$ and STM$_l$) were prepared for myofiber histochemistry using ATPase staining and the entire STM for gene expression analysis of myogenesis-related genes using RT-qPCR. In addition, various organs were removed and weighed. Data were analyzed using the MIXED model in SYSTAT.

Results: No effect of either IUC or dietary treatment was found in litter characteristics. Offspring of ARG sows displayed a greater muscle area in STM ($P < 0.01$) as a result of the greater myofiber hyperplasia ($P < 0.01$). The increase was more distinct in the STM$_l$ ($P < 0.05$) than in the STM$_d$ ($P = 0.131$). Offspring of OL sows were heavier at birth ($P < 0.01$), had a heavier STM ($P < 0.05$), liver ($P < 0.01$) and kidney ($P < 0.05$), but when expressed relative to birth weight, these differences were absent. In addition, IUC had an effect ($P < 0.05$) on the expression of one of the myogenesis-related genes investigated.

Conclusions: Independent from the extent of IUC, ARG improved BtW, muscle and organ weights and myofiber hyperplasia in offspring.

Keywords: Dietary supplement, Early gestation, Intra-uterine growth restriction, Myofiber hyperplasia, Neonate, Sow prolificacy

* Correspondence: giuseppe.bee@agroscope.admin.ch
[1]Agroscope Posieux, la tioleyre 4, 1725 Posieux, Switzerland
Full list of author information is available at the end of the article

Background

In the last decade, the selection for high prolificacy in modern sow herds has led to a marked increase in litter sizes. One consistent outcome of this strategy was the increasing number of less vital and less mature low birth weight (L-BtW) piglets [1, 2]. In these piglets, prenatal muscle development is impaired [3, 4] as evidenced by the lower myofiber number and the greater number of myofibers still expressing the fetal myosin heavy chain isoform at birth [4]. Compared with their heavier siblings, the lower prenatal myofiber hyperplasia observed in underprivileged piglets has negative consequences on postnatal growth efficiency [5] and lean meat deposition rate [6, 7].

The pig muscle develops in a biphasic manner [8] (Fig. 1). In a first wave from d 35-55 of gestation, an initial population of myoblasts fuse to form primary (P) myofibers. From d 55-90 of gestation, these primary myofibers serve as scaffold for the fusion of a second larger population of myoblasts, the so-called secondary (S) myofibers [9, 10]. As opposed to S myofiber development where crowding of the uterus at early gestation appears to be a compromising factor [11], P myofiber number is assumed to be a fixed genetic component, and its development is assumed to be unaffected by conditions occurring in utero [12]. However, a recent study showed that hyperplasia of P myofibers was greater in the *Semitendinosus* muscle (STM) of 75-day-old fetuses originating from sows fed a diet supplemented with L-arginine from d 14-28 of gestation [13]. Because supplementation of L-arginine occurred before the start of P myofiber formation, the authors hypothesized that this effect was an indirect effect. Earlier supplementation of L-arginine from d 0-14 is not advisable because of its detrimental effect on embryonic survival, which is likely due to reduced progesterone

secretion [14]. Arginine is a common substrate for nitric oxide and polyamine synthesis [15], both of which are key regulators of angiogenesis and placental growth [16]. Therefore, an increased dietary arginine supply might improve the fetal nutrient supply [17] and ultimately promote myofiber hyperplasia. This mode of action could be interesting especially for offspring from prolific sows suffering from intra-uterine growth retardation (IUGR) due to a crowded intra-uterine environment which impairs placental and, consequently, fetal development.

Based on the aforementioned association between dietary arginine supply, the extent of placental vascularization, the fetal nutrient supply and muscle development, it was hypothesized that supplementing L-arginine to an early gestational diet of the dams would promote hyperplasia leading to an increased number of myofibers in their offspring at birth. A second hypothesis tested the theory of whether L-arginine would be especially efficient in piglets suffering from IUGR. Two sow models were established for the present study, one 'non-crowded' and the other 'relatively crowded' simulated by intact (IN) prolific sows. The non-crowded intra-uterine environment was mimicked by using unilaterally oviduct ligated (OL) sows. In these sows, oocytes ovulated from the ovary ipsilateral to the ligated oviduct are prevented from being fertilized and entering the uterus. Therefore, the number of embryos and the extent of crowding in utero is markedly reduced in OL compared with IN sows [13, 17].

Methods

Animals and dietary treatments

The study was conducted as a 2 × 2 crossover design (two periods for each sow in their 5th and 6th parity). It involved five OL sows originating from a previous experiment of Pardo et al. [18] and five prolific IN sows.

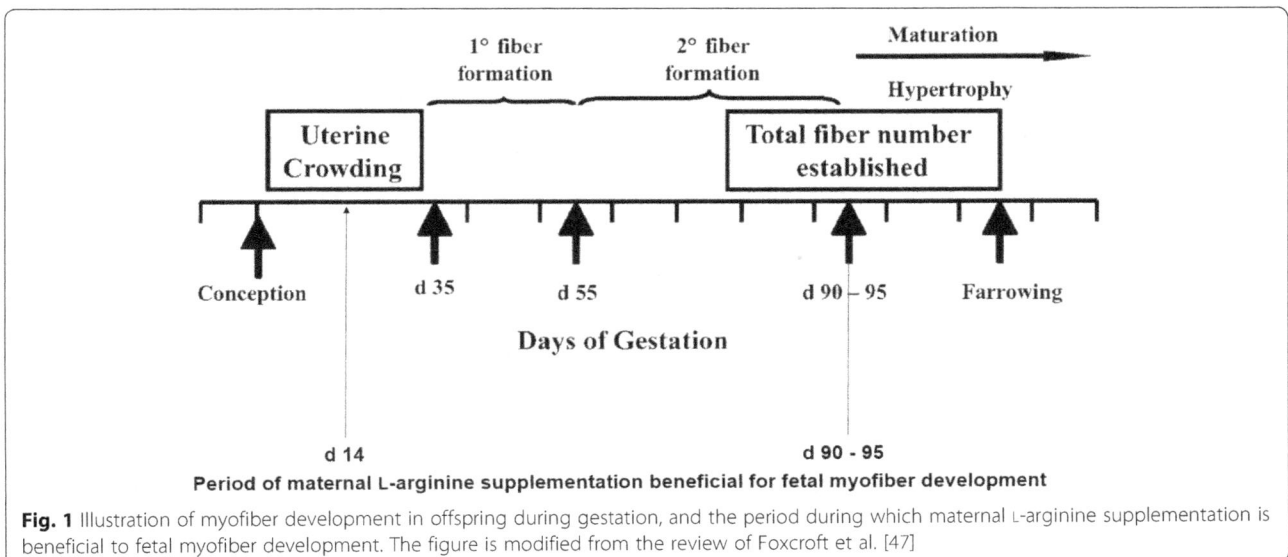

Fig. 1 Illustration of myofiber development in offspring during gestation, and the period during which maternal L-arginine supplementation is beneficial to fetal myofiber development. The figure is modified from the review of Foxcroft et al. [47]

The five IN sows were siblings to the five OL sows. Except from d 14 to 28 of gestation, the sows were reared from mating to farrowing with other multiparous sows in group pens equipped with an automatic feeder (Compident, Model 2000, Schauer, Prambachkirchen, Austria) and were offered 2.8 kg of a standard gestation diet daily (Table 1). From d 14 to 28 of gestation, the sows were kept in individual pens. They were randomly allotted to receive daily either 25 g L-arginine HCl (ARG), an amount based on the previous findings of Bérard and

Table 1 Ingredients and calculated nutrient composition of the gestation diet[1]

Item	Gestation diet
Ingredients, % as-fed	
Barley	25.0
Oat	20.0
Dried sugar beet pulp	14.3
Wheat bran	10.0
Soybean meal	5.7
Dried apple pomace	5.6
Dried whole maize plant	5.0
Animal fat (70% lard and 30 % tallow)	3.0
Linseed meal	2.0
Potato protein	2.0
Rapeseed meal	2.0
Molasses	2.0
Dicalcium phosphate	1.16
Calcium carbonate	0.75
Sodium chloride	0.56
Pellan[2]	0.40
Vitamin-mineral premix[3]	0.40
Amino acids[4], %	
L-lysine HCl	0.08
L-threonine	0.06
Calculated composition, % DM	
Dry matter, % of wet weight	88.35
Total ash	7.20
Ether extract	6.80
Crude protein	15.10
Digestible energy, MJ/kg DM	13.70

[1]In the ARG group, the pellets were top dressed with 25 g L-arginine HCl/ (sow·d), and in the Ctrl group, the pellets were top dressed with 43 g L-alanine/(sow·d)
[2]Binder that aids in pellet formation (Mikro-Technik, GmbH & Co. KG, Germany)
[3]Contains: vitamin A: 2,000,000 U/kg; vitamin D$_3$: 200,000 U/kg; vitamin E: 10 g/kg; I: 137.4 mg/kg; Mn: 5 g/kg; Cu: 1.75 g/kg; Zn: 1.4 g/kg; Se: 50 mg/kg
[4]Calculated amino acid composition, g/kg DM: alanine: 6.42; arginine: 8.51; aspartate plus asparagine: 12.70; cystine: 3.15; glutamate plus glutamine: 27.50; glycine: 6.54; histidine: 3.42; isoleucine: 5.74; leucine: 10.72; lysine: 7.90; methionine: 2.46; phenylalanine: 7.00; proline: 9.59; serine: 6.73; threonine: 6.30; tryptophan: 1.83; tyrosine: 5.03; valine: 7.38

Bee [13], or 43 g L-alanine (Ctrl) both as top dressing for 14 d. Alanine was used to compensate for the increased amount of nitrogen in the ARG treatment making the two dietary treatments isonitrogenous.

Data and tissue sample collection at farrowing
At the end of farrowing, litter characteristics including the number of piglets born alive and stillborn and their individual body weights were recorded. At the day of farrowing, two female and two male piglets per sow per parity were intentionally sacrificed, one with the lowest BtW (L: average ± standard deviation (SD): 1.24 ± 0.30 kg) and one with intermediate BtW (M: average ± STD: 1.49 ± 0.26 kg). Thus, the aim was to include a total of 80 piglets in the study. However, due to unforeseen events at farrowing, the final number was only 70 piglets, which were balanced accordingly between the extent of crowding (38 vs. 32, IN and OL), dietary treatment (36 vs. 34, Ctrl and ARG), BtW (33 vs. 37, L and M), sex (34 vs. 36, male and female), and parity (33 vs. 37, 5th and 6th). Pigs with a BtW of < 800 g were considered runts and were not included in the study. The selected piglets were anaesthetized using an isoflurane-oxygen mixture [4% vol/vol] and subsequently euthanized by exsanguination. Subsequently, the complete *Semitendinosus* muscle (STM) and *Psoas major* muscle (PM) were removed and weighed. The STM was split into the light and dark portions (STM$_l$ and STM$_d$). A section from each portion was removed from the middle of the muscle, snap-frozen in 2-methylbutane cooled in liquid nitrogen and subsequently stored at −80°C until histochemical analysis was performed. Subsequently, spleen, kidneys, heart, lungs, liver and brain were removed and weighed.

Histochemical analysis of the semitendinosus muscle
The P and S myofibers were differentiated histochemically using the protocol described previously by Bérard et al. [4]. Briefly, 10 μm cross-sections of the STM$_d$ and STM$_l$ were prepared and stained for the determination of myofibrillar ATPase activity after acid (pH 4.5) or alkaline (pH 10.3) preincubation. In the STM$_d$, the P myofibers stain dark, and the S myofibers light, using the acid preincubation condition, whereas the opposite occurs after basic preincubation. This differentiation was possible in the STM$_d$, whereas, in accordance with observations of the study by Bérard et al. [4], P and S myofibers could not be differentiated in the STM$_l$ of newborn pigs using this mATPase histochemistry assay. A sectional cut of the entire muscle area was stained using anti-slow myosin heavy chain monoclonal antibody (Novocastra lyophilized mouse monoclonal antibody myosin heavy chain (NCL-MHCs) diluted 1:20 in ultrapure water; Novocastra, Newcastle, UK). This allowed a clear determination of the STM$_d$ and STM$_l$ cross-sectional areas.

The number of P myofibers were determined in the mATPase sections after acid pre-incubation, where 750 P myofibers were counted in an area of 0.89 mm^2. The number of P myofibers, counted in the selected area, and the cross-sectional area of the STM$_d$ was used to estimate the total number of P myofibers. The number of S myofibers was determined in mATPase sections after alkaline preincubation. Four images were taken and analyzed using the analySIS software 5.0 (Soft Imagine System, Olympus Soft Imaging Solutions GmbH, Münster, Germany). At least 33 P myofibers were selected, and all the surrounding S myofibers were counted (> 1000 S myofibers). This permitted estimating the total S myofiber number as well as calculating the S:P myofiber ratio. The total number of myofibers (TNF) in the STM$_d$ was calculated from the respective estimated total number of P and S myofibers. In the STM$_l$, the number of myofibers was determined in mATPase sections after acid preincubation, where > 3000 myofibers were counted in an area of 1.47 mm^2. The number of counted myofibers, the according measurement area (1.47 mm^2) and the cross-sectional area of the STM$_l$ were used to estimate TNF.

Gene expression analysis of myogensis-related genes in the *Semitendinosus* muscle

Of the 70 piglets, samples from five individuals (1 L-BtW and 1 M-BtW male from ARG-IN sows, 2 L-BtW females from ALA-IN sows and 1 L-BtW male from ALA-IN sow) were of too poor quality for analysis of gene expression, thus in a total of 65 piglets, total mRNA was extracted from 50 mg STM by the phenol-chloroform extraction protocol involving homogenization with IKA® T10 basic Ultra-Turrax® (IKA, Staufen, Germany) and phase separation with peqGOLD TriFast™ (Peqlab, Erlangen, Germany). Centrifugation steps were carried out at 4 °C and 12,000 × g. Tissue was cut into pieces of 5 mg and homogenized for 20 s in 1 mL of TriFast TM reagent. After 5 min of incubation at room temperature, 200 µL of chloroform was added, shaken for 15 s and incubated at room temperature for 10 min. Subsequently, phases were separated by centrifugation, and the supernatant containing the mRNA was carefully removed and mixed with 500 µL of isopropanol, precooled at 4 °C. Following a 15 min incubation on ice, the tube was centrifuged, the supernatant discarded and the pellet washed in precooled (4 °C) 500 µL of ethanol. After a last centrifugation step at 7,500 × g for 8 min, the ethanol was discarded and the pellet was dried at room temperature. Finally, the mRNA was resuspended in 100 µL nuclease-free water under agitation at 55 °C. Quality and concentrations of sample mRNA were measured with a NanoDrop® ND-1000 (Thermo Scientific, Waltham, Massachusetts, USA). Primers for both target and reference genes are listed in Table 2. All primers were designed with Primer3 v.4.0.0 (http://primer3.wi.mit.edu/), tested for

dimerization with OligoAnalyzer® v.3.1 (Integrated DNA Technologies, Coralville, Iowa, USA), and subsequently purchased in SePOP quality dissolved in water (Eurogentec, Seraing, Belgium). Reverse transcription of complementary DNA (cDNA) was performed with ImProm-II™ Reverse Transcription System (Promega, Dübendorf, Switzerland). First, random hexamer primers were annealed to the mRNA (5 µL of DNAse-treated mRNA with 1 µL of random hexamers, incubation at 70 °C for 5 min. The mRNA was then subjected to the reverse transcription (using 20 µL reaction solutions with 4 µL of 5× buffer, 2.4 µL of 25 mmol/L MgCl$_2$, 1 µL of 10 mmol/L dNTPs, 1 µL of reverse transcriptase and 5.6 µL nuclease-free water, according to manufacturer's protocol). Finally, the freshly synthesized cDNA was diluted 12.5 × with nuclease-free water and stored at 4 °C until further use. Real time-qPCR was carried in 20 µL reaction solutions consisting of 10 µL of 2 × KAPA™ SYBR® FAST Master Mix (Kapa Biosystems, Woburn, Massachusetts, USA), 4.2 µL of water, 0.4 µL 10 µmol/L of both forward and reverse primers and 5 µL of cDNA sample. Amplification was carried out in a Rotor-Gene 6000 (Corbett Life Sciences, Sydney, Australia) under the following conditions: A 3 min hold at 95 °C followed by 40 cycles with 10 s of denaturation at 95 °C, 10 s of annealing at 60 °C: Insulin-like growth factor 2 (*IGF2*), Insulin-like growth factor binding protein 5 (*IGFBP5*), Myogenic factor 6 (*MYF6*), Myogenic differentiation 1 (*MYOD*), Protein kinase, AMP-activated, alpha 2 catalytic subunit (*PRKAA2*), TATA box binding protein (*TBP*), 20 s at 60 °C: Myogenin (*MYOG*) or 20 s at 62 °C: Myostatin (*MSTN*), Myogenic factor 5 (*MYF5*), Ribosomal protein L4 (*RPL4*) and 20 s extension time at 72 °C with the last 4 s being used for fluorescence measurements. The cycling was followed by acquisition of a melt curve (from 72 °C to 95 °C, in increments of 1 °C each 5 s) to control the specificity of the PCR product. Owing to the large number of samples, it was decided to amplify all samples for a given gene in one run. Two technical replicates were performed in two further runs (rather than measuring technical replicates in the same run). To reduce inter-run variation, the reactions in all three runs were prepared with the same mastermix.

The modified formula by Livak and Schmittgen [19] suitable for multiple reference genes was applied to gene expression data, and the relative quantification was performed with the qBase+ software (Biogazelle NV, Zwijnaarde, Belgium) [20].

Statistical analysis

Data were analyzed using the MIXED procedure of SYSTAT version 13 (Systat Software, San Jose, CA). All data were tested for normality of residuals. For litter traits, data were analyzed including IUC and dietary treatment (DIET), IUC × DIET interaction, parity, and

Table 2 Forward and reverse primers of myogenesis-related genes and reference genes[1]

	Accession no.	Forward primer (from 5′ to 3′)	Reverse primer (from 5′ to 3′)
Gene			
IGF2	NM_213883	TGGCATCGTGGAAGAGTG	AGGTGTCATAGCGGAAGAAC
IGFBP5	U41340	GTGTACCTGCCCAACTGTGA	AAGCTGTGGCACTGGAAGTC
MSTN	NM_214435	CCCGTCAAGACTCCTACAACA	CACATCAATGCTCTGCCAA
MYF5	Y17154	CCTGAATGCAACAGCCCT	CGGAGTTGCTGATCCGAT
MYF6	NM_001244672	CGCCATCAACTACATCGAGAGGT	ATCACGAGCCCCCTGGAAT
MYOD1	NM_001002824	GGTGACTCAGACGCATCCA	ATAGGTGCCGTCGTAGCAGT
MYOG	NM_001012406	CAACCAGGAGGAGCGAGAC	GAGGTGAGGGAGTGCAGATT
PRKAA2	NM_214266	CCCCTGAAACGAGCAACTATC	CACACTTCTTTCACAGCCTCAT
Reference genes			
RPL4	DQ845176	CAAGAGTAACTACAACCTTC	GAACTCTACGATGAATCTTC
TBP	DQ178129	GATGGACGTTCGGTTTAGG	AGCAGCACAGTACGAGCAA

[1]*IGF2* Insulin-like growth factor 2, *IGFBP5* Insulin-like growth factor binding protein 5, *MSTN* Myostatin, *MYF5* myogenic factor 5, *MYF6* myogenic factor 6, *MYOD1* Myogenic differentiation 1, *MYOG* Myogenin (myogenic factor 4), *PRKAA2* Protein kinase AMP-activated, alpha 2 catalytic subunit, *RPL4* (LOC100038029) Ribosomal protein L4, *TBP* TATA box binding protein

boar as fixed effect, offspring as random effect, and sow as the experimental unit. Data on absolute and relative STM and PM weight, absolute and relative organ weight, total area of STM, TNF of STM, P and S number of myofibers, the S:P ratio of STM_d, area of STM_d, STM_l and gene expression were analyzed including the main factors of IUC, DIET, sex, sow, parity, BtW (nested within sow), IUC × D and BtW × DIET interaction as fixed, and boar as random effect. For all analyses of the piglet data, sow was used as variable in a repeated measurement procedure. Birth weight was nested within sow as birth weight intervals (L and M) were not unique among sows, leading to the individual piglet being assigned either L or M depending on the dam. The final model was determined by performing backward elimination of non-significant three- and two-way interactions. Offspring was used as the observational unit.

Least squares means of the interaction means concerning the main factors IUC and DIET are presented in the result tables. As the effect of sex and BtW was rarely significant, these least square means were only reported in the text when $P < 0.05$. The PAIRWISE option using the Tukey adjustment was used to determine differences between interaction effects when two-way interactions occurred. Differences were considered statistically significant at $P < 0.05$ and as tendency at $0.05 < P < 0.10$.

Results

Sow performance and litter characteristics

The number of total born and born alive piglets, and the male to female ratio were not different between IN and OL sows and were also not affected by the dietary ARG supply (Table 3). Average litter BtW, average BtW of female and male newborns as well as BtW variability,

expressed as standard deviation (SD), also did not differ between IN and OL sows and between the ARG and Ctrl dietary treatments. Regarding stillborns, the five IN sows gave birth to one, two, two, four and five stillborn piglets, respectively, and only one OL sow gave birth to one stillborn piglet. Thus, the number of stillborn piglets were unevenly distributed: IN-Ctrl = 7; IN-ARG = 7; OL-Ctrl = 1; OL-ARG = 0. Due to the greatly unbalanced distribution between experimental treatment groups, statistical analysis of number and BtW of stillborn piglets was not performed.

Birth weight and absolute and relative muscle and organ weights

There were no significant effects of sow type and diet type on BtW of total born and born alive piglets (Table 3). Birth weight of males born alive from IN sows fed the ARG diet were heavier in tendency than those fed the Ctrl diet, whereas BtW of males born alive were lighter in tendency when originating from OL sows fed the ARG instead of the Ctrl diet (IUC × DIET interaction; $P = 0.087$). The average BtW of the selected offspring was lower than of the litter average because half of the piglets were of deliberately low BtW (Table 4). The piglets selected from the OL sows compared with those from the IN sows were heavier ($P = 0.008$) at birth, and had greater ($P \leq 0.033$) STM, PM, liver and kidney weights. The weight of the spleen tended ($P = 0.068$) to be greater in OL compared with IN offspring. When expressed per 100 g BtW, the relative weights of the muscles and organs did not differ. The selected offspring from ARG compared with Ctrl sows tended ($P < 0.10$) to have greater BtW, absolute STM weight, liver weight and a lower brain:liver weight ratio. Except for the relative heart weight, which was lower ($P = 0.030$) in ARG

Table 3 Reproductive characteristics of oviduct ligated (OL; non crowded) and intact (IN; relatively crowded) sows fed an unsupplemented (Ctrl) or L-arginine supplemented gestation diet (ARG) from d 14 to 28 of gestation[1]

| Item | IN | | OL | | SEM | P-value[2] | | |
	Ctrl	ARG	Ctrl	ARG		IUC	DIET	IUC × DIET
Observations	5	5	5	5				
Total born	15.5	15.3	10.5	11.9	1.71	0.255	0.722	0.657
Born alive	13.1	14.5	9.1	10.3	1.49	0.230	0.463	0.961
Male:female ratio	0.47	0.62	0.50	0.41	0.093	0.508	0.732	0.368
Birth weight, kg								
Total born	1.22	1.48	1.57	1.51	0.068	0.278	0.214	0.126
Born alive								
All	1.24	1.50	1.59	1.53	0.068	0.277	0.224	0.133
Male	1.31	1.46	1.74	1.63	0.079	0.224	0.511	0.087
Female	1.26	1.51	1.50	1.44	0.092	0.580	0.365	0.230
SD[c]	0.24	0.25	0.18	0.14	0.056	0.414	0.869	0.626

[1]Results are presented as least squares means of interactions between the two main factors extent of intra uterine crowding and dietary treatment and pooled SEM
[2]Probability values for the effects of extent of intra uterine crowding (IUC), dietary treatment (DIET), and interaction IUC × DIET
[3]Measure for variability in birth weight, expressed as standard deviation (SD).

Table 4 Birth weight, absolute and relative muscle and organ weights and brain:liver weight ratio of offspring born from oviduct ligated (OL; non crowded) and intact (IN; relatively crowded) fed an unsupplemented (Ctrl) or L-arginine supplemented diet (ARG) from d 14 to 28 of gestation[1]

| Item | IN | | OL | | SEM | IUC | P-value[2] | | | |
	Ctrl	ARG	Ctrl	ARG			DIET	BtW	IUC×DIET	DIET×BtW
Observations	19	19	17	15						
Birth weight, kg	1.06	1.21	1.58	1.59	0.128	0.008	0.068	0.564	0.137	< 0.001
Weights, g										
Semitendinosus muscle	2.08	2.48	3.47	3.60	0.417	0.025	0.074	0.571	0.426	0.009
Psoas major muscle	2.15	2.36	3.86	4.02	0.343	0.001	0.232	0.543	0.891	0.024
Heart	8.34[a]	10.54[b]	8.35[ab]	7.38[ab]	1.695	0.418	0.159	0.806	0.005	0.032
Liver	20.91	27.39	45.98	47.55	5.808	0.004	0.053	0.529	0.290	0.022
Spleen	1.05	1.00	1.45	1.50	0.164	0.063	0.938	0.918	0.537	0.052
Lungs	16.48	18.24	19.86	18.66	2.655	0.576	0.750	0.459	0.152	0.010
Kidneys	7.68	8.93	12.77	12.73	1.516	0.033	0.301	0.771	0.322	0.022
Brain	32.22	33.01	33.85	34.32	0.781	0.235	0.119	0.621	0.699	0.031
Brain:liver weight ratio	1.36	1.12	0.94	0.92	0.222	0.270	0.077	0.151	0.182	0.001
Weights, g/100 g BtW										
Semitendinosus muscle	2.09	2.08	2.19	2.28	0.095	0.328	0.415	0.716	0.341	0.511
Psoas major muscle	2.06	2.14	2.31	2.39	0.140	0.277	0.276	0.314	0.987	0.483
Heart	0.70	0.67	0.67	0.65	0.023	0.492	0.030	0.770	0.720	0.338
Liver	2.21	2.47	2.78	2.77	0.179	0.128	0.192	0.629	0.142	0.343
Spleen	0.10[b]	0.09[a]	0.09[ab]	0.09[ab]	0.006	0.665	0.176	0.734	0.018	0.556
Lung	1.47	1.39	1.31	1.27	0.109	0.363	0.250	0.235	0.762	0.655
Kidney	0.74	0.72	0.80	0.82	0.043	0.270	0.764	0.613	0.396	0.627
Brain	2.87	2.53	2.47	2.50	0.323	0.616	0.169	0.204	0.114	< 0.001

[1]Results are presented as least squares means of interactions between the two main factors extent of intra uterine crowding and dietary treatment and pooled SEM
[2]Probability values for the effects of extent of intra uterine crowding (IUC), dietary treatment (DIET), birth weight (BtW), and interactions IUC × DIET and DIET × BtW
[a,b]Within a row for the main factor treatment, least squares means without a common superscript differ (P < 0.05)

than Ctrl offspring, muscle and organ weights relative to BtW were not affected by the dietary treatment of the sow. There was an IUC × DIET interaction (P = 0.005) indicating that absolute heart and relative spleen weight were greater and lower, respectively, in IN-ARG compared with IN-Ctrl piglets. Birth weight had no effect on either absolute or relative STM, PM, organ weights and brain:liver weight ratio. There were a number of significant (P < 0.05) diet × BtW interactions. Compared with L-piglets, M-piglets born from sows fed the ARG, but not the Ctrl diet, had heavier STM (L-ARG = 2.679 vs. M-ARG = 3.401 g), PM (L-ARG = 2.840 vs. M-ARG = 3.547 g), hearts (L-ARG = 8.387 vs. M-ARG = 9.538 g), livers (L-ARG = 33.87 vs. M-ARG = 41.07 g), spleens (L-ARG = 1.07 vs. M-ARG = 1.42 g), lungs (L-ARG = 17.00 vs. M-ARG = 19.90 g) and kidneys (L-ARG = 9.66 vs. M-ARG = 12.00 g). On the other hand, L-piglets compared with M-piglets born from sows fed the ARG had higher relative brain weights (L-ARG = 2.80 vs. M-ARG = 2.25) and a higher brain:liver weight ratio (L-ARG = 1.14 vs. M-ARG = 0.90) (DIET × BtW interaction; P < 0.001). No such differentiation was observed between L-Ctrl vs. M-Ctrl. It is also noteworthy that the absolute brain weight did not differ between L-ARG (33.5 g) and M-ARG (33.8 g). Together, these results suggest that besides the obvious positive association between BtW of piglet and organ size, an early gestational ARG supplementation has slightly greater effect on organ development of offspring with intermediate compared with low BtW. Finally, males and females differed (P < 0.05) in absolute weights of liver (34.18 vs. 36.73 g), lung weight (18.80 vs. 17.82 g) and relative heart weight (0.69 vs. 0.66 g/100 g BtW).

Myofiber characteristics of the *Semitendinosus* muscle
Except for the greater (P = 0.078) area of the STM_d in OL compared with IN offspring, the extent of IUC did not affect total muscle area, muscle area of the STM_l nor STM_d and STM_l myofiber number of selected offspring (Table 5). By contrast, the areas of the STM_d and STM_l and, consequently, of the whole STM were larger (P ≤ 0.025) in offspring born from ARG sows compared with Ctrl sows. However, regarding the STM_d area, ARG supplementation had a greater impact on OL compared to IN offspring (IUC × DIET; P = 0.007). In addition, the STM tended to be larger in M piglets from ARG sows compared with piglets from the other BtW categories (STM; M-ARG = 8.573; L-ARG = 7.002; M-Ctrl = 6.996; L-Ctrl = 6.271 μm^2 × 10^7, and STM_l; M-ARG = 5.494; L-ARG = 4.301; M-Ctrl = 4.181; L-Ctrl = 3.875 μm^2 × 10^7; DIET × BtW interaction, P ≤ 0.061). Total myofiber number was 12% greater (P = 0.003) in the ARG group; this is primarily a result of a 13% greater (P = 0.022) TNF of the STM_l. In the STM_d, where

differentiation between P and S myofibers was possible, the number of S myofibers, but not P myofibers, was numerically (P = 0.124) greater in piglets of ARG sows. As a consequence, the S:P myofiber ratio tended (P = 0.051) to be greater in the STM_d of these piglets. While L- and M-piglets did not differ with respect to total muscle area and TNF of the whole STM, BtW had an effect (P < 0.05) on TNF in the STM_d (L = 188,925 vs. M = 208,792), the number of S myofibers (L = 182,927 vs. M = 202,776) and tended to have an effect (P = 0.069) on the number of P myofibers (L = 5,998 vs. M = 5,986) also in the STM_d. Due to the lack of an effect of BtW and of a DIET × BtW interaction in TNF, the positive effect of ARG supplementation on myofiber hyperplasia was similar for L and M piglets (L = 506,274 vs. M = 567,783 myofibers).

Gene expression in the *Semitendinosus* muscle
The level of *PRKAA2* expression in the STM was greater (P = 0.024) in IN compared with OL offspring (Table 6). In addition, ARG appeared to have a greater impact on the expression of this gene in IN than in OL offspring (IUC × DIET; P = 0.034). Furthermore, a IUC × DIET interaction effect (P = 0.013) was found in the expression of *IGF2*; however, Tukey's *post hoc* comparison did not identify individual differences between means. The expression of *MYOD1* tended (P = 0.056) to be lower in L- than M-pigs (L = 1.96 vs. M = 1.86), and a tendency (P = 0.091) to a DIET × BtW interaction effect on *PRKAA2* expression was found, where M-piglets from Ctrl sows displayed the lowest expression compared with the remaining piglets (M-Ctrl = 0.44; L-Ctrl = 0.53; M-ARG = 0.59; L-ARG = 0.60 relative expression level). Again, there were no differences in individual interaction means identified by the Tukey's *post hoc* comparison.

Discussion
The increased litter size of modern sow breeds is likely associated with increased incidents of IUC and consequently IUGR. Therefore, an increased proportion of underdeveloped piglets within the litters can be observed [21, 22]. These piglets display a lower postnatal survival rate [23], and a poorer growth performance compared with their larger littermates [24, 25]. These impairments can in part be attributed to the reduced myofiber hyperplasia mainly due to lower formation of S myofibers [8]. One approach to enhance hyperplasia is by supplementing L-arginine to the sows during early gestation, shown to be leading to increased formation of P myofibers, which serve as a scaffold for the development of the S myofibers [13]. To alter the intra-uterine environment, either unilateral ovary-hysterectomy or unilateral oviduct ligation of sows can be applied, where sows subjected to the latter procedure have shown to display minimal crowding and adequate placental development

Table 5 Myofiber characteristics in the *Semitendinosus* muscle of selected offspring born from oviduct ligated (OL; non crowded) and intact (IN; relatively crowded) sows fed an unsupplemented (Ctrl) or L-arginine supplemented diet (ARG) from d 14 to 28 of gestation[1]

Item	IN		OL		SEM	IUC	DIET	BtW	IUC×DIET	DIET×BtW
	Ctrl	ARG	Ctrl	ARG						
Observations	19	19	17	15						
Muscle area, $\mu m^2 \times 10^7$										
Total	5.762	6.780	7.505	8.795	0.9429	0.144	0.003	0.653	0.738	0.053
Dark portion	2.359[x]	2.278[x]	2.772[ax]	3.572[by]	0.2931	0.078	0.025	0.233	0.007	0.573
Light portion	3.151	4.118	4.906	5.677	0.8279	0.143	0.011	0.910	0.786	0.061
Myofiber number, N										
Total muscle	476,025	527,711	536,451	607,928	35,511.0	0.236	0.003	0.136	0.604	0.475
Dark portion										
Total myofibers	193,928	198,254	184,590	218,604	22,898.2	0.884	0.131	0.002	0.231	0.560
Primary myofibers (P)	6,105	5,840	5,868	6,156	638.8	0.970	0.974	0.069	0.422	0.575
Secondary myofibers (S)	187,823	192,414	178,722	212,448	22,455.9	0.883	0.124	0.002	0.231	0.559
S:P ratio[3]	30.8	33.2	30.6	34.4	2.86	0.912	0.051	0.356	0.643	0.337
Light portion	282,097	329,457	351,860	389,324	32,639.6	0.234	0.022	0.889	0.778	0.236

[1]Results are presented as least squares means of interactions between the two main factors extent of intra uterine crowding and dietary treatment and pooled SEM
[2]Probability values for the effects of extent of intra uterine crowding (IUC), dietary treatment (DIET), birth weight (BtW), and interactions IUC × DIET and DIET × BtW
[3]The secondary-to-primary myofiber ratio, calculated by dividing the number of secondary with the number of primary myofibers in the dark portion of the *Semitendinosus* muscle.
[a,b]Within a row for the main factor treatment, least squares means without a common superscript differ ($P < 0.05$)
[x,y]Within a row for the main factor treatment, least squares means without a common superscript differ ($0.05 < P < 0.10$)

[18]. Thus, compared with intact natural crowded sows as used in the current study, OL sows with average prolificacy are suitable models for investigating the consequences of IUC and IUGR in newborn piglets. Hence, intact prolific sows with naturally crowded uterus and OL sows with a "non-crowded uterus" were used in the present study to mimic two degrees of IUC and to investigate whether the positive effect of early gestational supplementation of the sows with L-arginine on myofiber hyperplasia observed in 75-day-old fetuses [13] would continue until birth and would differ depending on the extent of IUC.

Table 6 Relative expression of myogenesis-related genes in the *Semitendinosus* muscle of selected offspring born from oviduct ligated (OL; non crowded) and intact (IN; relatively crowded) sows fed an unsupplemented (Ctrl) or L-arginine supplemented diet (ARG) from d 14 to 28 of gestation[1]

Gene[3]	IN		OL		SEM	IUC	DIET	BtW	IUC×DIET	DIET×BtW
	Ctrl	ARG	Ctrl	ARG						
Observations	16	17	17	15						
IGF2	2.36	1.55	0.91	1.70	0.555	0.487	0.967	0.965	0.013	0.950
IGFBP5	5.63	5.65	2.32	4.62	2.190	0.489	0.233	0.275	0.290	0.297
MSTN[4]	1.60	1.64	0.46	0.98	1.754	0.351	0.232	0.105	0.241	0.110
MYF5[4]	0.35	0.42	0.24	0.28	1.694	0.631	0.531	0.895	0.974	0.159
MYF6	0.58	0.65	1.18	0.98	0.576	0.623	0.859	0.893	0.664	0.631
MYOD1[4]	2.36	1.84	0.68	0.48	2.143	0.223	0.340	0.056	0.885	0.418
MYOG	0.94	0.70	1.32	1.11	0.255	0.333	0.168	0.392	0.915	0.417
PRKAA2	0.72[x]	1.04[by]	0.26[x]	0.15[ax]	0.194	0.024	0.264	0.298	0.034	0.091

[1]Results are presented as least squares means of interactions between the two main factors extent of intra uterine crowding and dietary treatment and pooled SEM.
[2]Probability values for the effects of extent of intra uterine crowding (IUC), dietary treatment (DIET), birth weight (BtW), and interactions IUC × DIET and DIET × BtW
[3]*IGF2* Insulin-like growth factor 2, *IGFBP5* Insulin-like growth factor binding protein 5, *MSTN* Myostatin, *MYF5* myogenic factor 5, *MYF6* myogenic factor 6, *MYOD1* Myogenic differentiation 1, *MYOG* Myogenin (myogenic factor 4), *PRKAA2* Protein kinase AMP-activated, alpha 2 catalytic subunit
[4]Statistical evaluation was performed with log transformed data
[a,b]Within a row for the main factor treatment, least squares means without a common superscript differ ($P < 0.05$)
[x,y]Within a row for the main factor treatment, least squares means without a common superscript differ ($0.05 < P < 0.10$)

Sow performance and litter characteristics

Important causes of IUGR are impaired placental development, including suboptimal vascularization of the placenta, increased numbers of fetuses due to high prolificacy and uneven distribution of energy and dietary nutrients among the fetuses [26]. As indicated by the number of total and born alive piglets in combination with the corresponding BtW, IUGR occurred to a greater extent in IN sows compared with OL sows. A sow with a litter size of $\geq$ 15 is categorized as being high prolific [24, 27]. This was the case with the IN sows. On the contrary, OL sows had a litter size below the average herd level (9.7 vs. 12.3 live born). The OL sows compared with the IN sows gave birth to a smaller number of stillborn piglets, which was expected from previous findings associating increased litter size with a greater number of stillborn piglets [27]. Litter characteristics of IN sows were in agreement with results from a previous study where offspring developing in a crowded compared with an uncrowded intra-uterine environment exhibited phenotypes associated with IUGR [18]. In the present study, performance of OL sows with respect to litter size and average BtW was comparable to that observed in low to average prolific sows [22].

Effects of maternal dietary L-arginine supplementation on offspring have been previously investigated, but both dietary level, onset of supplementation during gestation and impact on sow and offspring traits have varied greatly among different studies as reviewed by Wu et al. [28]. Among reported effects of L-arginine are increased numbers of total born, born alive and total litter BtW in multiparous sows [29]. However, the sows included in the two former studies had an average litter size of 13 to 14 piglets, thus are to be categorized as average rather than highly prolific sows with no expected major impact of IUC and IUGR. In the present study, dietary supplementation of L-arginine did not significantly affect the sow reproductive performance. Nevertheless, numerical improvements of particularly the number of piglets born alive as well as a greater BtW of offspring in IN-ARG compared with IN-Ctrl sows were observed. However, the reproduction data must be interpreted with caution due to the low number of sows used in this study.

Muscle and organ weights of offspring

The greater BtW and weight of STM and PM of offspring from OL sows compared with IN sows were foreseen as it has been reviewed extensively that there is a high correlation between large litter size and extent of crowding as well as greater within-litter variation of BtW and greater number of low BtW offspring [2, 30, 31]. Organ weights of offspring were within the normal range of newborn piglets [18, 32]. The tendency for an increased absolute STM weight in offspring from ARG sows

suggests that, independent of IUC, L-arginine supplementation promotes muscle development of offspring, an observation that has to the best of the authors' knowledge not been shown in previous studies.

Besides reduced BtW, brain:liver weight ratio has been shown to positively correlate with uterine crowding [11] and therefore is an indicator of IUGR [33]. The brain-sparing mechanism ensures steady relative brain weight development when maternal nutrient supply does not match requirements for adequate fetal development [34]. The greater liver weight combined with the similar brain weight resulted in a lower brain:liver weight ratio in offspring from ARG sows compared with Ctrl sows and supports the hypothesis that L-arginine supplementation has the potential to alleviate the negative impact of IUGR. In addition, the absolute heart weight was only numerically greater in ARG compared with Ctrl, an observation that has also been related to low birth weight pigs at slaughter age [35]. Thus, with the exception of the brain:liver ratio, this observation would not support the positive effect of L-arginine on organ development under crowding conditions.

Due to the rather high level and intake of crude protein and supplementation of daily L-arginine, a great load of maternal arginine could be expected. However, as discussed above, no detrimental effects on either the sow or the offspring traits were observed, indicating extensive catabolism of arginine and that maternal intake of arginine was not excessive [36]. Thus, it was concluded that the level of supplemented arginine was within the safety margin eliciting no toxicity or adverse effects [37, 38], whereas higher crude protein level and arginine concentration of the diet might lead to impaired reproductive performance, antagonism between amino acids and toxicity of ammonia to sows and their fetuses [28].

Myofiber-related traits of the *Semitendinosus* muscle

It is widely accepted that the formation of S myofibers is highly dependent on fetal nutrient uptake [12]. Thus, a reduced nutrient availability in the placenta during mid to late gestation, is likely to result in a lower S myofiber hyperplasia. Previous studies showed that a lower number of myofibers was associated with reduced muscle mass at weaning and at slaughter age [6, 39] meaning an ongoing disadvantage during the entire fattening performance. In the present study, supplementing L-arginine to the sows during early gestation resulted in an increased TNF, with the greatest contribution derived from the second wave of myofiber formation as indicated by the greater S:P ratio in the STM_d and the greater TNF in the STM_l. In comparison, a similar previous study reported that early gestational supplementation with L-arginine enhanced prenatal hyperplasia, but, in contrast, the main increase of TNF originated from the formation of P myofibers [13]. However, in that study myofiber hyperplasia was assessed at d 75 of gestation, a time point where hyperplasia of S myofibers is still ongoing. It is worth mentioning that the increased hyperplasia observed in offspring from ARG sows was independent of BtW (L-ARG: 529,186 vs. M-ARG: 606,453; P = 0.136). This finding contradicts our hypothesis that underprivileged fetuses would preferentially benefit from ARG supplementation by an increased placental vascularization. Results of a recent study [40] suggested that supplementing the gestational diet with either L-arginine, ractopamine, or a combination of the two increased myofiber diameter, indicating that supplementation had an effect on muscle fiber hypertrophy rather than hyperplasia. Ultimately, this resulted in a lower proportion of low BtW piglets (< 0.8–1.2 kg BtW), and an increased proportion of medium and high BtW piglets (> 1.2–1.6 kg BtW) from L-arginine supplemented compared with unsupplemented sows [40]. However, interestingly and in contrast to the present study, a lower TNF was observed in offspring from supplemented compared with unsupplemented sows. One explanation for the discrepancy between the observation in our study and the study of Garbossa et al. [40] could be the mode of L-arginine supplementation, either during d 14-28 or 25-53 of gestation, respectively, and at either 0.89% or 1.00% of the diet, respectively. With these combined results in mind, it is relevant to reflect on the extent to which supplementary L-arginine is directing dietary energy for hyperplasia or hypertrophy in fetuses during the different phases of gestation. Furthermore, investigating the impact of gestational supplementation on hyperplasia especially in post-weaning pigs would be of great importance, as results on this matter are contradictory. For instance, supplementing L-carnitine daily to L-BtW piglets during the suckling period resulted in a third wave of hyperplasia [41] during the nursing period, but no difference in TNF was found between control and L-carnitine supplemented pigs at the age of slaughter [42]. Based on this contradiction, one could speculate that changes occurring during the prenatal period are more manifested and less prone to environmental changes in the postnatal period, thus, having permanent positive effects until the animal reaches slaughter weight.

In light of this discussion and based on earlier findings [43, 44], one can assume that maternal arginine concentration was increased in plasma of sows fed L-arginine during the gestational period from d 14-28.

Gene expression of myogenesis-related genes in the *Semitendinosus* muscle

Muscle development during the perinatal period requires the regulation of several myogenesis related genes

[45]. In the present study none of the genes analyzed were responding to the extent of IUC or dietary treatment, except for *PRKAA2*, the expression of which was up-regulated in the STM of IN offspring compared with OL offspring. *PRKAA2* is a known inhibitor of muscle protein synthesis [46], and its greater expression in IN offspring is consistent with their lower BtW and relative STM weight emphasizing the importance of this gene in pre-natal muscle development. In contrast to the increased TNF at birth elicited by maternal L-arginine supplementation, results from the gene expression analysis did not support a myogenic effect at the molecular level. As hyperplasia ceases around d 90 of gestation, genes involved in myogenesis including myofiber formation might no longer show a specific up-regulation, but would rather be expressed at a steady state at birth when the STM was sampled. However, although there was no direct effect of L-arginine supplementation on the gene expression of myogenesis-related genes, it appears that *IGF2* and *PRKAA2* expressions in neonate STM could be influenced by the dietary level of this particular amino acid in certain uterine environments because there was a significant interaction.

Conclusion

In conclusion, the present study further adds to the increasing evidence that intra-uterine crowding prevalent in high prolific sows critically affects the phenotype of the offspring. This was manifested by the low BtW in contrast to the moderate litter size and offspring BtW of the OL counterparts. Confirming our first hypothesis, L-arginine in the early gestational diet seems to reduce the negative impacts of IUGR, as shown by the increased hyperplasia, body weight and STM area of the offspring at birth. As muscle area increased more than TNF, L-arginine supplementation obviously not only enhanced prenatal myofiber hyperplasia but also hypertrophy. However, details concerning the way in which L-arginine affects myofiber development on cellular and molecular levels remain to be determined. The second hypothesis that L-arginine would be especially efficient in L-BtW piglets was not confirmed. Still, this feeding strategy could be of great benefit to especially L-BtW pigs as they are particularly vulnerable. Thus, L-arginine supplementation to sows in early gestation would potentially improve the survival rate during nursery, and the growth potential during fattening of L-BtW piglets.

Abbreviations

ARG: Arginine supplementation; BtW: Birth weight; Ctrl: Alanine supplementation; DIET: Dietary treatment; IGF2: Insulin-like growth factor 2; IGFBP5: Insulin-like growth factor binding protein 5; IN: Intact sows; IUC: Intra-uterine crowding; IUGR: Intra-uterine growth retardation; L: Low; M: Intermediate; MSTN: Myostatin; MYF5: Myogenic factor 5; MYF6: Myogenic factor 6; MYOD1: Myogenic differentiation 1; MYOG: Myogenin (myogenic factor 4); OL: Unilateral oviduct ligated; P: Primary; PM: *Psoas major;*

PRKAA2: Protein kinase AMP-activated alpha 2 catalytic subunit; RPL4 (LOC100038029): Ribosomal protein L4; S: Secondary; SD: Standard deviation;; STM: *Semitendinosus* muscle;; STM$_d$: *Semitendinosus* muscle dark portion; STM$_l$: *Semitendinosus* muscle light portion; TBP: TATA box binding protein TNF, total number of myofibers

Acknowledgements
The authors would like to thank Guy Maikoff and his staff for their excellent work in the piggery, and Dr. Paolo Silacci and his staff for their valuable support with our laboratory work.

Funding
No external funding was used for this study.

Authors' contributions
JGM analyzed and interpreted, CP performed the experimental work at slaughter and muscle histology, MK was a major contributor in reviewing the manuscript, GB had the idea for the experiment, planned and supervised the experiment and was a major contributor in writing the manuscript. All authors read and approved the final manuscript.

Competing interests
The authors declare that they have no competing interests.

Author details
[1]Agroscope Posieux, la tioleyre 4, 1725 Posieux, Switzerland. [2]ETH Zurich, Institute of Agricultural Sciences, Universitätsstrasse 2, 8092 Zurich, Switzerland.

References
1. Tuchscherer M, Puppe B, Tuchscherer A, Tiemann U. Early identification of neonates at risk: Traits of newborn piglets with respect to survival. Theriogenology. 2000;54(3):371–88. doi:10.1016/S0093-691X(00)00355-1.
2. Foxcroft GR, Bee G, Dixon W, Hahn M, Harding J, Patterson J, et al. Consequences of selection for litter size on piglet development. In: Wiseman J, Varley MA, McOrist S, Kemp B, editors. Paradigms in Pig Science. Nottingham: Nottingham Univ. Press; 2007. p. 207–29.
3. Rekiel A, Wiecek J, Batorska M, Kulisiewicz J. Effect of sow prolificacy and nutrition on preand postnatal growth of progeny – a review. Ann. Anim. Sci. 2014;14(1):3–15. doi:10.2478/aoas-2013-0060.
4. Bérard J, Pardo CE, Bethaz S, Kreuzer M, Bee G. Intra-uterine crowding decreases average birth weight and affects muscle fiber hyperplasia in piglets. J Anim Sci. 2010;88:3242–50. doi:10.2527/jas.2010-2867.
5. Dwyer CM, Fletcher JM, Stickland NC. Muscle cellularity and postnatal growth in the pig. J Anim Sci. 1993;71(12):3339–43. doi:10.2527/1993.71123339x.
6. Nissen PM, Jorgensen PF, Oksbjerg N. Within-litter variation in muscle fiber characteristics, pig performance, and meat quality traits. J Anim Sci. 2004; 82(2):414–21. doi:10.2527/2004.822414x.
7. Rehfeldt C, Kuhn G. Consequences of birth weight for postnatal growth performance and carcass quality in pigs as related to myogenesis. J Anim Sci. 2006;84(Suppl):E113–23. doi:10.2527/2006.8413_supplE113x.
8. Swatland HJ. Muscle growth in the fetal and neonatal pig. J Anim Sci. 1973; 37(2):536–45. doi:10.2527/jas1973.372536x.
9. Ashmore CR, Addis PB, Doerr L. Development of muscle fibers in the fetal pig. J Anim Sci. 1973;36(6):1088–93. doi:10.2527/jas1973.3661088x.
10. Beermann DH, Cassens RG, Hausman GJ. A second look at fiber type differentiation in porcine skeletal muscle. J Anim Sci. 1978;46(1):125–32. doi:10.2527/jas1978.461125x.
11. Town SC, Putman CT, Turchinsky NJ, Dixon WT, Foxcroft GR. Number of conceptuses in utero affects porcine fetal muscle development. Reproduction. 2004;128(4):443–54. doi:10.1530/rep.1.00069.
12. Wigmore PM, Stickland NC. Muscle development in large and small pig fetuses. J Anat. 1983;137(Pt 2):235–45.
13. Bérard J, Bee G. Effects of dietary L-arginine supplementation to gilts during early gestation on foetal survival, growth and myofiber formation. Animal. 2010;10:1680–7. doi:10.1017/S1751731110000881.

14. Li X, Bazer FW, Johnson GA, Burghardt RC, Erikson DW, Frank JW, et al. Dietary supplementation with 0.8% L-arginine between days 0 and 25 of gestation reduces litter size in gilts. J Nutr. 2010;140(6):1111–6. doi:10.3945/jn.110.121350.
15. Wu GY, Morris SM. Arginine metabolism: nitric oxide and beyond. Biochem. J. 1998;336:1–17. doi:10.1042/bj3360001.
16. Reynolds LP, Redmer DA. Utero-placental vascular development and placental function. J Anim Sci. 1995;73(6):1839–51. doi:10.2527/1995.7361839.
17. Hazeleger W, Ramaekers R, Smits C, Kemp B. Effect of Progenos on placenta and fetal development in pigs. J Anim Sci. 2007;85:98.
18. Pardo CE, Bérard J, Kreuzer M, Bee G. Intrauterine crowding in pigs impairs formation and growth of secondary myofibers. Animal. 2013;7(3):430–8. doi: 10.1017/S1751731112001802.
19. Livak KJ, Schmittgen TD. Analysis of relative gene expression data using real-time quantitative PCR and the 2(-Delta Delta C(T)) Method. Methods. 2001;25(4):402–8. doi:10.1006/meth.2001.1262.
20. Hellemans J, Mortier G, De Paepe A, Speleman F, Vandesompele J. qBase relative quantification framework and software for management and automated analysis of real-time quantitative PCR data. Genome Biol. 2007; 8(2):R19. doi:10.1186/gb-2007-8-2-r19.
21. Hermesch S, Luxford BG, Graser HU. Genetic parameters for lean meat yield, meat quality, reproduction and feed efficiency traits for Australian pigs: 3. Genetic parameters for reproduction traits and genetic correlations with production, carcase and meat quality traits. Livest Prod Sci. 2000;65(3):261–70. doi:10.1016/S0301-6226(00)00151-2.
22. Quesnel H, Brossard L, Valancogne A, Quiniou N. Influence of some sow characteristics on within-litter variation of piglet birth weight. Animal. 2008; 2(12):1842–9. doi:10.1017/S175173110800308X.
23. Roehe R, Kalm E. Estimation of genetic and environmental risk factors associated with pre-weaning mortality in piglets using generalized linear mixed models. Anim. Sci. 2000;70:227–40. doi:10.1017/S1357729800054692.
24. Quiniou N, Dagorn J, Gaudré D. Variation of piglets' birth weight and consequences on subsequent performance. Livest Prod Sci. 2002;78(1):63–70. doi:10.1016/S0301-6226(02)00181-1.
25. Beaulieu AD, Aalhus JL, Williams NH, Patience JF. Impact of piglet birth weight, birth order, and litter size on subsequent growth performance, carcass quality, muscle composition, and eating quality of pork. J Anim Sci. 2010;88(8):2767–78. doi:10.2527/jas.2009-2222.
26. Wootton R, McFadyen IR, Cooper JE. Measurement of placental blood flow in the pig and its relation to placental and fetal weight. Biol Neonate. 1977; 31(5-6):333–9. doi:10.1159/000240984.
27. Foxcroft GR, Dixon WT, Dyck MK, Novak S, Harding JC, Almeida FC. Prenatal programming of postnatal development in the pig. Soc Reprod Fertil Suppl. 2009;66:213–31.
28. Wu GY, Bazer FW, Satterfield MC, Li XL, Wang XQ, Johnson GA, et al. Impacts of arginine nutrition on embryonic and fetal development in mammals. Amino Acids. 2013;45(2):241–56. doi:10.1007/s00726-013-1515-z.
29. Gao K, Jiang Z, Lin Y, Zheng C, Zhou G, Chen F, et al. Dietary L-arginine supplementation enhances placental growth and reproductive performance in sows. Amino Acids. 2012;42(6):2207–14. doi:10.1007/s00726-011-0960-9.
30. Foxcroft GR, Town SC. Prenatal programming of postnatal performance - the unseen cause of variance. Advances in Pork Production. 2004;15:269.
31. Wu G, Bazer FW, Wallace JM, Spencer TE. Board-invited review: Intrauterine growth retardation: Implications for the animal sciences. J Anim Sci. 2006; 84(9):2316–37. doi:10.2527/jas.2006-156.
32. Pardo CE, Mueller S, Bérard J, Kreuzer M, Bee G. Importance of average litter weight and individual birth weight for performance, organ and myofiber characteristics of progeny. Livest Sci. 2013;157:330–8. doi:10.1016/j.livsci. 2013.06.015.
33. Town SC, Patterson JL, Pereira CZ, Gourley G, Foxcroft GR. Embryonic and fetal development in a commercial dam-line genotype. Anim. Reprod. Sci. 2005;85(3-4):301–16. doi:10.1016/j.anireprosci.2004.05.019.
34. DeLong GR. Effects of nutrition on brain development in humans. Am J Clin Nutr. 1993;57(2):286S–90S.
35. Rehfeldt C, Tuchscherer A, Hartung M, Kuhn G. A second look at the influence of birth weight on carcass and meat quality in pigs. Meat Sci. 2008;78(3):170–5. doi:10.1016/j.meatsci.2007.05.029.
36. Hou Y, Yao K, Yin Y, Wu G. Endogenous Synthesis of Amino Acids Limits Growth, Lactation, and Reproduction in Animals. Advances in Nutrition: An International Review Journal. 2016;7(2):331–42. doi:10.3845/an.115.010850.

37. Wu G, Bazer FW, Cudd TA, Jobgen WS, Kim SW, Lassala A, et al. Pharmacokinetics and safety of arginine supplementation in animals. J Nutr. 2007;137(6):1673S–1680.

38. Wu Z, Hou Y, Hu S, Bazer FW, Meininger CJ, McNeal CJ, et al. Catabolism and safety of supplemental l-arginine in animals. Amino acids. 2016:1–12. doi:10.1007/s00726-016-2245-9.

39. Rehfeldt C, Lang IS, Gors S, Hennig U, Kalbe C, Stabenow B, et al. Limited and excess dietary protein during gestation affects growth and compositional traits in gilts and impairs offspring fetal growth. J Anim Sci. 2011;89(2):329–41. doi:10.2527/jas.2010-2970.

40. Garbossa CA, Junior FM, Silveira H, Faria PB, Schinckel AP, Abreu ML, et al. Effects of ractopamine and arginine dietary supplementation for sows on growth performance and carcass quality of their progenies. J Anim Sci. 2015;93(6):2872–84. doi:10.2527/jas.2014-8824.

41. Lösel D, Kalbe C, Rehfeldt C. L-Carnitine supplementation during suckling intensifies the early postnatal skeletal myofiber formation in piglets of low birth weight. J Anim Sci. 2009;87(7):2216–26. doi:10.2527/jas.2008-1662.

42. Lösel D, Rehfeldt C. Effects of L-carnitine supplementation to suckling piglets on carcass and meat quality at market age. Animal. 2013;7(7):1191–8. doi:10.1017/S1751731113000268.

43. Mateo RD, Wu G, Bazer FW, Park JC, Shinzato I, Kim SW. Dietary L-arginine supplementation enhances the reproductive performance of gilts. J Nutr. 2007;137(3):652–6.

44. Li X, Bazer FW, Johnson GA, Burghardt RC, Frank JW, Dai Z, et al. Dietary supplementation with L-arginine between days 14 and 25 of gestation enhances embryonic development and survival in gilts. Amino Acids. 2014; 46(2):375–84. doi:10.1007/s00726-013-1626-6.

45. Pas MFW, Wit AAW, Priem J, Cagnazzo M, Davoli R, Russo V, et al. Transcriptome Expression Profiles in Prenatal Pigs in Relation to Myogenesis. J Muscle Res Cell Motil. 2005;26(2):157–65. doi:10.1007/s10974-005-7004-6.

46. Bolster DR, Crozier SJ, Kimball SR, Jefferson LS. AMP-activated Protein Kinase Suppresses Protein Synthesis in Rat Skeletal Muscle through Down-regulated Mammalian Target of Rapamycin (mTOR) Signaling. J Biol Chem. 2002;277(27):23977–80. doi:10.1074/jbc.C200171200.

47. Foxcroft GR, Dixon WT, Novak S, Putman CT, Town SC, Vinsky MDA. The biological basis for prenatal programming of postnatal performance in pigs. J Anim Sci. 2006;84(E. Suppl):E105–12. doi:10.2527/2006.8413_supplE105x.

13

Long-term effects of early antibiotic intervention on blood parameters, apparent nutrient digestibility, and fecal microbial fermentation profile in pigs with different dietary protein levels

Miao Yu, Chuanjian Zhang, Yuxiang Yang, Chunlong Mu, Yong Su, Kaifan Yu and Weiyun Zhu[*]

Abstract

Background: This study aimed to determine the effects of early antibiotic intervention (EAI) on subsequent blood parameters, apparent nutrient digestibility, and fecal fermentation profile in pigs with different dietary crude protein (CP) levels. Eighteen litters of piglets (total 212) were randomly allocated to 2 groups and were fed a creep feed diet with or without in-feed antibiotics (olaquindox, oxytetracycline calcium and kitasamycin) from postnatal d 7 to d 42. On d 42, the piglets within the control or antibiotic group were mixed, respectively, and then further randomly assigned to a normal- (20%, 18%, and 14% CP from d 42 to d 77, d 77 to d 120, and d 120 to d 185, respectively) or a low-CP diet (16%, 14%, and 10% CP from d 42 to d 77, d 77 to d 120, and d 120 to d 185, respectively), generating 4 groups. On d 77 (short-term) and d 185 (long-term), serum and fecal samples were obtained for blood parameters, microbial composition and microbial metabolism analysis.

Results: EAI increased ($P < 0.05$) albumin and glucose concentrations in low-CP diet on d 77, and increased ($P < 0.05$) urea concentration in normal-CP diet. On d 185, EAI increased ($P < 0.05$) globulin concentration in normal-CP diets, but decreased glucose concentration. For nutrient digestibility, EAI increased ($P < 0.05$) digestibility of CP on d 77. For fecal microbiota, the EAI as well as low-CP diet decreased ($P < 0.05$) E. coli count on d 77. For fecal metabolites, on d 77, EAI decreased ($P < 0.05$) total amines concentration but increased skatole concentration in low-CP diet. On d 185, the EAI increased ($P < 0.05$) putrescine and total amines concentrations in low-CP diets but reduced ($P < 0.05$) in the normal-CP diets. The low-CP diet decreased the concentrations of these compounds.

Conclusions: Collectively, these results indicate that EAI has short-term effects on the blood parameters and fecal microbial fermentation profile. The effects of EAI varied between CP levels, which was characterized by the significant alteration of glucose and putrescine concentration.

Keywords: Antibiotics long term effect, Antimicrobial, Blood parameters, Low protein diet, Metabolites, Microbiota

* Correspondence: zhuweiyun@njau.edu.cn
Jiangsu Key Laboratory of Gastrointestinal Nutrition and Animal Health,
Laboratory of Gastrointestinal Microbiology, College of Animal Science and
Technology, Nanjing Agricultural University, Nanjing, Jiangsu 210095, China

Background

In intensive swine husbandry systems, piglets commonly receive in-feed antibiotics during early life stage, mainly aiming to prevent outbreaks of respiratory and gastro-intestinal tract (GIT) infections, and promote growth [1]. However, their enormous and uncontrolled use has led to many resistant bacterial strains, causing detrimental effects in both humans and animals. In China, although the use of antibiotics in feed during weaning period of the pig is still a common practice, it has increasingly received safety concerns from consumers and thus the use of antibiotics in feed will be banned in the future. Thus, to better design alternatives to antibiotics for pigs, it is important to understand the post-term effect and the possible mode of action of the antibiotics. The immediate effects of antibiotic intervention on blood parameters [2], apparent nutrient digestibility [3], gut bacterial composition [4, 5] and microbial metabolism [6, 7] have been described, especially in newly weaned pigs. However, much less attention has been paid to antibiotics intervention in suckling and nursing stage and long-term (carry-over) effects on host metabolism and GIT microbiota after withdrawal of antibiotics. In pigs, early life antibiotic exposure affected pancreatic development and glucose metabolism 5 wk after antibiotic withdrawal [8]. A recent study found piglets that received an subcutaneous injection with 0.1 mL tulathromycin (dosage 2.5 mg/kg body weight) on d 4 after birth showed decreased microbial diversity and an altered microbial composition in jejunal digesta on d 176, but had limited effect on d 56 [9]. Using mice models, previous studies also have indicated that low-dose antibiotic exposure in early life had a long-term effect on host metabolism and adiposity [10, 11]. These results suggest that alteration of gut microbes at early life may have long-lasting effect on subsequent host metabolism and gut microbiota. However, it is still unclear whether the early use of antibiotics in suckling and nursing stage have long-term impact on subsequent metabolism and gut microbes in pigs.

In addition to the early use of antibiotics, dietary protein level could also affect the gut health of the animal [12]. In practice, low-protein amino acid-supplementation diets are used to decrease nitrogen excretion and feed costs. Our previous study found that low crude protein (low-CP) diet increased the counts of Bacteroidetes, *Clostridium* cluster IV and *Clostridium* cluster XIVa, decreased *Escherichia coli* counts, and decreased cadaverine, phenol and indole compounds concentrations [13, 14], suggesting a beneficial effect of low-CP diet. Antibiotics could decrease the nutrient utilization by gut microbiota and increase those available for body metabolism [15]. However, when fed low-CP diet, whether early antibiotic intervention affect body metabolism and fecal microbial fermentation profile in pigs remains unclear.

To test the hypothesis that early antibiotic intervention (EAI) (from d 7 to d 42) may have long-lasting effects on subsequent host metabolism and GIT microbes and microbial fermentation in pigs, the present study examined the blood parameters, apparent nutrient digestibility, and major bacterial taxonomic groups and microbial metabolism in feces on d 77 (short-term) and d 185 (long-term) in relation to different CP diets.

Methods

Animals and experimental design

A total of 18 litters crossbred (Duroc ×Landrace × Large White) newborn piglets (212 piglets in total) were used in this study. The piglets on d 7 while with the sow were randomly allotted to two groups (n = 9 litters) and offered creep feed (Additional file 1: Table S1) either without (control) or with a mixture of antibiotics (50 mg/kg oxytetracycline calcium, 50 mg/kg kitasamycin, and 50 mg/kg olaquindox) (antibiotic). This mixture of antibiotics is commonly used as a growth promoter for piglets in commercial farms in China in order to ensure healthy transition around weaning. Oxytetracycline calcium is a broad spectrum antibiotic with antibacterial activity against gram-negative and gram-positive bacteria, and kitasamycin exhibits activity mainly against gram-positive bacteria, whereas olaquindox exhibits antibacterial activity both gram-negative and gram-positive bacteria. The dosages for the antibiotics used were according to the dose limitation in Regulations of Feeding Drug Additives (announcement No. 168) approved by Ministry of Agriculture, China. On d 23, the sows were removed from the piglets, while the piglets remained in the environmental-controlled nursery with 1.8 × 2.5 m pens fitted with a hard plastic fully slotted floor, and fed the same creep feed until d 42, which is a typical feeding program in a commercial farms in China. On d 42, the piglets within the control or antibiotic group were mixed, respectively, and then randomly allotted to 1 of 2 dietary treatments (Low-crude protein, Low-CP; Normal-crude protein, Normal-CP, respectively) according to their equal average body weight (BW) and gender, respectively, which generated 4 groups (Control-Low CP, Con-LP; Control-normal CP, Con-NP; Antibiotic-Low CP, Ant-LP; Antibiotic-Normal CP, Ant-NP, respectively). There were 5 pens replicates per treatment group and 9 pigs per pen. The numbers of barrows and gilts were not equal on d 42, but spilt as evenly as possible across treatments (23 gilts and 22 barrows, 22 gilts and 23 barrows, 24 gilts and 21 barrows, and 23 gilts and 22 barrows for treatments Con-LP, Con-NP, Ant- LP, and Ant- NP, respectively).

The diets comprised a normal-CP diet (20%, 18%, and 14% CP during d 42 to d 77, d 77 to d 120, and d 120 to d 185, respectively, supplemented with lysine, methionine, threonine, and tyrptophan) and a low-CP diet (16%, 14%, and 10% CP during d 42 to d 77, d 77 to d 120, and d 120 to d 185, respectively, supplemented with four limiting AA mentioned above) (Table 1). The 2 diets were formulated to meet or exceed NRC [16] nutrient recommendations. Creep feed diets were fed as crumble, whereas normal- or low-CP diets were in a pelleted form. After d 42, the pigs were moved to a total confinement house with 2.5 × 3.0 m pens that had partial concrete slatted floors. During the experimental period, the pigs consumed feed ad libitum and feed residues were recorded daily. All pigs had free access to water via a low-pressure nipple drinker. The body weight (BW) of pigs was recorded at the beginning of the experiment and at the end of each period to determine average daily gain (ADG), average daily feed intake (ADFI), and the ratio of gain to feed (G:F).

Sample collection and preparation

The fecal samples were collected for 3 consecutive days, from d 75 to 77 and 183 to 185 of age, respectively, during the feeding period in the morning by rectal stimulation. Feces from each day were collected at least 3 pigs from each of the pen replicates and were pooled for each replicate to create composite samples. The feces were then frozen at −20 °C until the analysis of microbial metabolites and dominant bacterial community, which was completed within 2 weeks after sample collection.

On d 77 (28.4 ± 0.6 kg) and d 185 (102.0 ± 0.6 kg), when the experimental pigs were fasted for approximately 12 h, one median BW barrow from each pen was selected for blood sampling (n = 5). Blood samples (10 mL) were collected from the jugular vein and serum

Table 1 Ingredient and nutrient composition of experimental diets (%, as-fed basis)[a]

Age range, d: Ingredient, %	42 to 77		77 to 120		120 to 185	
	LP-16%	NP-20%	LP-14%	NP-18%	LP-10%	NP-14%
Corn	68.00	63.70	68.90	58.00	82.40	71.10
Soybean meal	18.00	21.00	18.80	29.38	8.71	19.13
Wheat bran	–	–	6.00	8.00	2.00	4.00
Whey power	4.30	4.30	–	–	–	–
Fishmeal	3.20	8.00	–	–	–	–
Soybean oil	2.70	0.70	2.56	1.60	2.50	2.00
Dicalcium phosphate	0.85	0.10	0.78	0.65	0.90	0.80
Calcium carbonate	0.50	0.36	0.89	0.89	0.82	0.90
Sodium chloride	0.30	0.30	0.30	0.30	0.30	0.30
Vitamin mixture[b]	0.20	0.20	0.20	0.20	0.20	0.20
Mineral mixture[b]	0.80	0.80	0.80	0.80	0.80	0.80
L-Lysine	0.65	0.33	0.49	0.17	0.78	0.48
DL-Methionine	0.20	0.09	0.10	–	0.20	0.11
L-Threonine	0.24	0.10	0.15	0.01	0.30	0.15
L-Tryptophane	0.06	0.02	0.03	–	0.09	0.03
Calculated content[c], %						
ME[d], MJ/kg	14.02	14.02	13.82	13.80	13.81	13.82
SID AA[e], %						
Lysine	1.23	1.23	0.99	0.98	0.98	0.99
Methionine + Cysteine	0.68	0.68	0.56	0.56	0.60	0.56
Threonine	0.73	0.73	0.59	0.60	0.61	0.60
Trptophane	0.20	0.20	0.17	0.20	0.18	0.17
Analyzed nutrient composition[f], %						
Crude protein	16.63	20.11	14.03	18.02	10.18	14.06

[a]The pigs of control and antibiotic groups were divided to 1 of 2 treatment groups (normal vs. low CP diet) after d 42. LP low protein diet, NP normal protein diet
[b]Provided per kilogram of diet: vitamin A, 2.85 million IU; vitamin D3, 0.6 million IU; vitamin E, 67.50 IU; vitamin K3, 750 mg; vitamin B1,750 mg; vitamin B2, 1500 mg; vitamin B6, 900 mg; vitamin B12, 7.5 mg; nicotinic acid, 7500 mg; folic acid, 300 mg; calcium pantothenate, 3750 mg; biotin, 37.5 mg; vitamin B4, 100 mg; antioxidants, 15 mg; Cu (CuSO4·5H2O), 60 mg; Fe (FeSO4·H2O), 51.6 mg; Co (CoCl2), 50 mg; Mn (MnSO4·H2O), 25 mg; Zn (ZnSO4·H2O), 21.2 mg; Zn (ZnO), 6 mg; I (KI), 3.9 mg; Se (Na2SeO3), 3.0 mg; Carrier (Sepiolite) 604.3 mg
[c]Calculate values according to NRC [16]
[d]ME = metabolizable energy
[e]SID AA = standardized ileal digestible AA
[f]Analytical results obtained according to AOAC [17]

was obtained by centrifuging at 3000×g for 10 min at 4 °C. Thereafter, samples were stored at −20 °C until chemical analysis.

Nutrient composition and apparent digestibility analysis

The dry matter (DM; procedure 930. 15; AOAC, 2007), crude protein (CP; N × 6.25; procedure 990.03; AOAC, 2007), crude fat (procedure 2003. 06; AOAC, 2007), ash (procedure 942.05; AOAC, 2007), contents of experimental diets and freeze-dried feces were analyzed according to the procedure of the Association of Official Analytical Chemists (AOAC) [17]. The acid insoluble ash (AIA) concentrations were analyzed by the method of ISO (method no.5985; ISO, 2003) [18]. Total tract apparent digestibility of nutrients was calculated based on the content of AIA as a marker in feed and feces, as previously described [19].

Serum biochemical parameters

Serum total protein, albumin, glucose, urea, cholesterol, triglyceride, high density lipoprotein-cholesterol (HDLC), and low density lipoprotein−cholesterol (LDLC) were measured by the enzymatic colorimetric method using an AU2700 autoanalyzer (Olympus, Tokyo, Japan). Globulin concentration of individual serum samples was calculated by subtracting the amount of albumin from the total protein concentration.

DNA extraction and quantitative real-time PCR

The total bacterial DNA was extracted from fecal samples (0.3 g) using the bead-beating and phenol-chloroform extraction methods [20]. The qPCR assay was performed on StepOnePlus™ Real-Time PCR System (Life Technologies, Califormia, USA) with SYBR Premix Ex Taq dye (Takala, Bio, Ostu, Japan). The sequences of the selected targets and primers are listed in Table 2. The reaction mixture

(20 μL) consist of 10.0 μL SYBR Green Supermix (Bio-Rad), 0.4 μL ROX Reference Dye, 2.0 μL of template DNA, 6.8 μL of double distilled water, and 0.4 μL of each primer set. The standard curves of each bacterial group were generated with 10-fold serial dilutions of the 16S rRNA genes amplified from the respective target strains. Quantitative real-time PCR amplification was performed using the following conditions: initial denaturation program (95 °C for 30 s), denaturation program repeated for 40 cycles (95 °C for 5 s, 60 °C for 30 s), followed by the melting curve program (60-95 °C with a heating rate of 0.1 °C/s). The melting curves were checked after amplification to ensure single product amplification of consistent melting temperature. Quantification was performed in triplicate, and the mean Ct value was calculated. The results were expressed as $\log_{10}$ 16S ribosomal DNA gene copies/g feces.

Measurement of fecal microbial metabolites

The short-chain fatty acid (SCFA), amines, ammonia, and phenolic and indolic compounds were selected as markers of GIT microbiota metabolism. The concentrations of SCFA were analyzed by gas chromatography as described previously [21], with slight modifications. Briefly, approximately 0.4 g of fecal samples were weighed into a 2-mL centrifuge tube, 1.6 mL of double distilled water was added. The mixture was vortexed for 10 min until the material was homogenized and then centrifuged at 13,000×g for 10 min at 4 °C. A portion of 1 mL of the clear supernatant was transferred into a new tube, and then added 0.2 mL 25% (w/v) metaphosphoric acid. After homogenization, the mixture was frozen at −20 °C and kept overnight to precipitate the proteins. After thawing, a portion of 100 μL internal standard (0.64% (w/v) crotonic acid solution) was added. The tubes were vortexed for 1 min and then centrifuged at 13,000×g for 10 min at 4 °C. The supernatant was filtered through a 0.22-μm syringe filter and then analyzed on an Aglient 7890B system with a flame ionization detector (Agilent Technologies Inc.). The following column conditions were used: nitrogen was used as the carrier gas with a flow rate of (17.68 mL/min); the oven, detector and injector port temperature were 130 °C, 250 °C, 220 °C, respectively. These acids were identified by their specific retention times and the concentrations determined and expressed as umol/g.

Amine concentrations in feces were determined by high-performance liquid chromatography (HPLC) with a method according to Yang et al. [22]. Briefly, 1.5 g of feces were treated with 3 mL of 5% trichloroacetic acid, homogenized for 10 min and then centrifuged at 3600×g for 10 min at 4 °C. The supernatant was mixed with an equal volume of n-hexane and vortexed for 5 min, the water phase (0.5 mL) was transferred into a new tube,

Table 2 Primers used for quantification in this study

Target	Primer sequence 5′→3′	Reference
Total bacteria	Forward: CGGTGAATACGTTCYCGG	[38]
	Reverse: GGWTACCTTGTTACGACTT	
Firmicutes	Forward: GGAGYATGTGGTTTAATTCGAAGCA	[39]
	Reverse: AGCTGACGACAACCATGCAC	
Bacteroidetes	Forward: GGARCATGTGGTTTAATTCGATGAT	[39]
	Reverse: AGCTGACGACAACCATGCAG	
Clostridium IV	Forward: GCACAAGCAGTGGAGT	[40]
	Reverse: CTTCCTCCGTTTTGTCAA	
Clostridium XIVa	Forward: CGGTACCTGACTAAGAAGC	[41]
	Reverse: AGTTTYATTCTTGCGAACG	
E.coli	Forward: CATGCCGCGTGTATGAAGAA	[42]
	Reverse: CGGGTAACGTCAATGAGCAAA	
Lactobacillus	Forward: AGCAGTAGGGAATCTTCCA	[43]
	Reverse: ATTCCACCGCTACACATG	

and then added with 1.5 mL saturated Na_2CO_3, 1 mL dansyl chloride, and 1 mL NaOH (2 mol/L). The mixed solution was heated at 60 °C for 45 min, and then added with 100 μL ammonia (2.8%) to stop the reaction. The mixture was kept in the water bath until the acetone was vaporized under nitrogen at 40 °C. Finally, the sample was extracted with 3 mL diethyl ether. The extracts were dried under nitrogen and then re-dissolved in acetonitrile. The mixture filtered through 0.22-μm syringe filter and then analyzed on an Aglient 1220 Infinity LC system with an UV detector (Agilent Technologies Inc.).

The ammonia concentration in feces was analyzed using UV spectrophotometer according to Chaney and Marbach [23]. Phenolic and indolic compounds concentration was determined by HPLC as previously described [24]. Briefly, 0.1 g of fecal sample was treated with 1 mL acetonitrile, homogenized for 10 min and then frozen at −20 °C for 20 min. Finally, the mixture was centrifuged at 3000×g for 10 min at 4 °C. The supernatant was filtered through a 0.22-μm syringe filter and analyzed for phenolic and indolic compounds (i.e., p-cresol, skatole, indole, and phenol) using HPLC with an UV detector (Agilent Technologies Inc.).

Statistical analysis

All data were analyzed using 2 × 2 factorial MIXED procedure of SAS for a randomized complete block design (SAS 9.2 Institute, Cary, NC, USA). The model included the fixed effect of antibiotic, protein level, associated two-interactions and the random errors of a pen or an individual pig. The ADG, ADFI, G:F, digestibility of nutrients,

and fecal fermentation profile were evaluated using the pen as the experimental unit. Blood parameters were assessed using the individual pig as the experimental unit. Differences were considered significant at $P \leq 0.05$, and tendency was declared with $0.05 < P < 0.10$. When a significant interaction among antibiotic and CP was observed, data were further analyzed by using a one-way ANOVA with Duncan's multiple range tests. Differences at $P < 0.05$ were identified significant.

Results

Growth performance and apparent nutrient digestibility

Growth performance of the pigs was shown in Table 3. For ADG, there was an interaction ($P < 0.05$) between EAI and dietary CP treatment from d 77 to d 185 and d 42 to d 185. The EAI increased ADG ($P < 0.05$) from d 42 to d 185 in the low-CP diets but not in the normal-CP diets. Meanwhile, low-CP diet tended to reduce the ADFI ($P = 0.08$) from d 42 to d 185, but G:F at each feeding period did not differ among treatments.

For apparent nutrient digestibility, no interactions between EAI and dietary CP treatment were observed (Table 4). On d 77, there was no difference of OM digestibility among treatments. The EAI significantly increased CP digestibility ($P < 0.05$) and tended to increase ($P = 0.06$) DM digestibility. The low-CP diet reduced CP and crude fat digestibility ($P < 0.05$). On d 185, digestibility of crude fat was decreased ($P < 0.001$) when pigs fed the low-CP diet, but DM, CP, and OM digestibility were not affected by treatment. Overall, these results indicated that EAI had short-term effect on nutrient digestibility.

Table 3 Effects of early antibiotic intervention on growth performance of pigs with different protein level diets[1]

Item	Low-CP		Normal- CP		SEM	P-value[2]		
	Con	Ant	Con	Ant		Ant	CP	Ant × CP
42 d BW, kg	12.23	12.20	12.28	12.25	0.07	0.851	0.743	0.981
ADG, kg/d								
d 42 to 77	0.50	0.52	0.51	0.51	0.02	0.686	0.972	0.786
d 77 to 185	0.78[b]	0.81[ab]	0.84[a]	0.83[a]	0.01	0.361	0.002	0.048
d 42 to 185	0.71[b]	0.74[a]	0.76[a]	0.75[a]	0.01	0.180	0.001	0.019
ADFI, kg/d								
d 42 to 77	0.72	0.72	0.78	0.78	0.02	0.945	0.140	0.993
d 77 to 185	2.24	2.36	2.44	2.54	0.06	0.349	0.119	0.891
d 42 to 185	1.85	1.94	2.02	2.09	0.05	0.368	0.089	0.895
G:F								
d 42 to 77	0.70	0.73	0.67	0.66	0.02	0.880	0.227	0.675
d 77 to 185	0.36	0.35	0.36	0.33	0.01	0.323	0.787	0.655
d 42 to 185	0.44	0.44	0.44	0.42	0.01	0.555	0.414	0.605

[a, b]Means in the same row with different superscripts differ ($P < 0.05$)
[1]The commercial creep feed with or without in-feed antibiotics (50 mg/kg olaquindox, 50 mg/kg oxytetracycline calcium, and 50 mg/kg kitasamycin) was fed to pig from d 7 to d 42. Thereafter, the Con and Ant group were further randomly assigned to provide a normal (20%, 18%, 14% CP from d 42 to d 77, d 77 to d 120, d 120 to d 185, respectively) or low CP diet (16%, 14%, 10% CP from d 42 to d 77, d 77 to d 120, d 120 to d 185, respectively), respectively
[2]The P values indicate main effects for antibiotic (Ant), protein level (CP) and their interaction (Ant x CP), respectively

Table 4 Effects of early antibiotic intervention on total tract apparent digestibility of pigs with different protein level diets[a]

Item, %	Low-CP		Normal-CP		SEM	P-value[2]		
	Con	Ant	Con	Ant		Ant	CP	Ant × CP
d 77								
DM	75.24	76.73	74.81	80.52	0.98	0.065	0.368	0.263
CP	60.00	64.84	64.54	73.45	1.75	0.032	0.040	0.498
OM	79.86	79.65	78.31	83.32	0.88	0.173	0.538	0.140
Crude fat	44.61	48.44	53.96	56.88	1.34	0.136	<0.001	0.831
d 185								
DM	77.12	72.61	79.24	71.82	1.67	0.103	0.843	0.666
CP	72.46	72.45	75.08	73.35	0.67	0.550	0.240	0.550
OM	82.93	81.67	82.14	80.54	0.51	0.198	0.372	0.867
Crude fat	50.52	48.48	57.72	56.74	1.34	0.320	<0.001	0.720

DM dry matter, *CP* crude protein, *OM* organic matter

[a]The commercial creep feed with or without in-feed antibiotics (50 mg/kg olaquindox, 50 mg/kg oxytetracycline calcium, and 50 mg/kg kitasamycin) was fed to pig from d 7 to d 42. Thereafter, the Con and Ant group were further randomly assigned to provide a normal (20%, 18%, 14% CP from d 42 to d 77, d 77 to d 120, d 120 to d 185, respectively) or low CP diet (16%, 14%, 10% CP from d 42 to d 77, d 77 to d 120, d 120 to d 185, respectively), respectively. [2]The *P* values indicate main effects for antibiotic (Ant), protein level (CP) and their interaction (Ant x CP), respectively

Serum biochemical measurements

To dissect whether EAI and dietary CP level affected the host metabolism, basic serum biochemistry was analyzed. On d 77, albumin, glucose, and urea concentrations in the serum revealed interactions ($P < 0.05$) between EAI and dietary CP treatments (Table 5). The EAI increased albumin and glucose concentrations in the low-CP diet, but increased urea concentration in the normal-CP diet. Meanwhile, EAI increased ($P < 0.05$) triglyceride concentration irrespective of dietary CP, and tended to increase concentration of total protein ($P = 0.08$). The low-CP diet reduced ($P < 0.05$) concentrations of total protein and LDLC irrespective of EAI. On d 185, EAI also interacted with dietary CP on the concentrations of globulin and glucose in the serum ($P < 0.05$). The EAI decreased glucose concentration in normal-CP diet, but increased globulin concentration. Furthermore, low-CP diet reduced ($P < 0.01$) concentrations of urea, triglyceride, and HDLC irrespective of EAI. In general, these results indicated that EAI had short-term effects on serum parameters, and these effects may be influenced by dietary CP levels.

Table 5 Effects of early antibiotic intervention on serum biochemical indexes of pigs with different CP level diets[1]

Item	Low-CP		Normal-CP		SEM	P-value[2]		
	Con	Ant	Con	Ant		Ant	CP	Ant × CP
d 77								
Total protein, g/L	62.00	66.21	63.61	70.79	1.67	0.082	0.031	0.643
Albumin, g/L	17.73[b]	23.20[a]	18.04[b]	19.45[b]	0.66	<0.001	0.122	0.049
Globulin, g/L	44.27	43.01	45.57	51.34	1.73	0.553	0.178	0.342
Glucose, mmol/L	4.39[c]	5.58[a]	5.44[ab]	5.26[b]	0.14	0.004	0.189	0.002
Urea, mmol/L	2.12[b]	2.16[b]	2.46[b]	3.26[a]	0.12	0.003	<0.001	0.003
Cholesterol, mmol/L	1.87	1.89	1.97	2.09	0.05	0.890	0.223	0.080
Triglyceride, mmol/L	0.45	0.60	0.63	0.75	0.03	0.039	0.145	0.323
HDLC, mmol/L	0.77	0.81	0.84	0.88	0.02	0.132	0.651	0.241
LDLC, mmol/L	1.03	0.99	1.04	1.14	0.04	0.430	0.045	0.603
d 185								
Total protein, g/L	66.62	65.14	65.26	69.34	1.45	0.378	0.342	0.072
Albumin, g/L	31.60	32.64	33.48	31.78	0.73	0.664	0.489	0.081
Globulin, g/L	35.02[ab]	32.50[b]	31.78[b]	37.56[a]	1.56	0.313	0.567	0.022
Glucose, mmol/L	4.54[b]	4.88[b]	5.54[a]	4.78[b]	0.12	0.245	0.021	0.011
Urea, mmol/L	4.38	4.12	5.02	6.04	0.26	0.352	0.004	0.134
Cholesterol, mmol/L	2.10	1.97	2.15	2.20	0.07	0.793	0.322	0.542
Triglyceride, mmol/L	0.46	0.45	0.61	0.57	0.03	0.684	0.020	0.743
HDLC, mmol/L	0.85	0.77	0.91	0.95	0.02	0.593	0.011	0.151
LDLC, mmol/L	0.99	0.99	1.01	0.91	0.04	0.572	0.712	0.574

HDLC, high density lipoprotein-cholesterol, *LDLC*, low density lipoprotein–cholesterol

[a-c]Means in the same row with different superscripts differ (P < 0.05)

[1]The commercial creep feed with or without in-feed antibiotics (50 mg/kg olaquindox, 50 mg/kg oxytetracycline calcium, and 50 mg/kg kitasamycin) was fed to pig from d 7 to d 42. Thereafter, the Con and Ant group were further randomly assigned to provide a normal (20%, 18%, 14% CP from d 42 to d 77, d 77 to d 120, d 120 to d 185, respectively) or low CP diet (16%, 14%, 10% CP from d 42 to d 77, d 77 to d 120, d 120 to d 185, respectively), respectively. [2]The *P* values indicate main effects for antibiotic (Ant), protein level (CP) and their interaction (Ant x CP), respectively

Fecal bacterial populations

In order to evaluate the effect of EAI on the quantitative change of bacterial community in the feces, real-time PCR was performed on some major bacterial groups. As shown in Additional file 1: Figure S1, dietary treatments had no effects on fecal major bacterial population on d 77 and d 185, such as total bacteria, Firmicutes, Bacteroidetes, *Clostridium* cluster IV, and *Clostridium* cluster XIVa. However, on d 77 (Fig. 1a), EAI significantly decreased ($P < 0.05$) the counts of *E. coli*, and tended to reduce the counts of *Lactobacillus* ($P = 0.09$). Meanwhile, the low-CP diet also reduced the count of *E. coli* ($P < 0.05$). On d 185 (Fig. 1b), pigs with low-CP diet tended to record a lower count of *Lactobacillus* ($P = 0.09$). Overall, these results suggest that the effect of EAI and dietary CP levels on bacterial groups in the fecal samples was limited except for *E. coli*.

Fermentation metabolites in fecal contents

To understand whether EAI affect fecal fermentation profiles with different CP diets, fermentation metabolites were determined. For SCFA (Fig. 1), EAI and low-CP diet did not affect fecal total SCFA, acetate, butyrate, isobutyrate, valerate, and isovalerate concentrations on d 77 (Fig. 1a), but the low-CP diet tended to reduce ($P = 0.088$) propionate concentration. On d 185 (Fig. 1b), EAI did not change the SCFA concentrations among groups in the present study. However, the low-CP diet decreased ($P < 0.05$) total SCFA, acetate, propionate, isobutyrate, and isovalerate concentrations.

Ammonia was produced through amino acids deamination by bacteria. On d 77, EAI did not affect the concentration of ammonia (data not shown). The low-CP diet significantly decreased ($P < 0.05$) ammonia concentration in feces. On d 185, dietary treatment did not affect ammonia concentration.

Amines are formed from the decarboxylation of proteins, amino acids, and other nitrogenous compounds in the GIT. As shown in Fig. 2, cadaverine and putrescine are the main amines. On d 77 (Fig. 2a), EAI significantly decreased ($P < 0.05$) the concentration of total amines, cadaverine, and increased ($P < 0.05$) the concentration of tryptamine. The low-CP diet significantly decreased ($P < 0.05$) cadaverine, tyramine, methylamine, and tryptamine concentrations. However, the concentrations of putrescine, spermidine, and spermine were not affected by dietary treatment. On d 185 (Fig. 2b), there

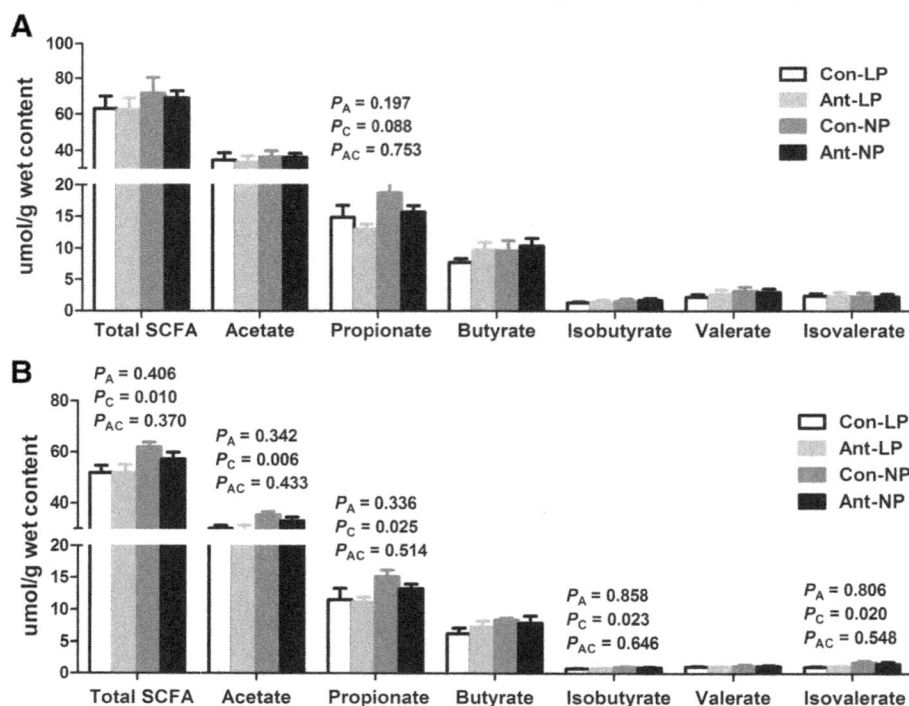

Fig. 1 Effects of EAI on SCFAs concentrations in the feces of pigs with different CP levels diets. (Con-LP, ▢; Ant-LP, ▨; Con-NP, ▨; Ant-NP, ■). **a** On d 77. **b** On d 185. EAI: early antibiotic intervention. Total SCFA, total short-chain fatty acid. The commercial creep feed with or without in-feed antibiotics (50 mg/kg olaquindox, 50 mg/kg oxytetracycline calcium, and 50 mg/kg kitasamycin) was fed to pig from d 7 to d 42. Thereafter, the control and antibiotic group were further randomly assigned to provide a normal (20%, 18%, 14% CP from d 42 to d 77, d 77 to d 120, d 120 to d 185, respectively) or low CP diet (16%, 14%, 10% CP from d 42 to d 77, d 77 to d 120, d 120 to d 185, respectively), respectively. The P values indicate main effects for antibiotic (A), protein level (C) and their interaction (AC), respectively

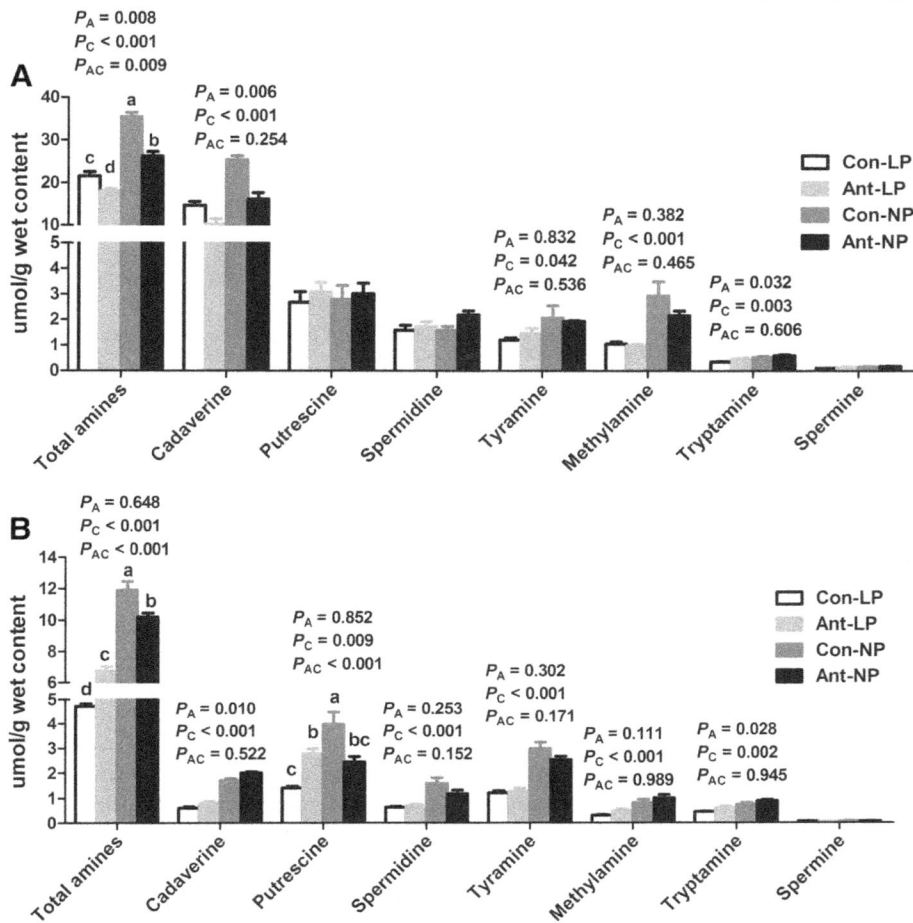

Fig. 2 Effects of EAI on amines concentrations in the feces of pigs with different CP levels diets. (Con-LP, ☐; Ant-LP, ▨; Con-NP, ▨; Ant-NP, ■). **a** On d 77. **b** On d 185. EAI: early antibiotic intervention. The commercial creep feed with or without in-feed antibiotics (50 mg/kg olaquindox, 50 mg/kg oxytetracycline calcium, and 50 mg/kg kitasamycin) was fed to pig from d 7 to d 42. Thereafter, the control and antibiotic group were further randomly assigned to provide a normal (20%, 18%, 14% CP from d 42 to d 77, d 77 to d 120, d 120 to d 185, respectively) or low CP diet (16, 14, 10% CP from d 42 to d 77, d 77 to d 120, d 120 to d 185, respectively), respectively. The P values indicate main effects for antibiotic (A), protein level (C) and their interaction (AC), respectively

was an interaction ($P < 0.001$) between EAI and dietary CP level for putrescine. The EAI reduced ($P < 0.05$) concentrations of putrescine in normal-CP diet, but increased its concentration of ($P < 0.05$) in low-CP diet. Additionally, EAI increased ($P < 0.05$) the concentrations of tryptamine and cadaverine. The low-CP diet decreased ($P < 0.01$) the concentrations of total amines, methylamine, tryptamine, putrescine, cadaverine, tyramine, and spermidine.

Phenolic and indolic compounds originate from the intestinal degradation of aromatic amino acids. As shown in Fig. 3, there was an interaction ($P < 0.05$) between EAI and dietary CP level for skatole concentration on d 77 (Fig. 3a). The EAI significantly increased ($P < 0.05$) concentration of skatole in low-CP diets, but not in normal-CP diet. Furthermore, pigs with low-CP diet reduced the concentrations of p-cresol and indole

($P < 0.05$). There was no dietary treatment effect on phenol concentration. On d 185 (Fig. 3b), the EAI did not affect the concentrations of phenol, p-cresol, and skatole, but tended to reduce ($P = 0.08$) indole concentration. The low-CP diet decreased ($P < 0.05$) the concentration of phenol and skatole but increased ($P < 0.05$) concentration of p-cresol. Collectively, these results indicated that EAI had a great impact on metabolites of protein fermentation in feces, but had little effect on carbohydrate fermentation. The low-CP diets reduced the protein fermentation products in feces on both d 77 and d 185.

Discussion

By employing a whole growth stage continuous feeding strategy, the current experiment enabled to evaluate the short- (77 d) and long-term (185 d) effect of EAI (from d 7 to d 42) on pig blood parameters, apparent nutrient

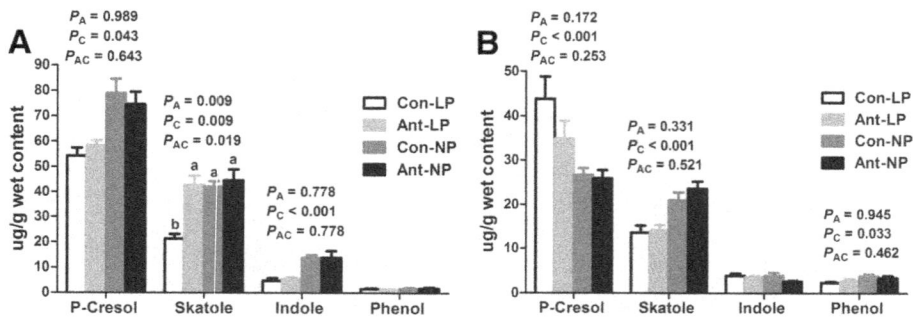

Fig. 3 Effects of EAI on phenolic and indole compounds concentrations in the feces of pigs with different CP levels diets. (Con-LP, ▢; Ant-LP, ▨; Con-NP, ▩; Ant-NP, ■). **a** On d 77. **b** On d 185. EAI: early antibiotic intervention. The commercial creep feed with or without in-feed antibiotics (50 mg/kg olaquindox, 50 mg/kg oxytetracycline calcium, and 50 mg/kg kitasamycin) was fed to pig from d 7 to d 42. Thereafter, the control and antibiotic group were further randomly assigned to provide a normal (20%, 18%, 14% CP from d 42 to d 77, d 77 to d 120, d 120 to d 185, respectively) or low CP diet (16, 14, 10% CP from d 42 to d 77, d 77 to d 120, d 120 to d 185, respectively), respectively. The P values indicate main effects for antibiotic (A), protein level (C) and their interaction (AC), respectively

digestibility, microbial metabolism, and bacterial counts in feces with different dietary CP level. Our results suggested that the EAI had short-term effects on blood parameters and microbial fermentation products in pigs, and that these effects were influenced by dietary CP levels.

In the intestine, there are differences in microbiome between gut segments and between fecal samples and intestinal samples. Along the hindgut fermentation in the pig (from cecum to distal colon), proportionally carbohydrate fermentation decreases while proteolytic fermentation increases. However, fecal samples are easy to access and easy for practice, and also can partly, though not exactly, reflect the general microbial metabolism in the gut. Thus, fecal samples are commonly used in many researches for host metabolism in humans and animals. In the present study, we intended to reflect the general gut metabolism through fecal samples analysis and the results showed that an alteration in microbial fermentation profiles of amino acid by dietary treatment, indicating that EAI and CP level affected the gut microbial composition and function. This approach by using fecal analysis can provide a feasible way to study the effects of EAI and CP on fermentation profiles of microbial amino acid metabolism.

Effects of early antibiotic intervention on growth performance and blood parameters with different protein levels after antibiotic withdrawal

At present, information on the effects of EAI on subsequent growth performance under different dietary CP levels is limited. Skinner et al. [25] reported that the use of antibiotics (therapeutic dose usage) in the nursery diets with normal-CP level appeared to reduce subsequent ADG of the pigs as compared to non-treated controls. In the present study, the antibiotics at low dosage did not have effect on subsequent ADG with normal-CP

diet, though increased the subsequent ADG under low-CP diet. Thus, the antibiotic effect on subsequent growth performance may vary depending on the antibiotic dosage and the diet CP level.

The level of serum parameters of animals reflects the body metabolic status according to their internal and external environments [26]. EAI exerted an effect on body metabolism on d 77, such as a significant increase of the concentrations of albumin and glucose, especially in low-CP diets, but not in normal-CP diets. A previous study also indicated that pigs received therapeutic amoxicillin exposure at birth through postnatal d 14 increased glucose concentration at postnatal d 49 [8]. Additionally, we found that the antibiotics effect on serum biochemistry was diminished on d 185. Thus, antibiotic intervention showed short-term effect on body metabolism, but had no long-term effect.

Effects of early antibiotic intervention on fermentation profiles of microbial amino acid metabolism with different protein levels after antibiotic withdrawal

Antibiotics are usually considered the most cost-effective way to reduce pathogenic bacteria [1]. Based on the microbiota analyses, we found that EAI had limited effect on the bacterial composition, only decreased the numbers of *E. coli* counts in the fecal samples irrespective of dietary CP levels on d 77, which suggested that EAI had limited long-term impact on fecal microbiota. Janczyk et al. [27] observed that piglets treated with amoxicillin (15 mg/kg body weight) by intramuscular administration at d 1 showed a decreased microbial diversity in colon on d 39. Collectively, these findings suggest that EAI less affected bacterial communities in feces.

EAI also affected profiles of biogenic amines in the pig feces. The significant decrease of cadaverine and total amines concentrations suggested a decreased amino acid decarboxylation after EAI treatment. High concentration of cadaverine can exert detrimental effects on the host, such as inducing DNA damage and oxidative stress [28]. Thus, the decrease of cadaverine and total amines concentration by EAI may exert a beneficial influence on the host. Coincidently, among the bacteria examined in the present study, *E. coli* species are a major protein-fermenting group [29]. Our previous study indicated that the cadaverine concentration was positively correlated with the abundance of *E. coli* [12]. Therefore, the decrease of cadaverine concentration by EAI may be due to the decrease in protein-fermenting bacteria counts (*E. coli*) and the subsequent reduction of the catabolism of precursor amino acid (lysine). EAI increased putrescine concentration in low-CP diet but decreased it in normal-CP diet. This results indicates that the effect of EAI on subsequent concentrations of amines may vary with CP level. Furthermore, we found that cadaverine concentration varied between d 77 and d 185. Cadaverine is derived from decarboxylation of lysine [30]. In the current study, the CP level and lysine concentration in the diet was much higher on d 77 than 185 d, which may lead to the variation in cadaverine concentration across ages.

On the other hand, EAI also changed aromatic amino acid metabolism by the microbes in the GIT. Tryptophan, one of the aromatic amino acid, can be decarboxylated into tryptamine by GIT microbes [30, 31]. In the present study, EAI increased tryptamine concentration in the feces irrespective dietary CP levels on d 77 and d 185, suggesting that EAI had a continuous effect on aromatic amino acid metabolism. Tryptamine can stimulate the secretion of serotonin by enterochromaffin cells, and then regulate intestinal motility [32].Thus, an increase of tryptamine may affect the function of intestine. Additionally, EAI also increased skatole concentration on d 77, especially in low-CP diets. Skatole is produced from tryptophan catabolism by GIT bacteria [30]. The increase of skatole concentration indicates that EAI also affected tryptophan catabolism in the gut. Collectively, combined with serum parameters and microbial fermentation profiles, the effects of EAI persisted to short-term after withdrawal, but weakened at finishing period of the pigs.

Low-CP diets affected host metabolism and fermentation profiles of microbial amino acid metabolism

In our study, low-CP dietary treatment affected the urea metabolism as reflected by the decreased urea concentration in serum both on d 77 and d 185, which has also been found in other reports [33, 34]. Urea is the main nitrogenous end product of amino acids catabolism in pigs. A decrease in serum urea level is indicative of a more efficient use of dietary nitrogen [35]. Thus, our study may indicate that low-CP diet affected nitrogen metabolism in pigs.

Corresponding to the decrease in urea concentration in serum, low-CP diet markedly affected the microbial metabolism of amino acids in the intestine, as shown by the decrease in concentration of most amines, suggesting decreased microbial decarboxylation. Furthermore, low-CP diet decreased some phenolic and indole compounds concentrations. Indole and skatole are products of tryptophan metabolism originating from dietary and endogenous protein [31]. Although the concentration of tryptophan in the normal-CP diet is equal to that of the low-CP diet in the current study, crystalline amino acid is quickly absorbed before reaching the hindgut, and then reduced the availability of substrates for microbial fermentation in the hindgut. Overall, these findings suggest that low-CP diets significantly reduced the microbial fermentation of protein or amino acids and the concentration of potentially harmful metabolites derived from microbial metabolism of feed-derived amino acids. Thus, the dietary nitrogen was used more efficiently with low level of potentially harmful metabolites in pigs fed low-CP amino acids-supplemented diets.

Low-CP diets altered the markers of microbial carbohydrate metabolism on d 185, such as decreased the concentrations of acetate, propionate, isobutyrate, isovalerate, and total SCFA, consistent with Le et al. [36], who observed that reducing dietary CP level from 15% to 12% led to a decrease in total SCFA, acetate, and propionate concentrations for finishing pigs. The concentration of SCFA produced mainly depends on the amount and composition of substrate and the type of microbes present in the GIT [37]. Thus, in the present study, a decrease of SCFAs may be related with the altered metabolism activity of gut microbiota after low-CP diet.

Conclusions

In conclusion, combined with the changes in blood metabolites, digestibility, and microbial metabolite, our results showed that EAI (from d 7 to d 42) showed short-term effects on pigs, especially on the body and microbial metabolism. The effects of EAI varied between CP levels, which was characterized by the significant alteration of glucose and putrescine concentration. The low-CP diet significantly decreased the amino acids fermentation products in feces.

Additional file

Additional file 1: Table S1. Ingredients and chemical composition of creep feed diets. Figure S1. Effects of early antibiotic intervention on bacterial abundance in the feces of pigs with different CP levels diets. (Con-LP, *White*; Ant-LP, *Light gray*; Con-NP, *Dark gray*; Ant-NP, *Black*). A: On d 77. B: On d 185. The commercial creep feed with or without in-feed antibiotics (50 mg/kg olaquindox, 50 mg/kg oxytetracycline calcium, and 50 mg/kg kitasamycin) was fed to pig from d 7 to d 42. Thereafter, the control and antibiotic group were further randomly assigned to provide a normal (20%, 18%, 14% CP from d 42 to d 77, d 77 to d 120, d 120 to d 185, respectively) or low CP diet (16%, 14%, 10% CP from d 42 to d 77, d 77 to d 120, d 120 to d 185, respectively), respectively. The *P* values indicate main effects for antibiotic (A), protein level (C) and their interaction (AC), respectively. (DOC 576 kb)

Abbreviations
ADFI: average daily feed intake; ADG: average daily gain; AIA: acid-insoluble ash; BCFA: branched-chain fatty acid; BW: body weight; CP: crude protein; DW: dry matter; EAI: early antibiotic intervention; G:F: the ratio of gain to feed; GIT: gastrointestinal bacteria; HDLC: high density lipoprotein cholesterol; LDLC: low density lipoprotein cholesterol; OM: organic matter; SCFA: short chain fatty acid

Acknowledgements
The authors sincerely thanks to Professor Sung Woo Kim of North Carolina State University for his critical discussion and reading during manuscript preparation.

Funding
This work was supported by National Key Basic Research Program of China (2013CB127300) and Natural Science Foundation of China (31430082). The funders had no role in the design of the study, collection, analysis, and interpretation of data and in writing of the manuscript.

Authors' contributions
The author' contributions are as follows: WYZ and YS were in charge of the whole trial; MY and WYZ wrote the manuscript; CLM and YXY for animal feeding and care; KFY, MY and CJZ assisted with sampling and laboratory analyses. All authors read and approved the final manuscript.

Competing interests
The authors declare that they have no competing interests.

References
1. Cromwell GL. Why and how antibiotics are used in swine production. Anim Biotechnol. 2002;13:7–27.
2. Puiman P, Stoll B, Mølbak L, de Bruijn A, Schierbeek H, Boye M, et al. Modulation of the gut microbiota with antibiotic treatment suppresses whole body urea production in neonatal pigs. Am J Physiol-Gastr L. 2013;304:300–10.
3. Yoon JH, Ingale SL, Kim JS, Kim KH, Lohakare J, Park YK, et al. Effects of dietary supplementation with antimicrobial peptide-P5 on growth performance, apparent total tract digestibility, faecal and intestinal microflora and intestinal morphology of weanling pigs. J Sci Food Agr. 2013;93:587–92.
4. Rettedal E, Vilain S, Lindblom S, Lehnert K, Scofield C, George S, et al. Alteration of the ileal microbiota of weanling piglets by the growth-promoting antibiotic chlortetracycline. Appl Environ Microb. 2009;75:5489–95.
5. Looft T, Johnson TA, Allen HK, Bayles DO, Alt DP, Stedtfeld RD, et al. In-feed antibiotic effects on the swine intestinal microbiome. P Natl Acad Sci USA. 2012;109:1691–6.
6. Dierick N, Vervaeke I, Decuypere J, Henderickx H. Influence of the gut flora and of some growth-promoting feed additives on nitrogen metabolism in pigs. I. Studies in vitro. Livest Prod Sci. 1986;14:161–76.
7. Bhandari S, Xu B, Nyachoti C, Giesting D, Krause D. Evaluation of alternatives to antibiotics using an K88 model of piglet diarrhea: effects on gut microbial ecology. J Anim Sci. 2008;86:836–47.
8. Li J, Yang K, Ju T, Ho T, McKay CA, Gao Y, et al. Early life antibiotic exposure affects pancreatic islet development and metabolic regulation. Sci Rep. 2017;7:41778..
9. Schokker D, Zhang J, Vastenhouw SA, Heilig HG, Smidt H, Rebel JM, et al. Long-lasting effects of early-life antibiotic treatment and routine animal handling on gut microbiota composition and immune system in pigs. PLoS One. 2015;10:e0116523.
10. Cho I, Yamanishi S, Cox L, Methé BA, Zavadil J, Li K, et al. Antibiotics in

early life alter the murine colonic microbiome and adiposity. Nature. 2012;488:621–6.
11. Cox LM, Yamanishi S, Sohn J, Alekseyenko AV, Leung JM, Cho I, et al. Altering the intestinal microbiota during a critical developmental window has lasting metabolic consequences. Cell. 2014;158:705–21.
12. Mu CL, Yang YX, Luo Z, Guan LL, Zhu WY. The colonic microbiome and epithelial transcriptome are altered in rats fed a high-protein diet compared with a normal-protein diet. J Nutr. 2016;146:474–83.
13. Luo Z, Li C, Cheng Y, Hang S, Zhu W. Effects of low dietary protein on the metabolites and microbial communities in the caecal digesta of piglets. Arch Anim Nutr. 2015;69:212–26.
14. Zhang CJ, Yu M, Yang YX, Mu CL, Su Y, Zhu WY. Effect of early antibiotic administration on cecal bacterial communities and their metabolic profiles in pigs fed diets with different protein levels. Anaerobe. 2016;42:188–96.
15. Mu CL, Yang YX, Yu KF, Yu M, Zhang CJ, Su Y, et al. Alteration of metabolomic markers of amino-acid metabolism in piglets with in-feed antibiotics. Amino Acids. 2017;49(4):771–81.
16. NRC. Nutrient requirements of swine. 11th ed. Washington, DC: Natl Acad Press; 2012.
17. AOAC. Official methods of analysis. 18th ed. Association of Official Analytical Chemists: Gaithersburg, MD; 2007.
18. ISO. Animal feeding stuffs-determination of ash insoluble in hydrochloric acid (ISO 5985). Geneva, Switzerland: International Organization for Standardization; 2003.
19. McCarthy J, Aherne F, Okai D. Use of HCl insoluble ash as an index material for determining apparent digestibility with pigs. Can J Anim Sci. 1974;54:107–9.
20. Zoetendal EG, Akkermans ADL, Vos WMD. Temperature gradient gel electrophoresis analysis of 16S rRNA from human fecal samples reveals stable and host-specific communities of active bacteria. Appl Environ Microb. 1998;64:3854–9.
21. Wang X, Mao S, Liu J, Zhang L, Cheng Y, Jin W, et al. Effect of gynosaponins on methane production and microbe numbers in a fungus-methanogen co-culture. J Anim Feed Sci. 2011;20:272–84.
22. Yang YX, Mu CL, Zhang JF, Zhu WY. Determination of biogenic amines in Digesta by high performance liquid chromatography with Precolumn Dansylation. Anal Lett. 2014;47:1290–8.
23. Chaney AL, Marbach EP. Modified reagents for determination of urea and ammonia. Clin Chem. 1962;8:130–2.
24. Jensen MT, Cox RP, Jensen BB. 3-Methylindole (skatole) and indole production by mixed populations of pig fecal bacteria. Appl Environ Microbiol. 1995;61:3180–4.
25. Skinner L, Levesque C, Wey D, Rudar M, Zhu J, Hooda S, et al. Impact of nursery feeding program on subsequent growth performance, carcass quality, meat quality, and physical and chemical body composition of growing-finishing pigs. J Anim Sci. 2014;92:1044–54.
26. Nicholson JK, Holmes E, Kinross JM, Darzi AW, Takats Z, Lindon JC. Metabolic phenotyping in clinical and surgical environments. Nature. 2012;491:384–92.
27. Janczyk P, Pieper R, Souffrant WB, Bimczok D, Rothkötter H-J, Smidt H. Parenteral long-acting amoxicillin reduces intestinal bacterial community diversity in piglets even 5 weeks after the administration. ISME J. 2007;1:180–3.
28. Holmes E, Li JV, Athanasiou T, Ashrafian H, Nicholson JK. Understanding the role of gut microbiome–host metabolic signal disruption in health and disease. Trends Microbiol. 2011;19:349–59.
29. Rist VT, Weiss E, Sauer N, Mosenthin R, Eklund M. Effect of dietary protein supply originating from soybean meal or casein on the intestinal microbiota of piglets. Anaerobe. 2014;25:72–9.
30. Davila A-M, Blachier F, Gotteland M, Andriamihaja M, Benetti P-H, Sanz Y, et al. Intestinal luminal nitrogen metabolism: role of the gut microbiota and consequences for the host. Pharmacol Res. 2013;68:95–107.
31. Rist V, Weiss E, Eklund M, Mosenthin R. Impact of dietary protein on microbiota composition and activity in the gastrointestinal tract of piglets in relation to gut health: a review. Animal. 2013;7:1067–78.
32. Yano JM, Yu K, Donaldson GP, Shastri GG, Ann P, Ma L, et al. Indigenous bacteria from the gut microbiota regulate host serotonin biosynthesis. Cell. 2015;161:264–76.
33. Nyachoti C, Omogbenigun F, Rademacher M, Blank G. Performance responses and indicators of gastrointestinal health in early-weaned pigs fed low-protein amino acid-supplemented diets. J Anim Sci. 2006;84:125–34.
34. Yue L, Qiao S. Effects of low-protein diets supplemented with crystalline amino acids on performance and intestinal development in piglets over

the first 2 weeks after weaning. Livest Sci. 2008;115:144–52.

35. Figueroa J, Lewis A, Miller PS, Fischer R, Gómez R, Diedrichsen R. Nitrogen metabolism and growth performance of gilts fed standard corn-soybean meal diets or low-crude protein, amino acid-supplemented diets. J Anim Sci. 2002;80:2911–9.

36. Le P, Aarnink A, Jongbloed A. Odour and ammonia emission from pig manure as affected by dietary crude protein level. Livest Sci. 2009;121:267–74.

37. van Beers-Schreurs HM, Nabuurs MJ, Vellenga L, Kalsbeek-van der Valk HJ, Wensing T, Breukink HJ. Weaning and the weanling diet influence the villous height and crypt depth in the small intestine of pigs and alter the concentrations of short-chain fatty acids in the large intestine and blood. J Nutr. 1998;128:947–53.

38. Suzuki MT, Taylor LT, DeLong EF. Quantitative analysis of small-subunit rRNA genes in mixed microbial populations via 5′-nuclease assays. Appl Environ Microbiol. 2000;66:4605–14.

39. Guo X, Xia X, Tang R, Zhou J, Zhao H, Wang K. Development of a real-time PCR method for Firmicutes and Bacteroidetes in faeces and its application to quantify intestinal population of obese and lean pigs. Lett Appl Microbiol. 2008;47:367–73.

40. Matsuki T, Watanabe K, Fujimoto J, Takada T, Tanaka R. Use of 16S rRNA gene-targeted group-specific primers for real-time PCR analysis of predominant bacteria in human feces. Appl Environ Microbiol. 2004;70:7220–8.

41. Bartosch S, Fite A, Macfarlane GT, McMurdo ME. Characterization of bacterial communities in feces from healthy elderly volunteers and hospitalized elderly patients by using real-time PCR and effects of antibiotic treatment on the fecal microbiota. Appl Environ Microbiol. 2004;70:3575–81.

42. Huijsdens XW, Linskens RK, Mak M, Meuwissen SG, Vandenbroucke-Grauls CM, Savelkoul PH. Quantification of bacteria adherent to gastrointestinal mucosa by real-time PCR. J Clin Microbiol. 2002;40:4423–7.

43. Khafipour E, Li S, Plaizier JC, Krause DO. Rumen microbiome composition determined using two nutritional models of subacute ruminal acidosis. Appl Environ Microbiol. 2009;75:7115–24.

14

Glucosamine supplementation during late gestation alters placental development and increases litter size

Jeffrey L. Vallet[1*], Jeremy R. Miles[1], Bradley A. Freking[1] and Shane Meyer[2]

Abstract

Background: During late gestation the placental epithelial interface becomes highly folded, which involves changes in stromal hyaluronan. Hyaluronan is composed of glucoronate and N-acetyl-glucosamine. We hypothesized that supplementing gestating dams with glucosamine during this time would support placental folded-epithelial-bilayer development and increase litter size. In Exp. 1, gilts were unilaterally hysterectomized-ovariectomized (UHO). UHO gilts were mated and then supplemented daily with 10 g glucosamine ($n = 16$) or glucose (control, $n = 17$) from d 85 of gestation until slaughter (d 105). At slaughter, the number of live fetuses was recorded and each live fetus and its placenta was weighed. Uterine wall samples adjacent to the largest and smallest fetuses within each litter were processed for histology. In Exp. 2, pregnant sows in a commercial sow farm were supplemented with either 10 g glucosamine or glucose daily from d 85 of gestation to farrowing. Total piglets born and born alive were recorded for each litter. In Exp. 3, the same commercial farm and same protocol were used except that the dose of glucosamine and glucose was doubled to 20 g/d.

Results: In Exp. 1, the number of live fetuses tended to be greater in glucosamine-treated UHO gilts ($P = 0.098$). Placental morphometry indicated that the width of the folded bilayer was greater ($P = 0.05$) in glucosamine-treated gilts. In Exp. 2, litter size did not differ between glucosamine- and glucose-treated sows. However in Exp. 3, the increased dose of glucosamine resulted in a significant treatment by parity interaction ($P \leq 0.01$), in which total piglets born and born alive were greater in glucosamine treated sows of later parity (5 and 6).

Conclusions: These results indicated that glucosamine supplementation increased the width of the folds of the placental bilayer and increased litter size in later parity, intact pregnant commercial sows.

Keywords: Fructose, Glucosamine, Hyaluronan, Swine, Uterine capacity

Background

Litter size contributes to the profitability of swine production, and is influenced by ovulation rate, fertilization rate, embryonic mortality and uterine capacity [1]. Fertilization rate and embryonic mortality are typically fixed rates that are independent of oocyte or embryo number [1, 2]. Thus, a greater ovulation rate results in increased embryos at d 30 of gestation [3]. However, greater ovulation rate does not result in greater litter size at farrowing [4], but does reduce birth weight [3]. In addition, conceptus losses under crowded uterine

conditions occur throughout gestation [3, 5]. Both phenomena are the result of reduced placental size caused by intrauterine crowding.

Recent studies suggest that the pig placenta compensates for reduced intrauterine space [6]. A component of this adaptation is likely to be increased depth of the microscopic folds of the placental epithelial bilayer [7]. However, Vallet and Freking [7] also reported that placenta of small fetuses lacked stromal tissue above the folded bilayer, especially during late gestation, potentially limiting the compensatory ability of the placenta. Hyaluronan is a major component of placental stroma [8, 9] and is composed of repeating units of N-acetyl glucosamine and glucuronate, which are both derivatives of glucose [10]. Interestingly, Glucose transporter (GLUT) 2,

* Correspondence: jeff.vallet@ars.usda.gov
[1]USDA, ARS, U.S. Meat Animal Research Center (USMARC), P.O. Box 166, Clay Center NE, Nebraska 68933, USA
Full list of author information is available at the end of the article

which is present at the fetal maternal interface [11], has greater transport capacity for glucosamine compared to glucose [12]. We hypothesized that supplementation of glucosamine in sow diets might preferentially promote placental stromal development, allowing increased placental epithelial bilayer fold development. This would increase uterine capacity of glucosamine-treated sows. The objective of these experiments was to test whether glucosamine supplementation alters placental fold development, uterine capacity and litter size.

Methods

Experiment 1

Gilts were unilaterally hysterectomized-ovariectomized (UHO) at 160 d of age. Gilts were anesthetized with sodium pentothal and anesthesia was maintained using fluothane. The UHO surgery involves removing one ovary and one uterine horn, and reduces the total intrauterine space by one-half, while the ovulation rate remains unaffected. The litter size of pigs after UHO is no longer affected by ovulation rate, and is considered to be a measure of one-half their uterine capacity [5]. Gilts were fed 2 kg/d of a diet that met NRC requirements for pigs, consisting of 70% ground corn, 25% soybean meal with the remainder made up of vitamin and mineral supplements, free lysine and soybean oil. Gilts were allowed to recover and were subsequently naturally mated to mature boars after at least one estrous cycle of normal length (17 to 23 d). Thirty-three pregnant UHO gilts were used in this experiment. Beginning at d 85 of gestation, gilts were fed in individual pens and received either 10 g glucosamine (Hard Eight Nutrition LLC, Henderson NV) or 10 g glucose (Pastry Chef Central Inc., Boca Raton FL) as a top dress on their daily feed. The dose of glucosamine was chosen to be similar to that routinely used for supplementation in humans, after accounting for differences in weight (1.5 to 3 g per day are recommended in adult humans). Supplementation on d 85 of gestation was chosen because secondary fold development begins at this time [13]. Secondary fold development requires significant stromal remodeling and hyaluronan turnover. Because hyaluronan is 50% glucosamine, we hypothesized that glucosamine supplementation would facilitate fold development by supporting hyaluronan turnover. At 105 d of gestation, gilts were humanely slaughtered and the remaining uterine horn and ovary were collected. Corpora lutea were counted, and the umbilical cord of each live fetus was exteriorized through a small antimesometrial hole in the uterine wall to minimize disruption of the placental vasculature. Then, blood samples were taken from each fetus. A fetus was considered alive if it had a visible pulse in the exposed umbilical cord. Blood samples were taken from the umbilical artery of each live fetus. The

remaining uterine horn was then opened completely and each live fetus was counted and weighed. The largest and smallest fetuses by weight were identified, their corresponding placentas were identified, and a uterine wall sample was collected that included these placentas. The smallest live fetus in each litter was chosen as this would be the most compromised fetus within the UHO litter. It was compared to the largest fetus in the litter as this would be the most uncompromised fetus in the litter. Thus, sampling the largest and smallest fetuses represented the full range of weight variation within each litter. Uterine samples were taken immediately adjacent but external to the amnion. Tissues were placed into cassettes and immersed in buffered formalin. Finally, a further sample of fetal placental tissue was collected and frozen in liquid nitrogen.

After formalin fixation, uterine wall samples were transferred to 70% ethanol in water, dehydrated through a series of increasing alcohol concentrations followed by xylene, and then embedded in paraffin. Tissues were then sectioned (10 μm), placed on slides, rehydrated and stained with hematoxylin and eosin, dehydrated and coverslipped. At least two sections were evaluated for each placental sample. Width of the folded bilayer, width of the stroma above the folds, and the interface length adjusted to constant placental length were measured using Bioquant (Bioquant image analysis corporation, Nashville, TN) as previously described [7], includes figure describing individual measures. Briefly, to obtain these measures, the area of a folded region within the placental interface was obtained by creating a closed polygon that extended from the base of the first fold to its top, across the top of several (3 to 5) adjacent folds, from the top to the bottom of the last fold, and across the bottom of the adjacent folds. Then, the length of the polygon along the long axis of the folded bilayer was measured by drawing a line through the center of the folded region within the polygon. The average width of the polygon (width of the folded bilayer) was calculated as the area of the polygon divided by its length. The width of the stroma was measured from the tip of each fold within the polygon to the adjacent edge of the stroma (border of the allantois). These measures were averaged for each slide to provide a single stromal width measure for each slide. Finally, the adjusted placental interface length was obtained by measuring (i.e., tracing) the length of the folded bilayer within the polygon, and dividing that length by the length of the polygon through the center. Each of the three measures was then averaged for the two slides for each placenta to provide a single measure of the width of folds, width of stroma and adjusted placental interface length for each placenta.

Fetal blood samples were allowed to clot, centrifuged, and serum was collected. Serum samples from each fetus were measured for glucose and fructose. Glucose was measured using the YSI 2700 Biochemistry Analyzer (YSI Life Sciences, Yellow Springs, OH) using instructions included in the manual. Fructose was measured using the procedure described by Zavy et al. [14].

Placental tissue samples were homogenized as described by Vallet et al. [9] and homogenates were measured for hyaluronan using a kit (Corgenix Inc., Broomfield CO). Homogenized placental samples were diluted 1:500 in PBS before assay.

Experiment 2

This trial took place at a commercial sow farm in Nebraska (Plymouth Ag Group, Diller, NE) in May–July, 2015. The farm farrowed approximately 255 sows over a two-week period in weekly batches. Sow parity in both weeks ranged from 2 to 8. Sows were managed (including diet) and bred by artificial insemination according to the normal protocols existing at the farm, and were then supplemented with either glucosamine (n = 128; 10 g/d) or glucose (n = 127; control, 10 g/d) as a top dress on their daily feed (~2 kg corn-soybean diet depending on body condition) beginning on d 85 of gestation. Care was taken to evenly distribute glucosamine or glucose treatment among parities. Top dress was delivered using plastic scoops previously calibrated to deliver the appropriate amounts of glucosamine or glucose. During gestation, sows were housed in individual gestation stalls according to the standard procedure for the farm, allowing for them to be dosed independently on a daily basis. At d 115 of gestation, sows received an injection of estrumate (cloprostenol; Merck Animal Health, Madison, NJ) to induce farrowing, which is also standard procedure for the management of farrowing on the farm. Number born, number born alive, number stillborn, number of mummified fetuses, and the number of piglets weaned were recorded for each litter. All live piglets were ear tagged at birth, and birth and weaning weights for each live piglet were also recorded for each litter. Piglets were crossfostered according to procedures used on the commercial farm, but it was not possible to record piglet movement except at weaning. Piglets were weaned at an average age of 19.6 d (range 11 to 28 d).

Experiment 3

This trial took place at the same commercial sow farm as described for Exp. 2, in May–July, 2016. The protocol used was the same as that described for Exp. 2, except that the dose of glucose and glucosamine was increased to 20 g/d, to account for the larger sows compared to the gilts in Exp. 1. In addition, only parity 3 to 7 sows were treated, no parity 2 sows were available within the management system at the time of the trial. In this experiment, 89 sows received glucosamine and 87 sows

received glucose. As in Exp. 2, piglets were ear tagged and weighed and then crossfostered after birth but it was not possible to record crossfostering until piglets were weighed at weaning.

Statistical analysis

Fetal number (uterine capacity) data from Exp. 1 was analyzed using PROC MIXED (SAS Inst. Inc., Cary, NC) with a model that included treatment. Fetal blood glucose and fructose were analyzed with a model that included treatment, fetal weight, placental weight, and treatment by fetal weight and treatment by placental weight interactions. Gilt within treatment was included as a random effect. Interaction terms that were not significant were sequentially dropped from the model to arrive at a final model. Placental morphometry and hyaluronan data were analyzed using a model that included effects of treatment, fetal size and the treatment by size interaction. Gilt within treatment was included as a random effect. Relationships between placental hyaluronan and fetal serum glucose and fructose for the corresponding largest and smallest fetuses were analyzed using a model that included treatment, fetal size, the treatment by fetal size interaction, fetal weight, fetal weight by treatment, fetal weight by fetal size, fetal weight by treatment by fetal size, placental weight, placental weight by treatment, placental weight by fetal size, and placental weight by treatment by fetal size. Nonsignificant effects were sequentially removed from the model, starting with complex interactions, until only significant effects remained.

Litter size data from Exp. 2 and 3 were analyzed using PROC MIXED with a model that included the effects of farrowing week, treatment, parity and the treatment by parity interaction. Orthogonal contrasts were used when necessary to further evaluate differences among treatment means. Birth and weaning weights were considered repeated measures of the birth dam and were therefore analyzed with a similar model to litter size data, including effects of farrowing week, treatment, parity and the treatment by parity interaction. Sow within week by treatment by parity interaction was included as a random effect. Finally, stillbirth rate and preweaning mortality were also considered a trait of the birth dam and analyzed using PROC GLIMMIX, treating alive or dead at birth and weaning as binary variables. Because it was not possible to record crossfostering, crossfostering was considered to be random and was not considered in the analysis of preweaning mortality. Thus, this analysis treats weaning weights and preweaning mortality as traits of the birth dam, not traits of the lactating dam. The model included effects of farrowing week, treatment, parity and treatment by parity, and the effect of sow within week by treatment by parity was included as a random effect.

Results

Experiment 1

Number of CL for glucosamine- and glucose-treated UHO gilts did not differ (16.2 ± 0.9 and 15.1 ± 0.9, respectively). There was a trend ($P = 0.098$) toward greater number of live fetuses in UHO gilts treated with glucosamine compared to glucose (8.4 ± 0.6 and 6.9 ± 0.6, respectively).

There were no treatment interaction effects on the relationships between fetal weight and fetal serum glucose and placental weight and fetal serum glucose, nor was there an overall effect of treatment on fetal serum glucose when interaction effects were removed from the model. However, both fetal weight and placental weight were associated ($P < 0.05$) with fetal serum glucose, and these relationships are illustrated in Fig. 1. The surface plot in Fig. 1 indicates that fetal serum glucose decreased with increasing fetal weight and increased with increasing placental weight (fetal glucose = – 0.0072 (fetal weight) + 0.002401 (placental weight)). The two effects balanced (no net effect on glucose concentrations) when placental weights were 30% of fetal weights (white line indicated in Fig. 1).

Fetal serum fructose relationships with fetal and placental weights were affected by treatment, and the effects are illustrated in Fig. 2. The final model included significant treatment by fetal weight ($P < 0.01$) and treatment by placental weight ($P < 0.05$) interaction terms. In glucosamine-treated gilts, fetal serum fructose was

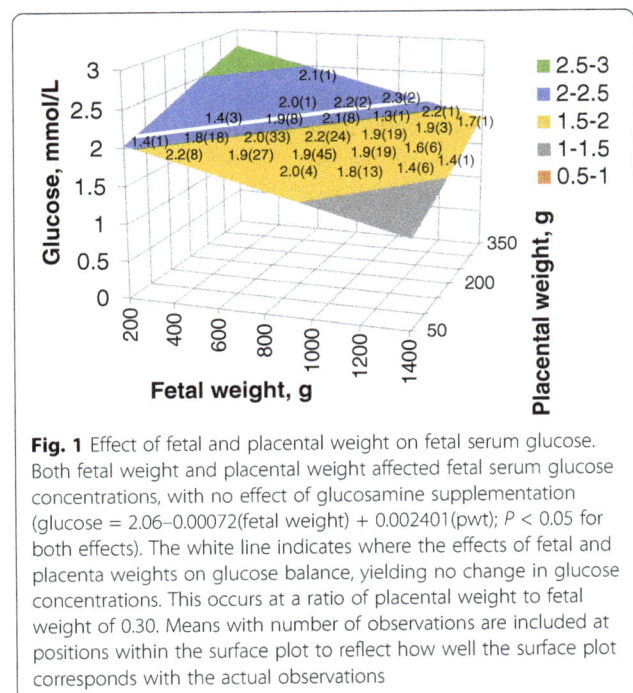

Fig. 1 Effect of fetal and placental weight on fetal serum glucose. Both fetal weight and placental weight affected fetal serum glucose concentrations, with no effect of glucosamine supplementation (glucose = 2.06–0.00072(fetal weight) + 0.002401(pwt); $P < 0.05$ for both effects). The white line indicates where the effects of fetal and placenta weights on glucose balance, yielding no change in glucose concentrations. This occurs at a ratio of placental weight to fetal weight of 0.30. Means with number of observations are included at positions within the surface plot to reflect how well the surface plot corresponds with the actual observations

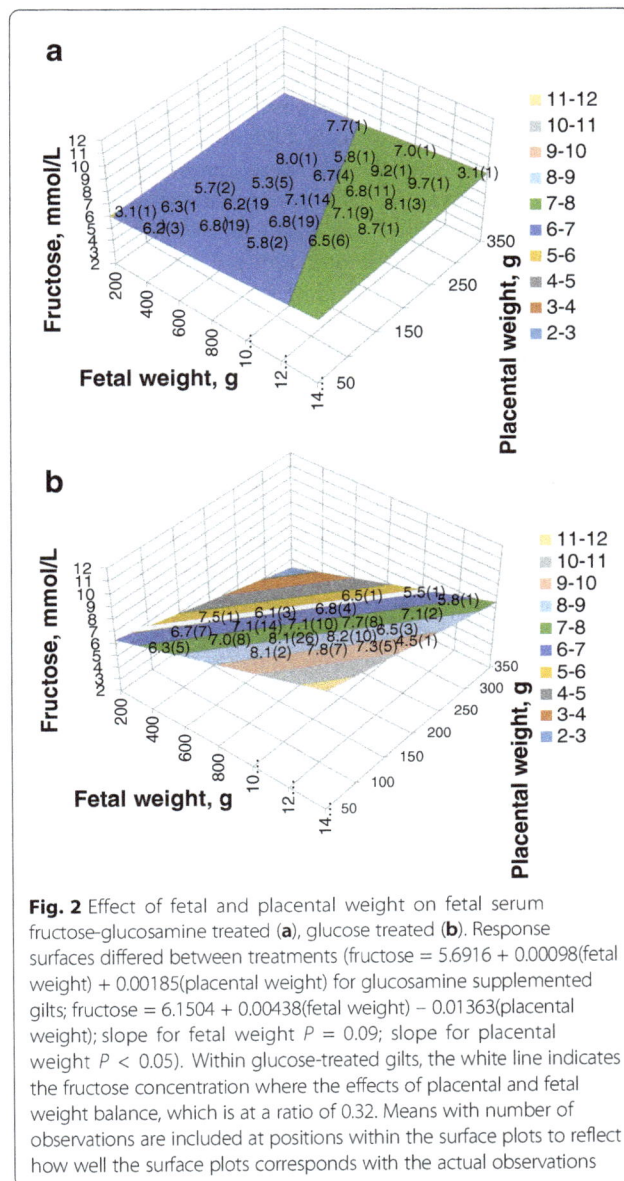

Fig. 2 Effect of fetal and placental weight on fetal serum fructose-glucosamine treated (**a**), glucose treated (**b**). Response surfaces differed between treatments (fructose = 5.6916 + 0.00098(fetal weight) + 0.00185(placental weight) for glucosamine supplemented gilts; fructose = 6.1504 + 0.00438(fetal weight) – 0.01363(placental weight); slope for fetal weight $P = 0.09$; slope for placental weight $P < 0.05$). Within glucose-treated gilts, the white line indicates the fructose concentration where the effects of placental and fetal weight balance, which is at a ratio of 0.32. Means with number of observations are included at positions within the surface plots to reflect how well the surface plots corresponds with the actual observations

positively related to both fetal weight and placental weight (fetal serum fructose = 5.6916 + 0.00098 (fetal weight) + 0.00185 (placental weight)). In glucose-treated gilts, fetal serum fructose was positively related to fetal weight and negatively related to placental weight (fetal serum fructose = 6.1504 + 0.00438 (fetal weight) – 0.01363 (placental weight)). Similar to relationships between fetal weight, placental weight and fetal serum glucose, the effects of fetal weight and placental weight on fetal serum fructose in glucose treated gilts balanced when placental weight was 32% of fetal weight (white line in Fig. 2a). Few live fetuses were present in the dataset where the placenta was greater than 30% of fetal weight at d 105 of gestation.

Fetal and placental weight, fetal blood glucose and fructose, and placental hyaluronan for the largest and smallest fetuses in each litter are summarized in Table 1. Hyaluronan data from the largest and smallest placentas from one gilt were unusually high (> 3 standard deviations above the mean), and were therefore deleted as outliers. There was no effect of treatment or treatment by fetal size interaction on these traits. Fetal and placental weights were significantly reduced ($P < 0.01$) in the smallest compared to the largest fetuses in the litter. Despite results from all fetuses within the litter indicating that glucose decreased with fetal weight, there was no significant effect of fetal size on fetal serum glucose in this reduced dataset. Similar to results from all fetuses in the litter, there was a significant decrease ($P < 0.05$) in fructose in the smallest fetus compared to the largest fetus in the litter. Placental hyaluronan was unaffected by fetal size as a category, or by fetal weight or placental weight as continuous variables. Further exploration of relationships between hyaluronan and fetal serum glucose and fructose using regression analysis indicated that the relationship between hyaluronan and serum fructose concentrations was unaffected by treatment or fetal size and that there was no overall relationship. The relationship between hyaluronan and fetal serum glucose did not differ between large and small fetuses, but did differ between treatments (treatment intercepts, $P < 0.05$; treatment slopes, $P = 0.06$; Fig. 3). Placental hyaluronan was positively related to glucose in glucosamine-treated gilts, and negatively related to glucose in glucose-treated gilts.

The results of placental morphometry are summarized in Table 2. Placental fold width was significantly greater ($P = 0.05$) in glucosamine-treated gilts compared to glucose-treated gilts. There was no effect of size of the fetus on placental fold width, and there was no treatment by size of fetus interaction. Stromal depth above the folded bilayer did not differ with treatment or size of fetus, but there tended to be a treatment by fetal size interaction ($P = 0.07$). This appeared to be due to greater

stromal depth in large fetuses from glucosamine-treated UHO gilts compared to the other three treatment by fetal size combinations ($P < 0.01$). The total width of the placenta was greater ($P < 0.05$) in glucosamine-treated gilts compared to glucose-treated gilts, and there was no effect of fetal size on total placental width nor was there a treatment by fetal size interaction. Finally, the length of the folded bilayer adjusted to a constant length of placenta was unaffected by treatment, fetal size or the treatment by fetal size interaction. Nevertheless, the adjusted length of the folded bilayer was significantly correlated ($r = 0.75$; $P < 0.01$) with the width of the folded bilayer.

Experiment 2

There was a treatment by parity interaction for number of stillborn piglets and stillbirth rate. No treatment by parity interaction was observed for any of the other traits measured in this experiment. The treatment by parity least squares means for number of stillborn piglets and stillbirth rate are presented in Table 3. For both number of stillborns and stillbirth rate, the interaction appeared to be due to greater stillbirth in glucosamine supplemented sows in later parities (parities 7 and 8).

Treatment least squares means for the other litter size and weight traits are presented in Table 4. There were no statistically significant effects of treatment on the number of total born, born alive, or mummies, or on birth weights, weaning weights, or preweaning mortality.

Significant parity effects (Table 5) were observed for the number of total born ($P < 0.05$) and live born piglets ($P < 0.05$), and for birth ($P < 0.01$) and weaning weights ($P < 0.05$). Number of mummies and preweaning mortality were not affected by parity. Total born and born alive increased gradually with increasing parity until parity 6, after which both litter size measures decreased. Birth weights were similar among parities until parity 6, after which they also decreased ($P < 0.05$). In contrast, average weaning weights increased progressively with increasing parity.

Table 1 Treatment effects on fetuses from Experiment 1

| Variable | Glucosamine[a] | | Glucose | |
	Large	Small	Large	Small
Fetal weight, g[b]	1017 ± 52	529 ± 52	1076 ± 49	639 ± 49
Placental weight, g	231 ± 13	108 ± 13	219 ± 13	129 ± 13
Serum glucose, mmol/L	1.8 ± 0.2	1.7 ± 0.2	1.9 ± 0.2	1.7 ± 0.2
Serum fructose, mmol/L[c]	6.7 ± 0.6	5.6 ± 0.6	7.7 ± 0.5	6.7 ± 0.5
Placental hyaluronan, mg/g tissue[d]	1.06 ± 0.11	0.87 ± 0.11	1.08 ± 0.10	1.18 ± 0.10

[a]Number of observations is 16 for glucosamine and 17 for glucose. Least squares means for fetal weight, placental weight, serum glucose and fructose and placental hyaluronan for the smallest and largest fetus in glucosamine and glucose-treated gilts from Exp. 1 are presented
[b]Effect of fetal size ($P < 0.01$)
[c]Effect of fetal size ($P < 0.05$)
[d]Placental hyaluronan results from one sow with very high hyaluronan (> 3 standard deviations above mean) were deleted from the analysis

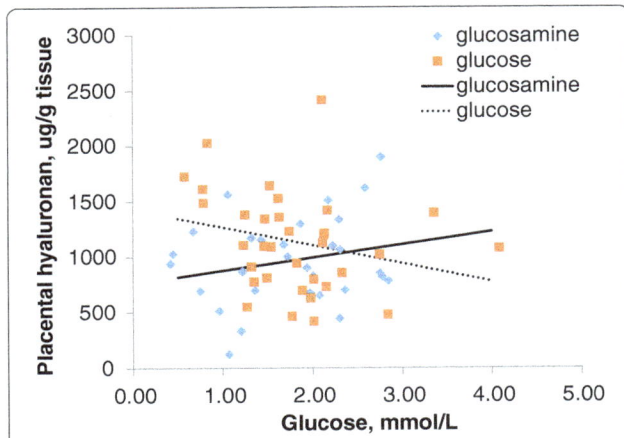

Fig. 3 Effect of glucose on hyaluronan by treatment. Heterogeneity of regression indicated that the two lines differed (glucosamine supplemented: hyaluronan = 760.44 + 115.77(glucose); glucose supplemented: hyaluronan = 1427−162.64; $P < 0.05$ for intercept; $P = 0.06$ for slope)

Experiment 3

There was an overall effect of treatment ($P < 0.05$; Total piglets born 17.9 ± 0.5 for glucosamine, 16.5 ± 0.5 for glucose; Number born alive 15.9 ± 0.4 for glucosamine, 14.6 ± 0.4 for glucose) and a treatment by parity interaction for total piglets born and piglets born alive ($P < 0.01$; Table 6). Orthogonal contrasts indicated that glucosamine treatment resulted in greater total piglets born and born alive in later parities (5 and 6) compared to early parities and parity 7. There were no effects of treatment on the number of stillbirths or mummies. Analysis of birth and weaning weights indicated that glucosamine had no effect on weights compared to glucose treatment. Finally, there was no difference in pre-weaning mortality between the two treatment groups.

Discussion

Results indicated that glucosamine supplementation tended to improve the number of live fetuses in UHO gilts (a measure of uterine capacity), altered the microscopic architecture of the developing pig placenta, and changed relationships between fetal size, fetal serum glucose and fructose, and placental hyaluronan. In Exp. 2, 10 g/d glucosamine supplementation of intact sows in a commercial herd did not result in a beneficial effect at the dose used in gilts. One possible contributing factor to the lack of treatment effect on the number of piglets born alive in Exp. 2 was an increased number of still-born piglets in late parity sows treated with glucosamine. Comparisons of total born, born alive and birth weights among parities confirmed that reductions in each occurred in parity 7 and 8 sows, so it is possible that some interaction between the reproductive competence of late parity sows and glucosamine supplementation may explain the increase in stillbirth incidence. Because the dose used in Exp. 2 may not have been sufficient in the larger sows to observe an effect on litter size, the trial was repeated in Exp. 3 using 20 g/d. Results of Exp. 3 indicate a substantial (+1.3 piglets born alive overall) gain in litter size compared to glucose treatment and the effect was greater in later parity sows.

Glucose and fructose are part of the pathway that results in glucosamine synthesis [15]. In Exp. 1, serum glucose and fructose concentrations were measured to determine whether they were altered by glucosamine supplementation. Results indicated significant relationships for both glucose and fructose with fetal and placental weights. Glucose concentrations were negatively related to fetal weight and positively related to placental weight. These relationships are consistent with the concept that within the conceptus, glucose originates from the placenta and is used by the fetus [16, 17]. These relationships were unaffected by treatment. In contrast, fetal and placental weight relationships with fructose concentrations were affected by treatment. Glucose treated gilts may be considered the "normal" fetal and placental weight relationships with fructose because a daily dose of 10 or 20 g glucose added to a diet of 2 kg/d of a corn-soy diet would not be expected to add much available glucose to the gestating gilt (the corn-soybean based daily diet contains at least 1 kg starch, which would be converted to glucose during digestion). Curiously, in

Table 2 Treatment effects on placental morphometry from Experiment 1

Variable	Glucosamine[a]		Glucose	
	Large	Small	Large	Small
Fold Width, µm[b]	778 ± 37	818 ± 37	716 ± 36	731 ± 36
Stromal Width, µm[c]	236 ± 28	142 ± 28	140 ± 27	146 ± 27
Total Width, µm[b]	1014 ± 47	961 ± 47	856 ± 46	876 ± 46
Interface length/unit placental length[c]	7470 ± 403	7546 ± 403	6969 ± 389	6727 ± 389

[a]Number of observations is 16 for glucosamine and 17 for glucose. Least squares means for bilayer fold width, stromal width above the folded bilayer, total placental width and placental bilayer interface length per unit placental length from Exp. 1 are presented
[b]Effect of treatment ($P ≤ 0.05$)
[c]Effect of treatment by fetal size ($P = 0.07$)

Table 3 Treatment by parity effect on number stillborn and stillbirth rate from Experiment 2

Parity[a]	Treatment			
	Glucosamine		Glucose	
	Number stillborn	Stillbirth rate	Number stillborn	Stillbirth rate
2	1.0 ± 0.3 (20)[b]	0.06 ± 0.02	0.8 ± 0.3 (20)	0.06 ± 0.02
3	1.4 ± 0.3 (30)	0.08 ± 0.02	1.0 ± 0.3 (30)	0.07 ± 0.01
4	0.9 ± 0.4 (15)	0.06 ± 0.02	1.3 ± 0.4 (16)	0.07 ± 0.02
5	1.8 ± 0.4 (14)	0.09 ± 0.02	1.8 ± 0.5 (11)	0.11 ± 0.03
6	1.8 ± 0.4 (14)	0.10 ± 0.03	2.2 ± 0.4 (15)	0.11 ± 0.03
7	2.1 ± 0.3 (23)	0.14 ± 0.03	1.5 ± 0.3 (23)	0.10 ± 0.02
8	2.6 ± 0.4 (12)	0.22 ± 0.04	1.2 ± 0.4 (12)	0.09 ± 0.03

[a]For both traits, the interaction contrast comparing the interaction between treatments and parities 2 through 6 combined versus parities 7 and 8 combined was statistically significant ($P < 0.05$), indicating that more stillborn piglets occurred in the glucosamine-treated parity 7 and 8 sows compared to earlier parity sows. Least squares means for the treatment by parity interaction for number of stillborn piglets and stillbirth rate from Exp. 2 are presented
[b]Numbers of observations are in parentheses

glucose-treated gilts at d 105 of gestation, fetal plasma fructose was positively related to fetal weights and negatively related to placental weights. Previous reports indicate that both the placenta and the fetus are sources of fructose during gestation [18, 19]. Results from glucose supplemented gilts suggest that at d 105 of gestation, the placenta may be a net consumer of fructose and the fetus may be a net producer of fructose. Glucosamine supplementation resulted in reducing the positive relationship with fetal weight, and converting the negative relationship with placental weight to a positive one, essentially stabilizing fructose concentrations as fetal and placental weights vary (Fig. 2). These changes could be consistent with the role of fructose in the synthesis of glucosamine [15]. Thus, Fig. 2 is consistent with the concept that providing exogenous glucosamine might reduce fructose use by the placenta and stabilize fructose concentrations.

Table 4 Treatment effects on litter traits from Experiment 2

Variable	Treatment[a]	
	Glucosamine	Glucose
Total born	15.6 ± 0.4 (128)	15.2 ± 0.4 (127)
Born alive	14.0 ± 0.3	13.8 ± 0.3
Mummies	0.56 ± 0.08	0.57 ± 0.08
Birth weights	1.37 ± 0.02 (1970)	1.35 ± 0.02 (1915)
Weaning weights	5.48 ± 0.06	5.43 ± 0.06
Preweaning mortality	0.17 ± 0.01	0.16 ± 0.01

[a]Differences between glucosamine and glucose supplementation were not statistically significant. Number of observations are in parentheses. Litter size trait, birth and weaning weight least squares means for glucosamine and glucose supplemented sows from Exp. 2 are presented

Hyaluronan is a major component of the placental stroma [7, 8]. Glucosamine is a major component of hyaluronan, the other component being glucuronic acid [9]. The folded bilayer interface undergoes extensive remodeling during late gestation (d 85 onward), characterized by increased stromal development between individual folds and development of secondary folds [7, 13]. These processes likely require both synthesis of new hyaluronan and turnover of existing hyaluronan. This is likely to require the synthesis of both glucuronic acid and glucosamine, which both originate from glucose [20], although fructose is an intermediate in glucosamine synthesis [15]. Paradoxically, glucosamine supplementation altered the relationship between serum glucose and placental hyaluronan, and no relationship between fructose and hyaluronan was demonstrated, despite fructose being an intermediate in glucosamine synthesis. However, no difference in placental hyaluronan occurred between treatments. It is possible that glucosamine altered the relationship between hyaluronan and glucose by improving turnover without affecting fructose, which may have made the placenta more efficient at glucose transport, resulting in a positive relationship. This possible mechanism would be supported by changes in the microscopic architecture of the placental folds.

Results indicated increased total thickness of the placenta, increased width of the folded bilayer, and increased stromal depth above the folded bilayer in placenta of large fetuses. These changes are all consistent with the concept that glucosamine encourages stromal development, and that this in turn provides the substrate needed to encourage folded bilayer development. Given our previous hypothesis [7], one would expect improvement in placental transport of nutrients. However, if nutrient transport was improved, one might expect an increase in fetal weight, especially for small fetuses. No change in the weight of the smallest fetus was observed. On the other hand, it is possible that there is a threshold weight of the smallest living fetus for fetal survival. If this is true, the weight of the smallest living fetus might be unchanged, but more fetuses would survive because more fetuses would be above the survival threshold as a result of improved transport. Thus, the trend in greater litter size in glucosamine-treated gilts observed in this experiment supports the hypothesis that the improvements in fold development resulted in improved nutrient transport that then increased litter size.

The UHO surgical procedure removes one ovary and one uterine horn, and the remaining ovary undergoes compensatory hypertrophy such that ovulation rate is unaffected. This results in the same number of available embryos in half the uterine space, and litter size in UHO gilts is considered to be a direct measure of one-half uterine capacity [5]. Results of experiment 1 indicated a

Table 5 Parity effects on litter traits from Experiment 2

| | Parity | | | | | | |
Variable	2	3	4	5	6	7	8
Total born[a]	14.7 ± 0.6(40)[b]	15.3 ± 0.5(60)	15.5 ± 0.7(31)	16.4 ± 0.8(25)	17.3 ± 0.7(29)	14.4 ± 0.6(46)	14.5 ± 0.8(24)
Born alive[a]	13.8 ± 0.6	14.1 ± 0.5	14.4 ± 0.6	14.6 ± 0.7	15.3 ± 0.7	12.6 ± 0.5	12.6 ± 0.7
Mummies	0.29 ± 0.14	0.59 ± 0.12	0.64 ± 0.16	0.53 ± 0.18	0.72 ± 0.17	0.69 ± 0.13	0.48 ± 0.18
Birth weights[c]	1.45 ± 0.04	1.45 ± 0.03	1.40 ± 0.04	1.39 ± 0.04	1.33 ± 0.04	1.34 ± 0.03	1.33 ± 0.05
Weaning weights[d]	5.29 ± 0.09	5.30 ± 0.08	5.34 ± 0.12	5.38 ± 0.12	5.57 ± 0.11	5.67 ± 0.09	5.64 ± 0.13
Preweaning mortality	0.14 ± 0.02	0.15 ± 0.02	0.17 ± 0.02	0.18 ± 0.02	0.16 ± 0.02	0.22 ± 0.02	0.19 ± 0.03

[a]Contrasts indicated a progressive increase to parity 6, followed by a precipitous decrease ($P < 0.05$). Litter size trait, birth and weaning weight least squares means for second through eighth parity sows from Exp. 2 are presented
[b]Number of observations are in parentheses
[c]Contrasts indicated no differences among parities 2 through 6, and a decrease in parities 7 and 8 ($P < 0.01$)
[d] Contrasts indicated a progressive increase from parity 2 to 8 ($P < 0.05$)

numerical increase in litter size of 1.4 live fetuses on 105 d of gestation, so it was conceivable that uterine capacity was increased by nearly 3 live fetuses. The UHO measure was uncomplicated by the incidence of stillbirth, because the measure was made at slaughter on d 105 of gestation. Previous selection for uterine capacity using the UHO model resulted in an increase in uterine capacity of approximately one fetus per uterine horn, and a significant increase in litter size in intact selected gilts, although the effect on litter size was less due to the influence of ovulation rate on litter size in intact gilts [21]. Nevertheless, given the results of Exp. 1, it seemed likely that glucosamine would increase litter size in intact pigs, thus warranting a larger test in intact sows.

Results of Exp. 2 indicated no significant effect of 10 g/d of glucosamine supplementation on litter size in sows ranging in parity from 2 to 8 in contrast to the results of Exp. 1 in which a trend toward improved litter size was obtained. In Exp. 2, we used the same dose of

glucosamine for sows as for the gilts in Exp. 1. However, because they were not as old, the gilts in Exp. 1 weighed less than the sows used in Exp. 2. To explore whether a greater dose would be effective, we performed a second commercial trial using 20 g/d glucosamine. Results of Exp. 3 indicated that a dose of 20 g/d in sows was effective in increasing both the total number of piglets born and the number born alive with greater increases occurring in later parity (5 and 6) sows. Analysis of birth and weaning weights indicated no change in piglet weights despite the increase in litter size of greater than 1 piglet per litter; therefore, the increase in litter size occurred without a depression in birth weights. Finally, there was no difference in preweaning mortality, indicating that the increase in litter size should have resulted in an increase in the number of weaned piglets.

An obvious complicating factor in Exp. 2 and 3 is the ovulation rate of the sows used, which we were not able to measure because of the commercial setting. Previous

Table 6 Treatment by parity effects on litter traits from experiment 3

Treatment	Total born[a]	Born alive[a]	Stillborns	Mummies	Birth weight	Weaning weight	Preweaning mortality
Glucosamine							
Parity 3	17.9 ± 0.7	15.9 ± 0.7	1.3 ± 0.3	0.8 ± 0.2	1.40 ± 0.04	5.0 ± 0.1	0.11 ± 0.02
Parity 4	16.8 ± 0.9	14.9 ± 0.8	1.3 ± 0.3	0.5 ± 0.2	1.41 ± 0.05	4.9 ± 0.1	0.15 ± 0.02
Parity 5	20.0 ± 1.2	17.2 ± 1.0	1.7 ± 0.5	1.1 ± 0.3	1.26 ± 0.06	5.1 ± 0.2	0.16 ± 0.03
Parity 6	18.7 ± 1.2	17.2 ± 1.1	0.5 ± 0.5	1.0 ± 0.3	1.32 ± 0.07	5.1 ± 0.2	0.17 ± 0.03
Parity 7	16.2 ± 1.0	14.3 ± 0.9	1.5 ± 0.4	0.3 ± 0.2	1.37 ± 0.06	5.2 ± 0.2	0.11 ± 0.02
Glucose							
Parity 3	17.6 ± 0.8	16.0 ± 0.7	1.0 ± 0.3	0.6 ± 0.2	1.36 ± 0.04	5.1 ± 0.1	0.11 ± 0.03
Parity 4	17.9 ± 0.9	15.8 ± 0.8	1.5 ± 0.4	0.6 ± 0.2	1.41 ± 0.05	4.8 ± 0.2	0.18 ± 0.03
Parity 5	15.8 ± 1.1	14.7 ± 1.0	0.9 ± 0.4	0.2 ± 0.3	1.42 ± 0.06	4.8 ± 0.2	0.12 ± 0.03
Parity 6	15.1 ± 1.1	12.1 ± 1.0	2.4 ± 0.4	0.6 ± 0.3	1.30 ± 0.06	5.0 ± 0.2	0.24 ± 0.04
Parity 7	15.8 ± 1.1	14.3 ± 0.9	1.1 ± 0.4	0.5 ± 0.3	1.41 ± 0.06	4.8 ± 0.2	0.08 ± 0.02

[a]A treatment by parity interaction was present. Orthogonal contrasts indicated that total born and born alive were greater in later parities (5 and 6) in glucosamine treated sows compared to glucose treated sows, but were not different in early parities or in parity 7. Litter size trait, birth and weaning weight least squares means for glucosamine- and glucose-supplemented sows for each parity from Exp. 3 are presented

observations of ovulation rates in commercial sows have indicated very high ovulation rates, particularly in later parity sows [22]. High ovulation rates in parity 5 and 6 sows may explain the interaction effect observed. If similar ovulation rates to those previously reported occurred in the parity 5 and 6 sows used in our commercial trials, they are likely to have been sufficient to provide a good test of uterine capacity in intact females. Ovulation rates in earlier parity animals may not have been sufficient, resulting in the interaction observed. The commercial farm in this experiment obtains maternal line gilts from DNA genetics (Columbus, NE; formerly Danbred USA), but we could find no published estimates of ovulation rates in these maternal line gilts. However, there are published litter size estimates for Danish maternal line gilts, demonstrating excellent genetic progress in litter size selection [23], and it is likely that high ovulation rates contributed to the reported increase in litter size due to selection. Thus, ovulation rates were likely to be high enough in the parity 5 and 6 sows to put greater emphasis on uterine capacity as a determining factor for litter size. Exception to this may have been the late parity sows in Exp. 2 and 3 (parity 7 and 8). In Exp. 2 these sows clearly had reduced numbers of total born and born alive piglets, and one contributing factor to this decrease could have been reduced ovulation rates. We could not find any published reports of ovulation rates specifically in sows in later parities for commercial herds, primarily because previous reports combined ovulation rate estimates of parities 4 or greater [22]. Thus, whether ovulation rate decreased in parity 7 and 8 sows under some conditions, resulting in lower litter size, remains unknown.

The increase in the number of stillborn piglets and stillbirth rates in parity 7 and 8 sows supplemented with glucosamine in Exp. 2 was an unexpected result. Results of Exp. 1 indicated that glucosamine supplementation increased the depth of the folded bilayer in the pig placenta regardless of the size of the fetus, and because of this the total width of the placenta was increased. Van Rens and Van der Lende [24] implicated a thicker placenta in prolongation of individual piglet birth intervals, and suggested that placental thickness may contribute to stillbirth rate due to the well-known relationship between piglet birth intervals and stillbirth [25]. Van Rens and Van der Lende [24] suggested that thicker placentas may present a greater barrier to delivery of the piglet during farrowing, increasing birth intervals. However, the effective width of the placenta is likely to be the width of the stroma above the folded bilayer, which we have reported to be reduced in small fetuses compared to large fetuses [7]. Results of Exp. 1 confirm that the stroma above the folds is greater in placenta of large fetuses in gilts supplemented with glucosamine. The

incidence of larger fetuses, and therefore thicker placenta, might be expected to increase with decreasing litter size. It is possible that the reduced litter size in parity 7 and 8 sows observed in Exp. 2 resulted in thicker placenta in these sows upon glucosamine supplementation, which could have prolonged birth intervals (which we did not measure), and increased stillbirth. Whatever the mechanism for the increase in stillbirth rate, results from Exp. 2 suggest that glucosamine supplementation may be detrimental in parity 7 and 8 sows due to increased stillbirth rate.

Conclusions

Maternal supplementation with glucosamine altered fetal serum glucose and fructose dynamics and increased the depth of the placental folded bilayer in placenta of UHO gilts. Results are consistent with the concept that glucosamine supplementation may have stabilized fetal fructose concentrations as fetal and placental weights vary among conceptuses. The placental epithelial bilayer fold-depth changes occurred with a trend toward increased uterine capacity of about 1.4 fetuses per uterine horn measured in UHO gilts. In commercial trials, glucosamine supplementation in intact sows resulted in a slight numerical but nonsignificant increase in the number of piglets born and born alive at a dose of 10 g/d glucosamine and a significant increase in litter size of greater than 1 piglet per litter at a dose of 20 g/d. The increased litter size occurred in later parity 5 and 6 sows. Thus, it is likely that 10 g/d glucosamine was not sufficient in commercial sows due to their larger size compared to the gilts in Exp. 1. The increase in litter size from a dose of 20 g/d occurred without reductions in birth or weaning weights and with no reduction in preweaning mortality. Our results further indicated an increase in the number of stillborn piglets and stillbirth rate in glucosamine-treated late parity sows (parity 7 and 8) in Exp. 2, which was unexpected. This may have been due to the effect of glucosamine supplementation on placental thickness, which could have occurred due to the reduced litter sizes in these late parity sows. These results suggest that glucosamine supplementation at a dose of 20 g/d could be useful in increasing litter size in commercial parity 5 and 6 sows, resulting in increased number of piglets weaned. Because the mechanism of the increase is likely to be based on improved placental function and uterine capacity, an increase would only be expected if ovulation rates are sufficient.

Abbreviations

GLUT: Glucose transporter; UHO: Unilaterally hysterectomized ovariectomized; USMARC: US Meat Animal Research Center

Acknowledgements

The authors wish to acknowledge with thanks the efforts of L. Parnell in preparing this manuscript, N. Felder, J. Wood, E. Wright and T. McCain in Exp. 1 and the swine staff at Plymouth Ag group without whose help Exp. 2 and

3 would not have been accomplished. In addition, we would like to acknowledge the efforts of T. DeVries in Exp. 2 and J. Bornschlegl in Exp. 3.

[1]Mention of trade names or commercial products in this article is solely for the purpose of providing specific information and does not imply recommendation or endorsement by the USDA.

[3]The U.S. Department of Agriculture (USDA) prohibits discrimination in all its programs and activities on the basis of race, color, national origin, age, disability, and where applicable, sex, marital status, familial status, parental status, religion, sexual orientation, genetic information, political beliefs, reprisal, or because all or part of an individual's income is derived from any public assistance program. (Not all prohibited bases apply to all programs.) Persons with disabilities who require alternative means for communication of program information (Braille, large print, audiotape, etc.) should contact USDA's TARGET Center at (202) 720-2600 (voice and TDD). To file a complaint of discrimination, write to USDA, Director, Office of Civil Rights, 1400 Independence Avenue, S.W., Washington, D.C. 20250-9410, or call (800) 795-3272 (voice) or (202) 720-6382 (TDD). USDA is an equal opportunity provider and employer.

Funding

This work was partially supported by the Nebraska Pork Producers Council [grant number 14–238].

Authors' contributions

All authors read and approved the final manuscript.

Authors' information

Not applicable.

Competing interests

Not applicable.

Author details

[1]USDA, ARS, U.S. Meat Animal Research Center (USMARC), P.O. Box 166, Clay Center NE, Nebraska 68933, USA. [2]Plymouth Ag Group, Diller NE, Nebraska 68342, USA.

References

1. Bennett GL, Leymaster KA. Integration of ovulation rate, potential embryonic viability and uterine capacity into a model of litter size in swine. J Anim Sci. 1989;67(5):1230–41.
2. Polge C. Fertilization in the pig and horse. J Reprod Fertil. 1978;54(2):461–70.
3. Freking BA, Leymaster KA, Vallet JL, Christenson RK. Number of fetuses and conceptus growth throughout gestation in lines of pigs selected for ovulation rate or uterine capacity. J Anim Sci. 2007;85(9):2093–103.
4. Johnson RK, Nielsen MK, Casey DS. Responses in ovulation rate, embryonic survival, and litter traits in swine to 14 generations of selection to increase litter size. J Anim Sci. 1999;77(3):541–57.
5. Christenson RK, Leymaster KA, Young LD. Justification of unilateral hysterectomy-ovariectomy as a model to evaluate uterine capacity in swine. J Anim Sci. 1987;65(3):738–44.
6. Vallet JL, McNeel AK, Johnson G, Bazer FW. Triennial reproduction symposium: limitations in uterine and conceptus physiology that lead to fetal losses. J Anim Sci. 2013;91(7):3030–40.
7. Vallet JL, Freking BA. Differences in placental structure during gestation associated with large and small pig fetuses. J Anim Sci. 2007;85(12):3267–75.
8. Steele VS, Froseth JA. Effect of gestational age on the biochemical composition of porcine placental glycosaminoglycans. Proc Soc Exp Biol Med. 1980;165(3):480–5.
9. Vallet JL, Miles JR, Freking BA. Effect of fetal size on fetal placental hyaluronan and hyaluronoglucosaminidases throughout gestation in the pig. Anim Reprod Sci. 2010;118(2–4):297–309.
10. Toole BP. Hyaluronan is not just a goo! J Clin Invest. 2000;106(3):335–6.
11. Bazer FW, Gao H, Johnson GA, Wu G, Bailey DW, Burghardt RC. Select nutrients and glucose transporters in pig uteri and conceptuses. Soc Reprod Fertil Suppl. 2009;66:335–6.
12. Uldry M, Ibberson M, Hosokawa M, Thorens B. GLUT2 is a high affinity glucosamine transporter. FEBS Lett. 2002;524(1–3):199–203.
13. Friess AE, Sinowatz F, Skolek-Winnisch R, Träutner W. Structure of the epitheliochorial porcine placenta. Bibl Anat. 1982;22:140–3.
14. Zavy MT, Clark WR, Sharp DC, Roberts RM, Bazer FW. Comparison of glucose, fructose, ascorbic acid and glucosephosphate isomerase enzymatic activity in uterine flushings from nonpregnant and pregnant gilts and pony mares. Biol Reprod. 1982;27(5):1147–58.
15. Milewski S. Glucosamine-6-phosphate synthase—the multi-facets enzyme. Biochim Biophys Acta. 2002;1597(2):173–92.
16. Ford SP, Reynolds LP, Ferrell CL. Blood flow, steroid secretion and nutrient uptake of the gravid uterus during the periparturient period in sows. J Anim Sci. 1984;59(4):1085–91.
17. Père M-C, Etienne M. Uterine blood flow in sows: effects of pregnancy stage and litter size. Reprod Nutr Dev. 2000;40(4):369–82.
18. Père MC. Maternal and fetal blood levels of glucose, lactate, fructose, and insulin in the conscious pig. J Anim Sci. 1995;73(10):2994–9.
19. Rama F, Castellano MA, Germino NI, Micucci M, Ohanian C. Histochemical location of ketose-reductase in the placenta and fetal tissues. J Anat. 1973; 114(Pt 1):109–13.
20. Egger S, Chaikuad A, Kavanagh Kathryn L, Oppermann U, Nidetzky B. UDP-glucose dehydrogenase: structure and function of a potential drug target. Biochem Soc Trans. 2010;38(5):1378–85.
21. Christenson RK, Leymaster KA. Correlated responses in gravid uterine, farrowing and weaning traits to selection of pigs for ovulation rate or uterine capacity. World Congr Genet Appl Lvstk Prod (Proc. 7th). 2002: Comm. No. 08–25.
22. Town SC, Patterson JL, Pereira CZ, Gourley G, Foxcroft GR. Embryonic and fetal development in a commercial dam-line genotype. Anim Reprod Sci. 2005;85(3–4):301–16.
23. Nielsen B, Su G, Lund MS, Madsen P. Selection for increased number of piglets at d 5 after farrowing has increased litter size and reduced piglet mortality. J Anim Sci. 2013;91(6):2575–82.
24. van Rens BTTM, van der Lende T. Parturition in gilts: duration of farrowing, birth intervals and placenta expulsion in relation to maternal, piglet and placental traits. Theriogenology. 2004;62(1–2):331–52.
25. Vallet JL, Miles JR, Brown-Brandl TM, Nienaber JA. Proportion of the litter farrowed, litter size, and progesterone and estradiol effects on piglet birth intervals and stillbirths. Anim Reprod Sci. 2010;119(1–2):68–75.
26. Federation of Animal Science Societies (FASS). Guide for the Care and Use of Agricultural Animals in Research and Teaching, Third Edition. Champaign, IL: Federation of Animal Science Societies 2010.

15

Betaine affects muscle lipid metabolism via regulating the fatty acid uptake and oxidation in finishing pig

Sisi Li, Haichao Wang, Xinxia Wang, Yizhen Wang and Jie Feng[*]

Abstract

Background: Betaine affects fat metabolism in animals, but the specific mechanism is still not clear. The purpose of this study was to investigate possible mechanisms of betaine in altering lipid metabolism in muscle tissue in finishing pigs.

Methods: A total of 120 crossbred gilts (Landrace × Yorkshire × Duroc) with an average initial body weight of 70. 1 kg were randomly allotted to three dietary treatments. The treatments included a corn–soybean meal basal diet supplemented with 0, 1250 or 2500 mg/kg betaine. The feeding experiment lasted 42 d.

Results: Betaine addition to the diet significantly increased the concentration of free fatty acids (FFA) in muscle ($P < 0.05$). Furthermore, the levels of serum cholesterol and high-density lipoprotein cholesterol were decreased ($P < 0.05$) and total cholesterol content was increased in muscle ($P < 0.05$) of betaine fed pigs. Experiments on genes involved in fatty acid transport showed that betaine increased expression of *lipoprotein lipase(LPL)*, *fatty acid translocase/cluster of differentiation (FAT/CD36)*, *fatty acid binding protein (FABP3)* and *fatty acid transport protein (FATP1)* ($P < 0.05$). The abundance of fatty acid transport protein and fatty acid binding protein were also increased by betaine ($P < 0.05$). As for the key factors involved in fatty acid oxidation, although betaine supplementation didn't affect the level of carnitine and malonyl-CoA, betaine increased mRNA and protein abundance of carnitine palmitransferase-1(CPT1) and phosphorylated-AMPK ($P < 0.05$).

Conclusions: The results suggested that betaine may promoted muscle fatty acid uptake via up-regulating the genes related to fatty acid transporter including *FAT/CD36*, *FATP1* and *FABP3*. On the other hand, betaine activated AMPK and up-regulated genes related to fatty acid oxidation including *PPARα* and *CPT1*. The underlying mechanism regulating fatty acid metabolism in pigs supplemented with betaine is associated with the up-regulation of genes involved in fatty acid transport and fatty acid oxidation.

Keywords: Betaine, Fatty acid intake, Fatty acid oxidation, Muscle, Pig

Background

Betaine is a derivative of the amino acid glycine with three chemically reactive methyl groups. Betaine is distributed widely in animals, plants and microorganisms, and it is also a metabolite of choline oxidation in animals [1]. The principal physiologic role of betaine is as a methyl group donor [2], which means betaine participates in many important biochemical pathways, including methionine-homocysteine cycle and the biosynthesis of many compounds such as carnitine, creatine and phospholipids. Since carnitine is required for transport of long chain fatty acids into mitochondria [3], scientists have paid much attention to effects of betaine on energy metabolism especially lipid metabolism in animals. Studies showed that dietary betaine supplementation affected energy partitioning in pigs [4, 5] and it's also widely reported that betaine promotes animal growth and decreases carcass fat percentage in finishing pigs [6–10]. Further investigations found that betaine supplementation could decrease hepatic triglyceride accumulation [11, 12] and prevent fatty liver in rats fed high-fat-diets

* Correspondence: fengj@zju.edu.cn
Key Laboratory of Animal Nutrition & Feed, Zhejiang Province, College of Animal Science, Zhejiang University, Hangzhou, China

[13, 14]. The intramuscular fat content in the longissimus muscle was increased when pigs were fed betaine [15, 16]. Madeira et al. [17] reported that betaine might be involved in the differential regulation of some key genes of lipid metabolism in muscle and subcutaneous adipose tissue. However, studies on the mechanism of betaine affecting lipid metabolism in muscle are lacking. Therefore, the objective of the present study was to investigate possible mechanisms of betaine in altering lipid metabolism in muscle tissue of finishing pigs.

Methods

Animals and treatments

The experiment protocol used in this study was approved by the Institutional Animal Care and Use Committee of Zhejiang University. A total of 120 crossbred gilts (Landrace × Yorkshire × Duroc) with an average initial body weight of 70.1 kg (SD 0.70 kg) were randomly allotted to three dietary treatments. Each treatment consisted of four pens replicates with 10 gilts per pen. The treatment diets included a corn–soybean meal basal (Table 1) supplemented with 0, 1250 mg/kg (Low Betaine) or 2500 mg/kg (High Betaine) betaine (provided by Healthy Husbandry Sci-tech Co., Ltd. Hangzhou, China) respectively at the expense of corn. The basal diet was formulated to meet or exceed the nutrient requirements of finishing pigs [18]. Chemical analyses of the basal diet were carried out according to the methods of AOAC [19]. The feeding experiment lasted 42 d after a 7-day adaptation period. All pigs were housed in a curtain-sided pig barn with concrete slotted floors. Feed and water were provided for ad libitum consumption throughout the experiment.

Table 1 Nutrition formulation of basic diet

Ingredients	%	Nutrient	%
Corn	67.83	Digestible energy, MJ/kg[a]	13.42
Soybean meal	23	Dry matter	87.09
Rapeseed meal	3	Crude protein	17.02
Wheat midding	3	Crude fat	3.98
CaHPO₄	1.5	Calcium	0.85
Limestone	1.0	Phosphorus	0.64
Salt	0.3	Lysine	0.92
Lysine	0.10	Met	0.27
Trace element premix[b]	0.25		
Vintamin premix[c]	0.02		

[a]All of the data were analyzed value except digestible energy which was calculated using swine NRC(2012) values
[b]Provided the following amounts per kilogram of diet: Fe (FeSO4·7H2O), 50 mg; Cu (CuSO4·5H2O), 5 mg; Mn (MnSO4·H2O), 5 mg; Zn (ZnSO4·7H2O), 50 mg; I (KI), 0.35 mg; Se (NaSe2O3), 0.15 mg
[c]Provided the following amounts per kilogram of diet: vitamin A, 3000 IU; vitamin D3, 610 IU; vitamin E, 20 IU; vitamin B2, 5 mg; vitamin B12, 0.021 mg; biotin, 0.1 mg; pantothenic acid, 10 mg; nicotinic acid, 15 mg

Sample collection

At the end of the trial, eighteen pigs (six from each dietary treatment) weighing about 111.8 kg (SD 2.08 kg) were selected to collect tissue samples. Following an overnight fast, pigs were stunned by electrical shock and bleeding. Individual blood samples were collected at slaughter during exsanguinations. After collection of blood, samples were kept at room temperature for 2 h and then centrifuged for 10 min at 3000×g at 4 °C. Serum was collected and frozen at −80 °C until subsequent analyses. Samples of longissimus muscle between the 6th and 7th rib were obtained on the left side of the carcass within 5 min after slaughter, and then snap frozen in liquid nitrogen and stored at −80 °C until subsequent analyses.

Analysis of lipid metabolites in serum

Serum concentration of high-density lipoprotein cholesterol (HDLC), total cholesterol (TC), free fatty acid (FFA) and triglyceride were measured with commercial assay kits (Nanjing Jiancheng Bio-engineering Institute, Code No. A112−2, A111−2, A042−1 and A110−2, respectively, Nanjing, China) following the manufacturer's instructions.

Muscle lipid metabolites analysis

A 10% muscle homogenate was prepared with a mixture of chloroform and formaldehyde (a volume ratio of 2:1). Then extracted at room temperature for 24 h [20]. The organic solvent layer was taken and the level of triglyceride in muscle was measured with commercial assay kit (Nanjing Jiancheng Bio-engineering Institute, A110−2, Nanjing, China). Before the levels of TC and FFA in muscle were measured by the kits (Nanjing Jiancheng Bio-engineering Institute, Code No. A112−2, A111−2, A042−1 and A110−2, respectively, Nanjing, China), muscle tissue was made homogenate with physiological saline. The concentrations of carnitine and malonyl-CoA were measured using ELISA kits (Biovol Technologies, Code No.50R−E.3088P & 50R−E.3035P, Shanghai, China) for porcine assay according to the instructions.

RT-PCR analysis

Total RNA was extracted from frozen porcine muscle tissue using the Trizol reagent as described by the manufacturer (Invitrogen). The RNA concentration and purity were determined by the NanoDrop ND-2000 spectrophotometer (Thermofisher, USA) and its integrity was confirmed by agarose gel electrophoresis. The cDNA synthesis was performed in a 10-μL reaction volume containing 2 μg total RNA using the SYBR PrimeScript™ RT-PCR kit with gDNA Eraser (Code No. RR047A, TaKaRa, Dalian, China). Genomic DNA is eliminated by treatment for 2 min at 42 °C with gDNA

Eraser, which has potent DNA degrading activity. Then a reverse-transcription reaction reagent is added that includes a component that completely inhibits DNA degradation activity, and the reverse-transcription reaction proceeds for 15 min at 37 °C. The abundance of the target genes was measured by quantitative real-time PCR, performed with the ABI Stepone Plus™ RT-PCR system (ABI Biotechnology, USA) using SYBR Premix Ex Taq™ (Tli RNaseH Plus) RT-RCR kit (TaKaRa, Dalian, China). Primers for the selected genes were synthesized commercially by Invitrogen (Shanghai, China), shown in Table 2. The reaction protocol comprised a cycle of 95 °C for 1 min, 40 cycles of 95 °C for 10 s and 64 °C for 25 s. The expression of the target genes were normalized by the endogenous housekeeping gene (β-actin) [21, 22]. Each sample was analyzed in triplicate and the PCR amplification efficiency was close to 100%. The gene expression was calculated by using the comparative ($2^{-\Delta\Delta Ct}$) method [23].

Western blot analysis

Protein form muscle samples was extracted by T-PER Tissue Protein Extraction Reagent containing protease inhibitor cocktail (Thermo Pierce, Code No.78510, USA), and quantified with BCA protein assay kit (Beyotime, Code No.P0010, Shanghai, China) according to kit instructions. Proteins were separated on SDS - PAGE gels (12%), and then electrophoretically transferred onto immobilon-P polyvinylidene fluoride membranes (PVDF membrane, Millipore, Code No. IPVH00010, America).

Membranes were blocked 1 h in Tris-buffered saline containing 5% nonfat-dried milk at room temperature. Membranes were then incubated overnight at 4 °C in blocking buffer containing primary antibodies (as shown in Table 3). A goat anti-rabbit IgG (H + L) Secondary antibody (Thermo Pierce, Code NO.31210, USA) with 1/5000 dilution was used in the detection of specific proteins. For loading control, β-actin antibody was used as control. In addition, the relative expression of p-AMPK was normalized with AMPK. Finally, Super Signal West Dura Extended Duration Substrate (Thermo Pierce, Code No. 34075, USA) was used to visualize the protein bands. Band intensities were determined by using BandScan 5.0 software.

The relative expressions of target proteins = (The optical density of target proteins/The optical density of β-actin).

Statistical analyses

Results were presented as means and standard deviations. Statistical analysis was performed by one-way analysis of variance (ANOVA) and the Duncan method was used to put up multiple comparison with the statistical software SPSS 19.0. In all analyses, the level of significant difference was set at $P < 0.05$.

Results

Betaine on serum lipid metabolites

As shown in Fig. 1, there was no significant difference in the levels of serum FFA and triglyceride in the pigs fed

Table 2 Primers of target genes for RT-PCR

Genes	GenBank accession	Primers sequences(5′ to 3′)	Product size, bp	Annealing temperature, °C
β-actin	XM_003124280.3	CCTGCGGCATCCACGAAAC TGTCGGCGATGCCTGGGTA	123	63
AMPKα2	AY159788.1	GGTCTGGTTCCTCAACACCTCA GGCTCTCCGCAGTGACAGAAT	90	63
PPARγ	NM_214379	GTGGAGACCGCCCAGGTTTG GGGAGGACTCTGGGTGGTTCA	108	64
LPL	NM_214286.1	CCCTATACAAGAGGGAACCGGAT CCGCCATCCAGTCGATAAACGT	138	63
CPT1	NM_001007191.1	GGACGAGGAGTCTCACCACTATGAC TCTTGAACGCGATGAGGGTGA	128	63
FATP1	NM_001083931.1	CCCTCTGCGTCGCTTTGATG GCTGCGGTCCCGGAAATACA	151	63
FAT/CD36	NM_001044622.1	CTGGTGCTGTCATTGGAGCAGT CTGTCTGTAAACTTCCGTGCCTGTT	160	63
FABP3	NM_001099931.1	CCAACATGACCAAGCCTACCACA ACAAGTTTGCCTCCATCCAGTGT	176	63
PPARα	NM_001044526.1	GGCTTACGGCAATGGCTTCA CGGTCTCCGCACCAAATGA	168	64

Table 3 The primary antibodies for Western blot

Primary antibody	Order numbers	Dilution	Size, kDa
Anti-Cardiac FABP	abca ab45966	1:1500	15
Anti-FATP1	abcam ab81875	1:2000	65
Anti-CPT1B	abcam ab104662	1:2000	88
Anti-Phospho-AMPK	Cell Signaling Technology 2535	1:1000	62
Anti-AMPKα	Cell Signaling Technology 5832	1:1000	62
β-actin (C4)	Santa Cruz SC-47778	1:1500	43

betaine compared with control group. Additionally, the concentration of HDLC and TC were significantly lower in the betaine treated pigs ($P < 0.05$).

Betaine on muscle lipid metabolites
The level of FFA and TC were markedly higher in muscle when pigs were fed betaine ($P < 0.05$, Fig. 2). Compared to the control group, the level of triglyceride in muscle was not affected by betaine addition ($P > 0.05$).

Key factors involved in muscle FFA intake
As shown in Fig. 3, the gene expression of *FAT/CD36*, *FATP1* and *PPARγ* ($P < 0.05$) were higher in betaine-fed groups than control group. The addition of 2500 mg/kg betaine markedly up-regulated the gene expression of *FABP3* and *LPL* ($P < 0.05$). In addition, the abundance of fatty acid transport protein and fatty acid binding protein were significantly increased by betaine supplementation ($P < 0.05$, Fig. 4).

Key factors involved in muscle FA oxidation
Betaine supplementation did not affect carnitine or malonyl-CoA in muscle compared to the control group ($P > 0.05$, Fig. 5).

The gene expression of *AMPKα2*, *PPARα* and *CPT1* were significantly higher in pigs fed with betaine than the control group. ($P < 0.05$, Fig. 6). Furthermore, betaine supplementation markedly increased the abundance of phosphorylated-AMPK and CPT1 in muscle ($P < 0.05$, Fig. 7).

Discussion
Fatty acid metabolism in muscle includes uptake, synthesis and oxidation [24–26], but the synthesis is at a slow rate [27]. The main source of fatty acid in muscle tissue includes transport from plasma and hydrolysis from chylomicron and very-low-density-lipoprotein (VLDL) with LPL. Our study found that the concentration of FFA was significantly increased in muscle when pigs

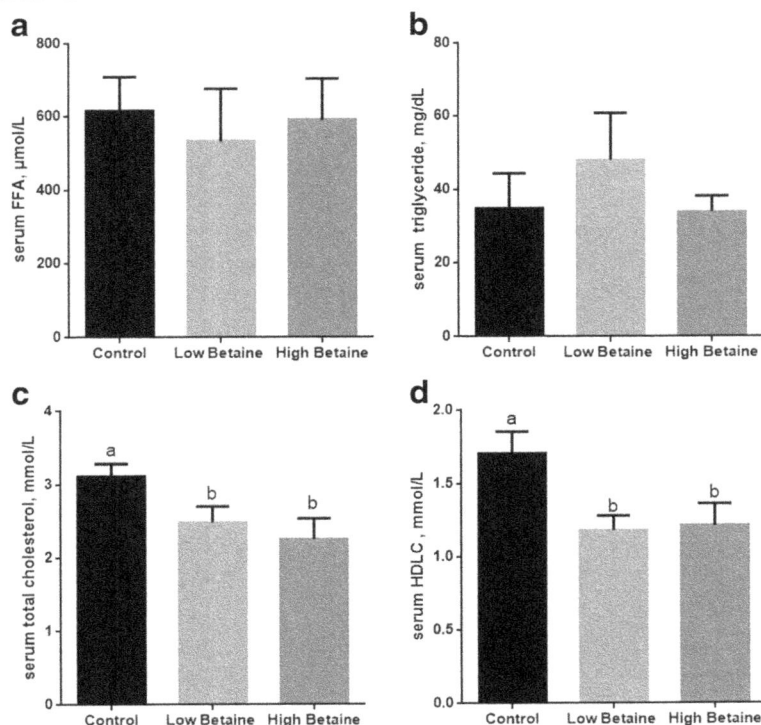

Fig. 1 Effect of betaine supplementation on serum parameters of lipid metabolism. The levels of serum free fatty acid (FFA, **a**), triglyceride (**b**), total cholesterol (**c**) and high-density lipoprotein cholesterol (HDLC, **d**). [a,b]Values without common superscript letters differ significantly ($P < 0.05$). Low betaine and high betaine represent 1250 mg/kg and 2500 mg/kg betaine addition, respectively

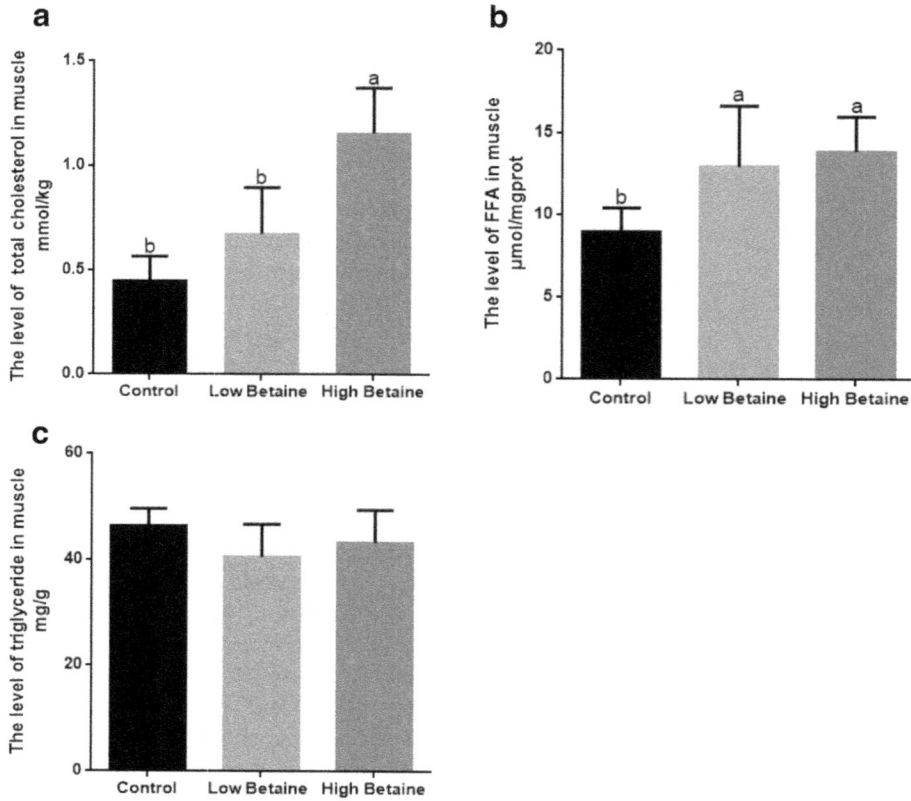

Fig. 2 Effect of betaine supplementation on total cholesterol, FFA and triglyceride in muscle. The levels of total cholesterol (**a**), free fatty acid (FFA, **b**) and triglyceride (**c**) in muscle. [a,b]Values without common superscript letters differ significantly (*P* < 0.05). Low betaine and high betaine represent 1250 mg/kg and 2500 mg/kg betaine addition, respectively

were fed betaine, similarly to the studies carried out by Yang et al. [28] and Fernández-Fígares et al. [29]. We speculated that the transport of FFA and/or the hydrolysis may be enhanced. More experiments were carried out regarding factors involved in fatty acid transport in muscle tissue. It is widely recognized that long chain fatty acid (LCFA) cross the plasma membrane via a protein-mediated mechanism. A number of fatty acid transporters have been identified, including fatty acid translocase/cluster of differentiation (FAT/CD36) and fatty acid transport proteins (FATP1) [30]. We found that betaine supplementation up-regulated gene expression for

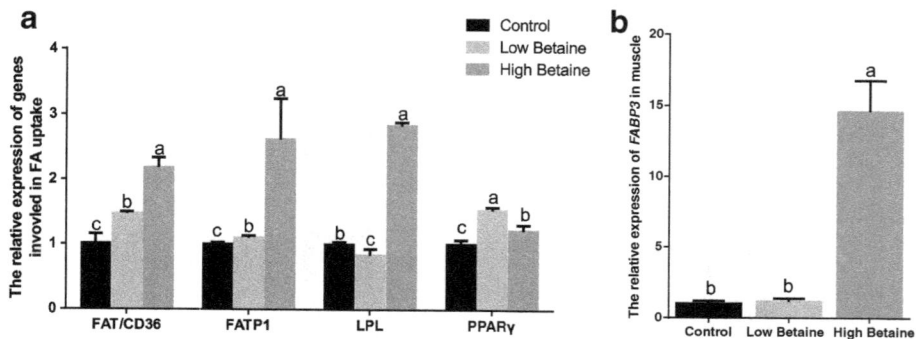

Fig. 3 The relative gene expression of key factors involved fatty acid uptake in muscle. mRNA expression was performed by RT-PCR and β-actin was chosen as reference gene. **a**The relative expression of *FAT/CD36*, *FATP1*, *LPL* and *PPARγ* in muscle, (**b**) The relative expression of *FABP3* in muscle. [a,b]Values without common superscript letters differ significantly (*P* < 0.05). Low betaine and high betaine represent 1250 mg/kg and 2500 mg/kg betaine addition, respectively

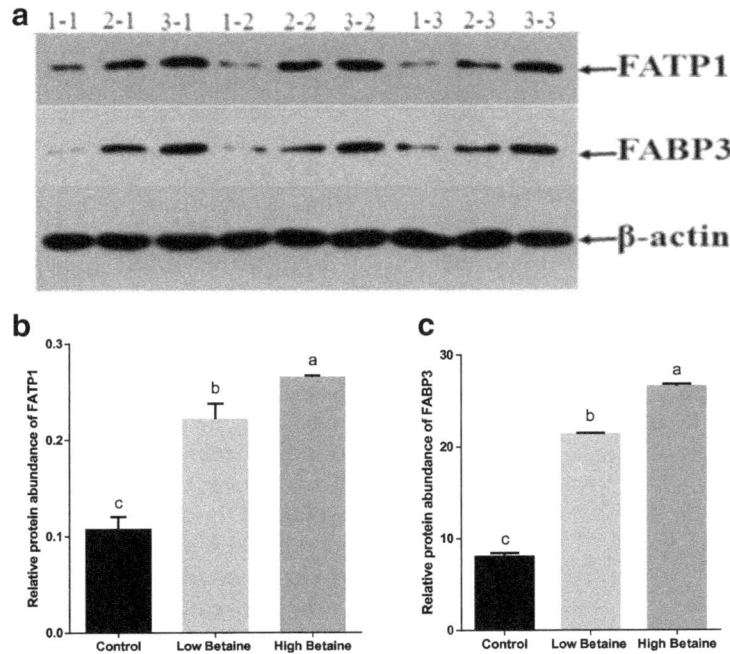

Fig. 4 The relative protein abundance of FATP1 and FABP3 in muscle. Western blot results were shown in **a** (The control group: 1–1, 1–2, 1–3; Low betaine group: 2–1, 2–2, 2–3; High betaine group: 3–1, 3–2, 3–3). Data were normalized with β-actin as shown in **b**, **c**. [a,b]Values without common superscript letters differ significantly ($P < 0.05$). Low betaine and high betaine represent 1250 mg/kg and 2500 mg/kg betaine addition, respectively

FATP1 and *FAT/CD36*. Experiments in vitro have shown that over expression of *FATP1* increased the uptake of LCFA in cells [31] and studies in vivo documented that muscle-specific over-expression of *FAT/CD36* enhanced cellular fatty acid uptake in mice [32]. FABP3, another important protein in fatty acid transportation, plays a role in transporting fatty acid from the sarcolemma to their intracellular sites of metabolism [33]. In muscle cells, the intracellular transport of LCFAs is facilitated to a great extent by FABP3 [34] Additionally, FABP3 is confirmed to be

associated with intramuscular fat in pigs [35]. Our studies showed that feeding betaine up-regulated the protein abundance of FABP3. In addition, the gene expression of *FABP3* was enhanced when pigs were fed with 2500 mg/kg betaine but no difference was found with 1250 mg/kg-betaine addition. The possible reason maybe that *FABP3* expression is translationally rather than transcriptionally regulated [36]. In summary, betaine may promote the uptake of fatty acids in muscle via regulating the expression of *FAT/CD36*, *FATP1* and *FABP3*. As mentioned above,

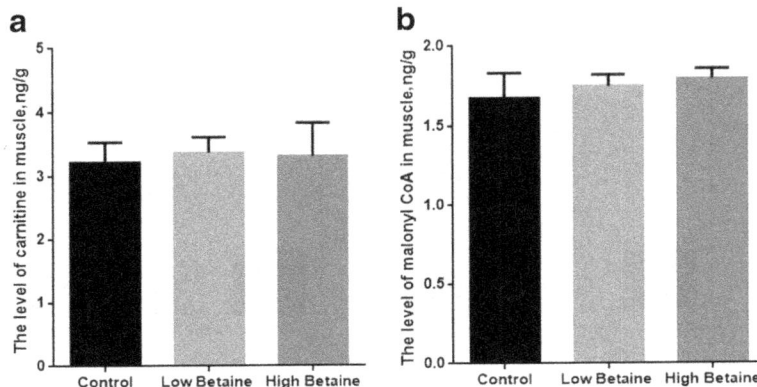

Fig. 5 Effect of betaine supplementation on the level of carnitine(**a**) and malonyl-CoA (**b**) in muscle. Low betaine and high betaine represent 1250 mg/kg and 2500 mg/kg betaine addition, respectively

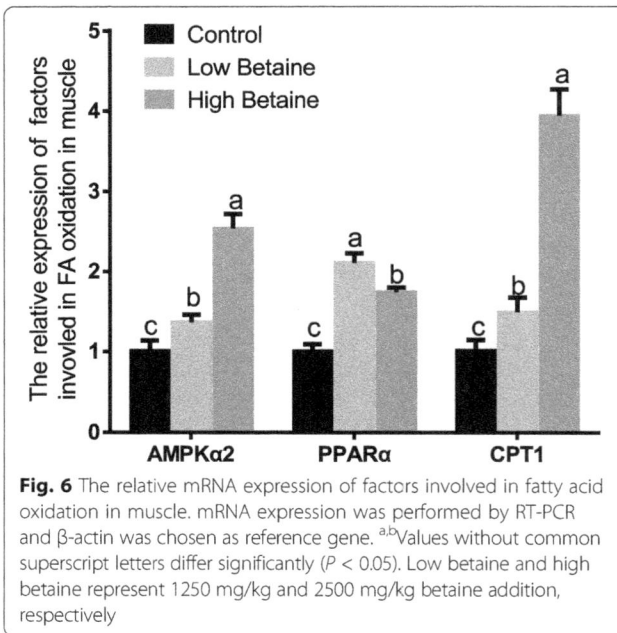

Fig. 6 The relative mRNA expression of factors involved in fatty acid oxidation in muscle. mRNA expression was performed by RT-PCR and β-actin was chosen as reference gene. [a,b]Values without common superscript letters differ significantly ($P < 0.05$). Low betaine and high betaine represent 1250 mg/kg and 2500 mg/kg betaine addition, respectively

LPL is the principal enzyme that hydrolyzes circulating triglycerides and it also can increase lipid uptake [37]. The results showed a significant increase in the gene expression of *LPL* with the addition of 2500 mg/kg betaine, which indicates betaine might enhance lipid uptake as well as chylomicron hydrolysis. The nuclear receptor PPARγ is a central regulator of adipose tissue development and an important modulator of expression in adipocytes [38]. To date, only a limited number of genes are known to be direct targets of PPARγ in adipose tissue. The majority of these encode proteins with direct links to lipid metabolism including LPL, FATP and FAT/CD36 [39, 40]. In present study, the gene expression of *PPARγ* was significantly higher in betaine-fed groups than the control group. We found that the effect of betaine on PPARγ was similar to its downstream target genes. All these results were similar to Albuquerque [41] and imply that betaine may facilitate fatty acids uptake in muscle via affecting key factors involved in FFA uptake, and the specific regulation mechanism needs more research.

The concentration of FFA in muscle tissue resultes from the balance of transport and oxidation. As a methyl donor, betaine participates in the biosynthesis of carnitine and because of this, betaine may be related to

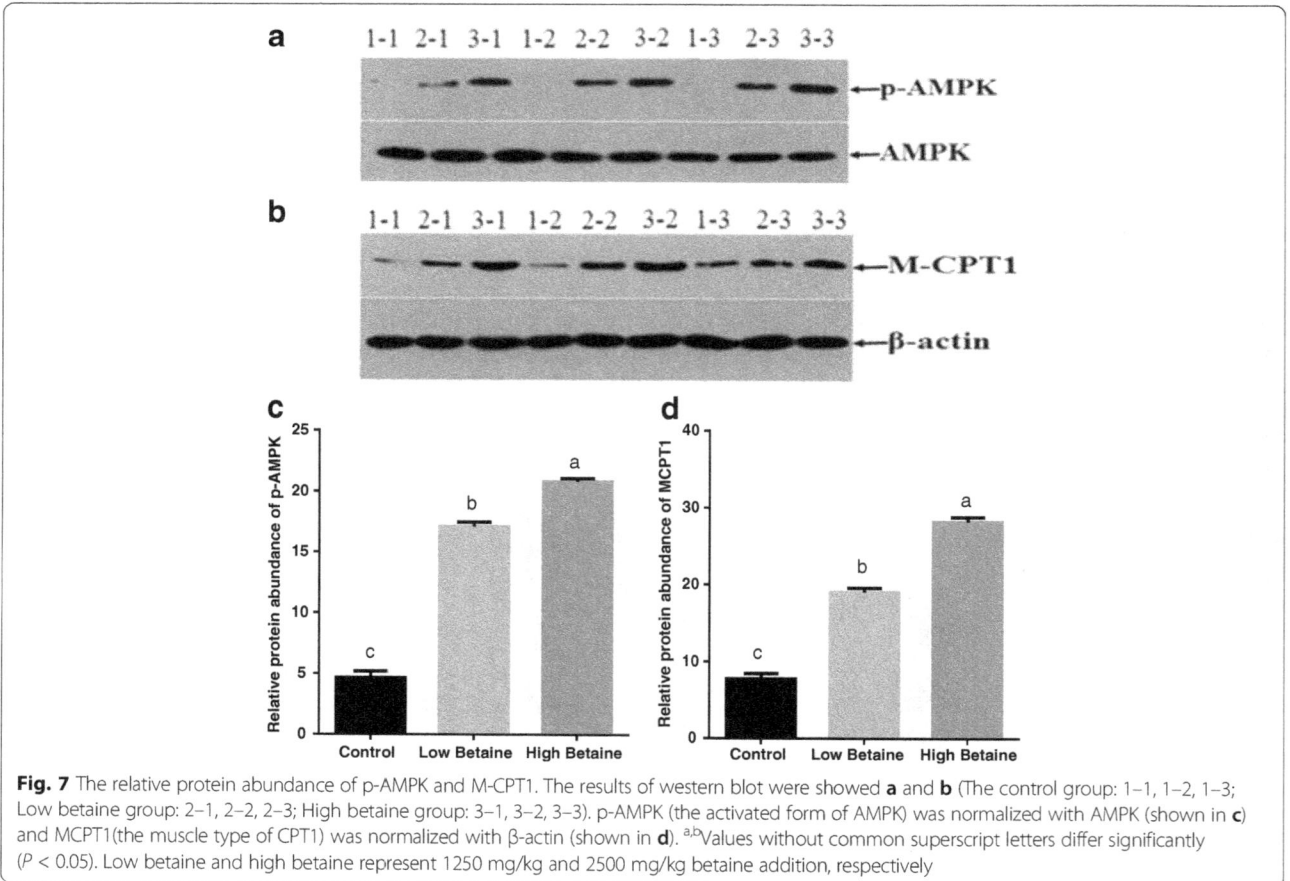

Fig. 7 The relative protein abundance of p-AMPK and M-CPT1. The results of western blot were showed **a** and **b** (The control group: 1–1, 1–2, 1–3; Low betaine group: 2–1, 2–2, 2–3; High betaine group: 3–1, 3–2, 3–3). p-AMPK (the activated form of AMPK) was normalized with AMPK (shown in **c**) and MCPT1(the muscle type of CPT1) was normalized with β-actin (shown in **d**). [a,b]Values without common superscript letters differ significantly ($P < 0.05$). Low betaine and high betaine represent 1250 mg/kg and 2500 mg/kg betaine addition, respectively

fatty acid β-oxidation. LCFAs are first transformed into acyl CoA, then transferred into mitochondria after combining with carnitine where it is oxidized. Carnitine palmitoyl transferase I (CPT1) is the rate-limiting enzyme that controls the step of combination and malony-CoA is an allosteric inhibitor of CPT1 [42]. Whereas the synthesis of malonyl-CoA is catalyzed by acetyl-CoA carboxylase (ACC), the activity of the ACC is regulated by phosphorylation of AMPK [43]. Hence, AMPK-ACC-CPT1 is an important signaling pathway to regulate fatty acid β-oxidation in mitochondria. Cai et al. [44] found that gestational dietary betaine supplementation down-regulated expression of *ACC* in neonatal piglets and Pekkinen et al. [11] found betaine supplementation had an impact on carnitine metabolism in high-fat-fed mice. Our experiment didn't find significant changes in muscle concentrations of malonyl-CoA or carnitine. The different results might be related to the different experiment condition and the mechanism needs to be further investigated. Increased gene expression and protein expression of *CPT1* were up-regulated with betaine addition, which implied betaine may enhance fatty acid β-oxidation in muscle tissue. However, others have shown betaine supplementation reduced the activity of CPT1 and mRNA abundance, and further increased IMF in finishing pigs [Duroc × (Seghers × Seghers)] [15]. We speculate that the effect of betaine addition on CPT1 might be influenced by breed and muscle type. In order to get a better understanding, we further analyzed effects of betaine on AMP-activated protein kinase (AMPK) and PPARα, which are both upstream regulatory factors of CPT1. AMPK is a crucial energy sensor for cells, which can promote the catabolism of fatty acids by enhancing their uptake into mitochondria and their consequent breakdown by beta-oxidation [45]. It was reported that activated AMPK in muscle enhances the gene expression of *PPARα* and *CPT1* [46], and CPT1 also seems to be a target of PPARα [47]. In the current experiment, the gene expression of both *PPARα* and *AMPK* were higher in betaine-fed groups as well as protein expression of p-AMPK (the activated form of AMPK). Similar to our previous report in rat liver [12], it can be inferred that betaine affected fatty acid oxidation in muscle via activating AMPK and up-regulated *PPARα* and *CPT1* gene expression.

The effect of betaine supplementation on cholesterol metabolism was of interest. The present study showed that betaine supplementation decreased the concentration of serum cholesterol and HDLC and increased cholesterol level in muscle, which was consistent with the studies by Albuquerque et al. [41] and Yang et al. [20]. However, Matthews et al. [48] and Martins et al. [49] reported that betaine supplemented pigs presented higher serum cholesterol. The efficacy of betaine in regulating the concentration of cholesterol in pigs shows variable results and seems to depend on both animal and dietary factors. Although the results were inconsistent, it seems to indicate that betaine might affect cholesterol partitioning or maybe enhances the transport of cholesterol in pigs, and more research is needed to clarify the specific mechanism.

Conclusions

In present study, betaine supplementation increased the level of free fatty acids in muscle, which may have resulted due to a change in the balance of fatty acid uptake and oxidation. Betaine may promote fatty acid uptake via increasing the expression of fatty acid transporters including FAT/CD36, FATP1 and FABP3 in muscle. Additionally, betaine activated AMPK and up-regulated *PPARα* and *CPT1*, and may enhance fatty acid oxidation in muscle. Fatty acid accretion in muscle represents a balance between uptake and oxidation, and it seems that the effect of betaine on uptake was stronger than oxidation.

Abbreviations
CPT1: Carnitine palmitoyl transferase 1; FABP3: Fatty acid binding protein; FAT/CD36: Fatty acid translocase/cluster of differentiation; FATP1: Fatty acid transporter protein1; FFA: Free fatty acid; HDLC: High-density lipoprotein cholesterol; LPL: Lipoprotein lipase; TC: Total cholesterol

Acknowledgements
We would like to thank Healthy Husbandry Sci-tech Co., Ltd. (Hangzhou, China) for their offer with betaine.

Funding
This work was funded and supported by National Basic Research Program of China (No.2012CB124705), Zhejiang Provincial Key Research and Development Program (2015C03006) and Provincial Key S&T Special Projects (2015C02022).

Authors' contributions
YW and JF designed the study. SL and HW were involved in performing the experiment and data interpretation. SL drafted the manuscript and XW revised the manuscript. All authors read and approved the final version of the manuscript.

Competing interests
The authors declare that they have no completing interests.

References
1. Eklund M, Bauer E, Wamatu J, Mosenthin R. Potential nutritional and physiological functions of betaine in livestock. Nutr Res Rev. 2005; 18(01):31–48.
2. Craig SAS. Betaine in human nutrition. Am J Clin Nutr. 2004;80(3):539–49.
3. Wray-Cahen D, Fernández-Fígares I, Virtanen E, Steele NC, Caperna TJ. Betaine improves growth, but does not induce whole body or hepatic palmitate oxidation in swine (Sus Scrofa Domestica). Comp Biochem Physiol A Mol Integr Physiol. 2004;137(1):131–40.
4. Fernández-Fígares I, Wray-Cahen D, Steele NC, Campbell RG, Hall DD, Virtanen E, et al. Effect of dietary betaine on nutrient utilization and

partitioning in the young growing feed-restricted pig. J Anim Sci. 2002; 80(2):421–8.

5. Schrama JW, Heetkamp MJW, Simmins PH, Gerrits WJJ. Dietary betaine supplementation affects energy metabolism of pigs. J Anim Sci. 2003;81(5): 1202–9.

6. Wang YZ, Xu ZR, Feng J. The effect of betaine and DL-methionine on growth performance and carcass characteristics in meat ducks. Anim Feed Sci Technol. 2004;116(1):151–9.

7. Feng J, Liu X, Wang YZ, Xu ZR. Effects of betaine on performance, carcass characteristics and hepatic betaine-homocysteine methyltransferase activity in finishing barrows. Asian-Australas J Anim Sci. 2006;19(3):402.

8. Huang QC, Xu ZR, Han XY, Li WF. Changes in hormones, growth factor and lipid metabolism in finishing pigs fed betaine. Livest Sci. 2006;105(1):78–85.

9. Huang QC, Xu ZR, Han XY, Li WF. Effect of betaine on growth hormone pulsatile secretion and serum metabolites in finishing pigs. J Anim Physiol Anim Nutr (Berl). 2007;91(3–4):85–90.

10. Nakev J, Popova T, Vasileva V. Influence of dietary betaine supplementation on the growth performance and carcass characteristics in male and female growing-finishing pigs. Bulg J Agric Sci. 2009;15(3):263–8.

11. Pekkinen J, Olli K, Huotari A, Tiihonen K, Keski-Rahkonen P, Lehtonen M, et al. Betaine supplementation causes increase in carnitine metabolites in the muscle and liver of mice fed a high-fat diet as studied by nontargeted LC-MS metabolomics approach. Mol Nutr Food Res. 2013;57(11):1959–68.

12. Xu L, Huang D, Hu Q, Wu J, Wang Y, Feng J. Betaine alleviates hepatic lipid accumulation via enhancing hepatic lipid export and fatty acid oxidation in rats fed with a high-fat diet. Br J Nutr. 2015;113(12):1835–43.

13. Zhang W, Wang L, Wang L, Li X, Zhang H, Luo L, et al. Betaine protects against high-fat-diet-induced liver injury by inhibition of high-mobility group box 1 and toll-like receptor 4 expression in rats. Dig Dis Sci. 2013; 58(11):3198–206.

14. Deminice R, Da Silva RP, Lamarre SG, Kelly KB, Jacobs RL, Brosnan ME, et al. Betaine supplementation prevents fatty liver induced by a high-fat diet: effects on one-carbon metabolism. Amino Acids. 2015;47(4):839–46.

15. Huang QC, Han XY, Xu ZR, Yang XY, Chen T, Zheng XT. Betaine suppresses carnitine palmitoyltransferase I in skeletal muscle but not in liver of finishing pigs. Livest Sci. 2009;126(1):130–5.

16. Martins JM, Neves JA, Freitas A, Tirapicos JL. Effect of long-term betaine supplementation on chemical and physical characteristics of three muscles from the Alentejano pig. J Sci Food Agric. 2012;92(10):2122–7.

17. Madeira MS, Rolo EA, Alfaia CM, Pires VR, Luxton R, Doran O, et al. Influence of betaine and arginine supplementation of reduced protein diets on fatty acid composition and gene expression in the muscle and subcutaneous adipose tissue of cross-bred pigs. Br J Nutr. 2016;115(06):937–50.

18. National Research Council. Nutrient requirements of swine: eleventh revised edition. Washington, D.C.: The National Academies Press; 2012.

19. Association of Official Analytical Chemists (AOAC). Van Nostrand's Encyclopedia of Chemistry. Hoboken: John Wiley & Sons, Inc. 2005.

20. Zeng T, Xie K, Zhang C, Yu L, Zhu Z. Determination the level of triglycerides in liver with chloroform/ methanol homogenate. J Hyg Res. 2008;37(5):550–1. (in Chinese)

21. Nygard A-B, Jørgensen CB, Cirera S, Fredholm M. Selection of reference genes for gene expression studies in pig tissues using SYBR green qPCR. BMC Mol Biol. 2007;8(1):67.

22. Erkens T, Van Poucke M, Vandesompele J, Goossens K, Van Zeveren A, Peelman LJ. Development of a new set of reference genes for normalization of real-time PT-PCR data of porcine backfat and longissimus dorsi muscle, and evaluation with PPARGC1A. BMC Biotechnol. 2006;6:41.

23. Livak KJ, Schmittgen TD. Analysis of relative gene expression data using real-time quantitative PCR and the 2−ΔΔCT method. Methods. 2001;25(4): 402–8.

24. Lopaschuk GD, Ussher JR, Folmes CDL, Jaswal JS, Stanley WC. Myocardial fatty acid metabolism in health and disease. Physiol Rev. 2010;90(1):207–58.

25. Kelley DE, Goodpaster B, Wing RR, Simoneau J-A. Skeletal muscle fatty acid metabolism in association with insulin resistance, obesity, and weight loss. Am J Physiol-Endoc M. 1999;277(6):E1130–41.

26. Clarke SD. Regulation of fatty acid synthase gene expression: an approach for reducing fat accumulation. J Anim Sci. 1993;71(7):1957.

27. Christensen K. In vitro studies on the synthesis of intramuscular fat in the longissimus dorsh muscle of pigs. Livest Prod Sci. 1975;2(1):59–68.

28. Yang HS, Lee JI, Joo ST, Park GB. Effects of dietary glycine betaine on growth and pork quality of finishing pigs. Asian-australas J Anim Sci. 2009;22(5):706–11.

29. Fernández-Fígares I, Lachica M, Martín A, Nieto R, Gonzalez-valero L, Rodriguez-Lopez JM, et al. Impact of dietary betaine and conjugated linoleic acid on insulin sensitivity, protein and fat metabolism of obese pigs. Animal. 2012;6(07):1058–67.

30. Nickerson JG, Alkhateeb H, Benton CR, Lally J, Nickerson J, Han XX, et al. Greater transport efficiencies of the membrane fatty acid transporters FAT/CD36 and FATP4 compared with FABPpm and FATP1 and differential effects on fatty acid esterification and oxidation in rat skeletal muscle. J Biol Chem. 2009;284(24):16522–30.

31. Sebastián D, Guitart M, García-Martínez C, Mauvazin C, Orellana-Gavalda JM, Serra D, et al. Novel role of FATP1 in mitochondrial fatty acid oxidation in skeletal muscle cells. J Lipid Res. 2009;50(9):1789–99.

32. Ibrahimi A, Bonen A, Blinn WD, Hajri T, Li X, Zhong K, et al. Muscle-specific overexpression of FAT/CD36 enhances fatty acid oxidation by contracting muscle, reduces plasma triglycerides and fatty acids, and increases plasma glucose and insulin. J Biol Chem. 1999;274(38):26761–6.

33. Hertzel AV, Bernlohr DA. The mammalian fatty acid-binding protein multigene family: molecular and genetic insights into function. Trends Endocrinol Metab. 2000;11(5):175–80.

34. Glatz JFC, Schaap FG, Binas B, Bonen A, Van Der Vusse GJ, Luiken JJFP. Cytoplasmic fatty acid-binding protein facilitates fatty acid utilization by skeletal muscle. Acta Physiol Scand. 2003;178(4):367–71.

35. Cho KH, Kim MJ, Jeon GJ, Chung HY. Association of genetic variants for FABP3 gene with back fat thickness and intramuscular fat content in pig. Mol Biol Rep. 2011;38(3):2161–6.

36. Gerbens F, Verburg FJ, Van Moerkerk HTB, Engel B, Buist W, Veerkamp JH, et al. Associations of heart and adipocyte fatty acid-binding protein gene expression with intramuscular fat content in pigs. J Anim Sci. 2001;79(2):347–54.

37. Yagyu H, Chen G, Yokoyama M, Hirata K, Augustus A, Kako Y, et al. Lipoprotein lipase (LpL) on the surface of cardiomyocytes increases lipid uptake and produces a cardiomyopathy. J Clin Invest. 2003;111(3):419–26.

38. Walczak R, Tontonoz P. PPARadigms and PPARadoxes: expanding roles for PPARγ in the control of lipid metabolism. J Lipid Res. 2002;43(2):177–86.

39. Rosen ED, Walkey CJ, Puigserver P, Spiegelman BM. Transcriptional regulation of adipogenesis. Genes Dev. 2000;14(11):1293–307.

40. Ahmadian M, Suh JM, Hah N, Liddle C, Atkins AR, Downes M, et al. PPAR [gamma] signaling and metabolism: the good, the bad and the future. Nat Med. 2013;99(5):557–66.

41. Albuquerque A, Neves JA, Redondeiro M, Laranjo M, Felix MR, Freitas A, et al. Long term betaine supplementation regulates genes involved in lipid and cholesterol metabolism of two muscles from an obese pig breed. Meat Sci. 2017;124:25–33.

42. Saha AK, Ruderman NB. Malonyl-CoA and AMP-activated protein kinase: an expanding partnership. Mol Cell Biochem. 2003;253(1):65–70.

43. Xue B, Kahn BB. AMPK integrates nutrient and hormonal signals to regulate food intake and energy balance through effects in the hypothalamus and peripheral tissues. J Physiol Lond. 2006;574(1):73–83.

44. Cai D, Wang J, Jia Y, Liu H, Yuan M, Dong H, et al. Gestational dietary betaine supplementation suppresses hepatic expression of lipogenic genes in neonatal piglets through epigenetic and glucocorticoid receptor-dependent mechanisms. Biochim Biophys Acta. 2016;1861(1):41–50.

45. Hardie DG, Ross FA, Hawley SA. AMPK: a nutrient and energy sensor that maintains energy homeostasis. Nat Rev Mol Cell Biol. 2012;13(4):251–62.

46. Lee WJ, Kim M, Park HS, Kim HS, Jeon MJ, Oh KS, et al. AMPK activation increases fatty acid oxidation in skeletal muscle by activating PPARα and PGC-1. Biochem Biophys Res Commun. 2006;340(1):291–5.

47. Ferré P. The biology of peroxisome proliferator-activated receptors. Diabetes. 2004;53(suppl 1):S43–50.

48. Matthews JO, Southern LL, Higbie AD, Persica MA, Bidner TD. Effects of betaine on growth, carcass characteristics, pork quality, and plasma metabolites of finishing pigs. J Anim Sci. 2001;79(3):722–8.

49. Martins JM, Neves JA, Freitas A, Tirapicos JL. Research article Betaine supplementation affects the cholesterol but not the lipid profile of pigs. Eur J Lipid Sci Technol. 2010;112:295–303.

16

Effects of pituitary-specific overexpression of FSHα/β on reproductive traits in transgenic boars

Wenting Li[1,2†], Yujun Quan[1,3†], Mengmeng Zhang[1], Kejun Wang[1,2], Muzhen Zhu[1], Ye Chen[4], Qiuyan Li[5*] and Keliang Wu[1*]

Abstract

Background: Follicle-stimulating hormone (FSH) is a gonadotropin synthesized and secreted by the pituitary gland. FSH stimulates follicle development and maturation in females. It also plays an important role in spermatogenesis in males, including humans and mice. However, the effects of FSH on male pigs are largely unknown. In this study, we generated transgenic pigs to investigate the effects of FSHα/β overexpression on reproductive traits in boars.

Results: After five transgenic F_0 founders were crossed with wide-type pigs, 193 F_1 animals were obtained. Of these, 96 were confirmed as transgenic. *FSHα* and *FSHβ* mRNAs were detected only in pituitary tissue. Transgenic boars exhibited significantly higher levels of *FSHα* and *FSHβ* mRNA, serum FSH, and serum testosterone, compared to full-sib non-transgenic boars. Significant increases in testis weight, vas deferens diameter, seminiferous tubule diameter, and the number of Leydig cells were observed, suggesting that the exogenous FSHα/β affects reproductive traits. Finally, transgenic and non-transgenic boars had similar growth performance and biochemical profiles.

Conclusions: Pituitary-specific overexpression of *FSHα/β* genes is likely to impact reproductive traits positively, as indicated by enhancements in serum testosterone level, testis weight, the development of vas deferens, seminiferous tubules, and Leydig cells in transgenic boars. A high level of serum FSH induces secretion of serum testosterone, possibly by boosting the number of Leydig cells, which presumably increases the libido and the frequency of sexual activity in transgenic boars. Our study provides a preliminary foundation for the genetic improvement of reproductive traits in male pigs.

Keywords: Boar, FSHα/β, Reproductive traits, Transgene

Background

Follicle-stimulating hormone (FSH) is a gonadotropin and glycoprotein polypeptide hormone with a mass of 35.5 kDa [1]. As a member of the glycoprotein hormone superfamily, it consists of two subunits (α and β) that combine non-covalently to form an active heterodimer, as is also the case for luteinizing hormone (LH), thyroid-stimulating hormone (TSH), and human chorionic gonadotropin (hCG) [1]. The synthesis and secretion of FSHα and FSHβ is regulated by gonadotropin-releasing hormone (GnRH). FSHβ is also regulated by inhibin, leptin, and activins derived from brain, pituitary, placenta, and other tissues [2–4]. In females, FSH plays a key role in antral follicle development and stimulates preovulatory follicular growth in cooperation with LH [5, 6]. In males, FSH is required for the mitotic division of germ cells, and together with testosterone, is involved in spermatocyte maturation and spermatogenesis [7].

Transgenic mouse models incorporating human *FSHα* and *FSHβ* genes have been used to study the effect of FSH on reproductive function [8]. In transgenic mice carrying a 10 kb human *FSHβ* construct, the inserted gene is highly and specifically expressed in pituitary tissue and the mice exhibit normal fertility [9, 10]. *FSH*-null

---

* Correspondence: liqiuyan@cau.edu.cn; liangkwu@cau.edu.cn
†Equal contributors
⁵State Key Laboratory for Agrobiotechnology, China Agricultural University, Beijing 100193, China
¹College of Animal Science and Technology, China Agricultural University, Beijing 100193, China
Full list of author information is available at the end of the article

(knockout) male mice are fertile and sire normal-sized litters, although they show reductions in epididymal sperm number, sperm motility, and testicle size, while female knockouts are infertile [5]. *FSHβ* has been verified to be an important gene controlling litter size in Chinese Erhualian pigs, one of the most prolific pig breeds in the world [11]. In transgenic mice exhibiting pituitary-specific overexpression of the Chinese Erhualian *FSH* gene, ovulation rate and litter size increase markedly [12].

F_0 transgenic pigs, in which *FSHα/β* expression is pituitary-specific, were generated previously [13]. In this study, we obtained 193 F_1 transgenic animals derived from five F_0 founders crossed to wild-type Large White pigs. Integration of the exogenous *FSHα/β* genes and their expression were confirmed. Since genetic improvements are more efficiently transferred by males than by females in pig breeding, we focused on the effects of FSHα/β on reproductive traits in boars. As is typical in reproductive trait studies, multiple traits were assessed, including semen volume, sperm quality parameters, sperm per ejaculate, epididymis weight, reproductive tract weight, and seminiferous tubule diameters (Animal QTL database) [14]. Hormone assays and histological analyses were performed to investigate the effects of exogenous FSH expression on the reproductive traits of male offspring. In addition, the health status of transgenic pigs was evaluated based on growth and various biochemical criteria. The results are directly relevant to strategies for improving the fecundity of multiparous mammals.

Methods
Generation of transgenic pigs
BAC DNA used for the production of transgenic animals in this study was described previously [12]. BAC clones for *FSHα* (BAC412H8) and *FSHβ* (BAC183O11) were isolated from a BAC library constructed using genomic DNA from a male Erhualian pig [15]. The LoxP-neo-LoxP cassette was introduced into two BAC constructs (*FSHα* and *FSHβ*) by homologous recombination (Fig. 1). BAC DNAs were linearized with *NotI* and co-transfected into fetal fibroblast cells. Positive cells were used as donors to produce transgenic founder pigs following standard procedures [16]. Transgenic F_0 pigs were mated with non-transgenic Large White pigs to produce F_1 pigs.

Identification of transgenic pigs and detection of gene expression
Transgenic pigs were identified by PCR and Southern blot using genomic DNA extracted from ear tissue. Three pairs of primers, *FSHα*-5-453-F/R (453 bp product), *FSHβ*-5-737-F/R (737 bp product) [13], and *Neo*-382-F/R (382 bp product), were used to amplify *FSHα*, *FSHβ*, and *Neo*, respectively. PCR products were digested with *AvaII*

and *PstI* prior to gel electrophoresis. The primers *Neo*-382-F/R (forward 5′-GTTGTCACTGAAGCGGGAAG-3′ and reverse 5′-CACAGTCGATGAATCCAGAAAA-3′) were used to generate a digoxigenin (DIG)-labeled probe for the Southern blot assay (Roche Diagnostics, Mannheim, Germany). All primers were synthesized by the Sangon Company (Shanghai, China).

Three F_1 transgenic (Tg) boars and three non-transgenic (NTg) full-sib boars were slaughtered at approximately 300 d of age. Tissue samples from hypothalamus, pituitary, testis, epididymis, vas deferens, seminal vesicle, prostate, Cowper's gland, heart, liver, spleen, lung, kidney, and pancreas were collected, rapidly frozen in liquid nitrogen, and stored at −80 °C. Tissue-specific expression of the *FSHα* and *FSHβ* transgenes was determined by reverse transcription PCR (RT-PCR) and quantified by real-time PCR [17]. Total RNA was extracted using an animal total RNA extraction kit according to the manufacturer's instructions (Tiangen, Beijing, China). cDNA synthesis was performed with 1 µg total RNA following the protocol accompanying the FastQuant RT Kit (Tiangen, Beijing, China). *GADPH* expression was used for normalization. The specific primers used for quantifying expression were: *FSH-α* (forward: 5′-GGGTGCCCCAATCTATCAGTG-3′, reverse: 5′-GTGGCATTCGGTGTGGTTCTC-3′), *FSH-β* (forward: 5′-CACCCCAAGATGAAGTCGCTG-3′, reverse: 5′-GCCAGGTACTTTCACGGTCTCG-3′), and *GADPH* (forward: 5′-GTTTGTGATGGGCGTGAAC-3′, reverse: 5′-ATGGACCTGGGTCATGAGT-3′).

Phenotype measurements
Body weight
Body weight of 20 F_1 pigs (10 Tg and 10 NTg half-sib individuals) was recorded at the ages of 1 d (birth weight), 10 d and 21 d (weaning weight), 60 d, 90 d, and 150 d.

Serum biochemistry
Serum was separated from blood samples obtained from F_1 pigs (5 Tg and 5 of NTg half-sib individuals) at 300 d, 307 d, and 315 d. The following compounds were measured: glucose (GLU), urea (UREA), creatinine (CREA), blood urea nitrogen/creatinine (BUN/CREA), phosphorus (PHOS), calcium (CA), total protein (TP), albumin (ALB), globulin (GLB), alanine aminotransferase (ALT), alkaline phosphatase (ALKP), γ-glutamyl transpeptidase (GGT), cholesterol (CHOL), triglyceride (TRIG), amylase (AMYL), lipase (LIPA), and creatine kinase (CK). All assays were conducted at Beijing Tianzewanwu Veterinary Hospital, China.

Hormone assays
Serum from three pairs of randomly chosen Tg and NTg full-sib boars was collected 3 times within one week at ~300 d of age. Levels of FSH, LH, testosterone, and

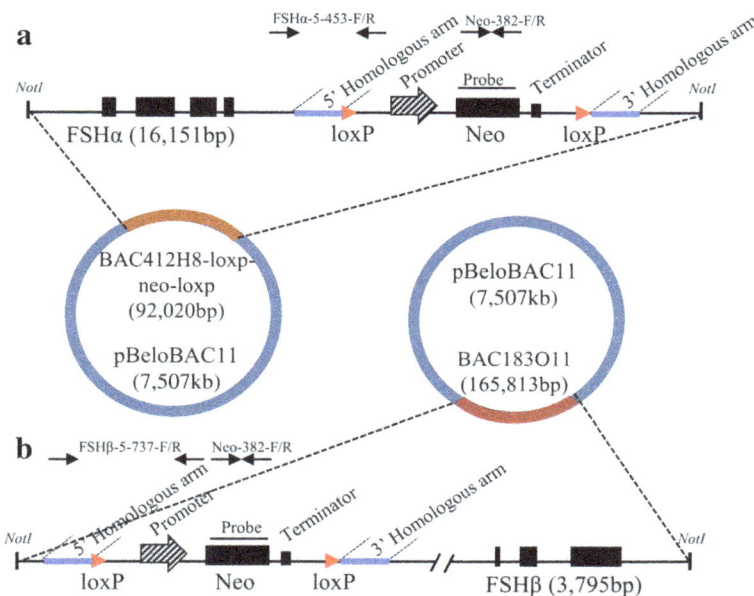

Fig. 1 Schematic view of *FSHa* and *FSHβ* expression vectors. The vectors include the complete DNA sequences of the *FSHa* and *FSHβ* genes, along with the *Neo* gene and its promoter and terminator. Solid boxes represent exons. The red arrows represent *LoxP*, and the homologous arms are represented in blue. PCR primers (*FSHa*-5-453-F/R, *FSHβ*-5-737-F/R and *Neo*-382-F/R) are represented by black arrows. Southern blot probes are indicated by the label "probe". *NotI* was the restriction enzyme cutting site

estradiol (E2) were measured in triplicate using a standard radioimmunoassay. Assays were conducted at the Beijing North Institute of Biological Technology, China.

Assessment of sperm quality

Semen collection and quality assessments were performed as described [18]. Briefly, semen was collected from five pairs of Tg and NTg half-sib boars at an approximate age of 300 d. Three successive collections were performed at 7-day intervals. Semen volume was measured using graduated semen collection jars. Sperm concentration and motility were analyzed using the Sperm Quality Analyzer (Beijing, China). Total sperm number per ejaculate was calculated using the formula: sperm concentration × semen volume. The fraction of sperm exhibiting teratospermia, intact acrosomes, and normal mitochondrial function was assessed using methods described previously [19]. Seminal plasma quality was assessed by measuring levels of zinc, fructose, neutral α-glucosidase (NAG), and acid phosphatase (ACP), using a ChemWell BRED Analyzer (Guangdong, China) at the Beijing North Institute of Biological Technology.

Histological analysis

After slaughter, testes and epididymis were isolated and weighed. Testes tissue and vas deferens was fixed in 4% paraformaldehyde, embedded in paraffin, and sectioned. Tissue sections were stained with hematoxylin-eosin (H&E) and observed with a light microscope (Nikon, Japan). The diameters of vas deferens and seminiferous tubules were measured in ~30 fields. Leydig cells were counted in ~10 fields for each pig at 200× magnification and the average value was calculated.

Statistical analysis

Student's *t*-test was performed using SPSS Statistics (IBM Corporation, USA). All values are presented as mean ± standard error (SEM). $P < 0.05$ was the threshold for statistical significance.

Results

Transgenic pigs exhibiting pituitary-specific overexpression of the *FSHα/β* genes were generated using the BAC DNAs (*FSHα* and *FSHβ*) shown in Fig. 1. Five F_0 transgenic animals (two boars and three sows), in which both BACs were intact, were identified by PCR and Southern blot analysis, as described by Bi [13].

Integration and expression of exogenous *FSH*

Five founders were crossed with wild-type Large White pigs to obtain 193 F_1 progenies, of which nearly half (43 boars and 53 sows) were positive for the exogenous *FSHα*, *FSHβ* and *Neo* genes, as determined by PCR (Fig. 2a). The *Neo* gene was also detected by Southern blot in all 96 F_1 pigs (Fig. 2b). These data confirm that the integrated *FSHα*, *FSHβ* and *Neo* genes were transmitted to both male and female F_1 pigs with the expected Mendelian ratio.

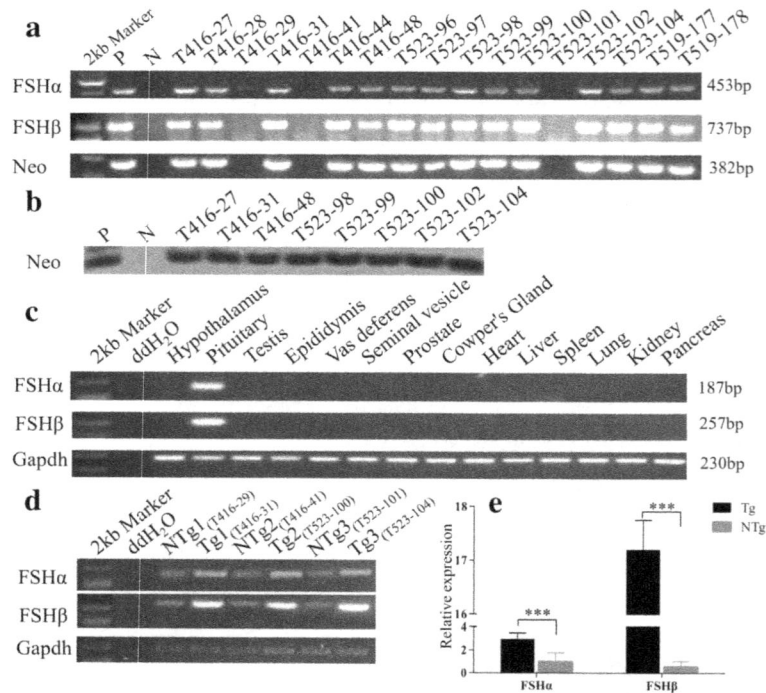

Fig. 2 Identification of exogenous *FSHα/β* insertion and expression analysis. (**a**) Identification of F₁ transgenic pigs by PCR using DNA obtained from ear tissue. P, a single F₀ transgenic pig as positive control; N, a single non-transgenic Large White pig as negative control; T416–27, 28, 29, 31, 41, T523–96, 97, 98, 99, 100, 101, 102, 104, and T519–177, 178 are identifiers for F₁ transgenic pigs. (**b**) Southern blot for transgenic pig identification. The *Neo* gene in transgenic pigs was detected using the probe shown in Fig. 1. DNAs were digested with *AvaII* and *PstI* to generate a target fragment of 463 bp. (**c**) RT-PCR analysis of *FSHα* and *FSHβ* from pituitary and 13 other tissues. *GADPH* was used as a control. (**d**) RT-PCR analysis of *FSHα* and *FSHβ* expression using mRNA from the pituitaries of six transgenic pigs. (**e**). *FSHα* and *FSHβ* mRNA expression levels in the pituitaries of Tg and NTg pigs analyzed using qPCR. Relative expression was calculated relative to *β-actin* (reference gene). Values are expressed as means ± SEM. ***, $P < 0.001$

To determine whether the exogenous *FSHα* and *FSHβ* genes in the F₁ transgenic pigs were expressed in a tissue specific manner, *FSH* mRNA from pituitary gland and 13 other tissues was subjected to RT-PCR. *FSHα* and *FSHβ* expression was observed only in pituitary tissue (Fig. 2c). Because this experiment does not distinguish between contributions made by exogenous and endogenous *FSH* genes, *FSHα* and *FSHβ* expression in the pituitary glands of three pairs of full-sib transgenic and non-transgenic boars was compared by RT-PCR (Fig. 2d), and total *FSH* mRNA expression was quantified in the same samples using qPCR (Fig. 2e). As expected, mRNA levels of both *FSHα* and *FSHβ* were significantly higher in transgenic animals ($P < 0.001$).

Serum concentrations of FSH, LH, testosterone, and E2

To examine the effects of FSHα/β overexpression on hormone levels, FSH, LH, testosterone and E2 levels were compared in full-sib transgenic and non-transgenic boars at an approximate age of 300 d (Fig. 3). Serum levels of FSH were significantly higher in transgenic animals (2.25 ± 0.18 mIU/mL vs. 1.75 ± 0.20 mIU/mL,

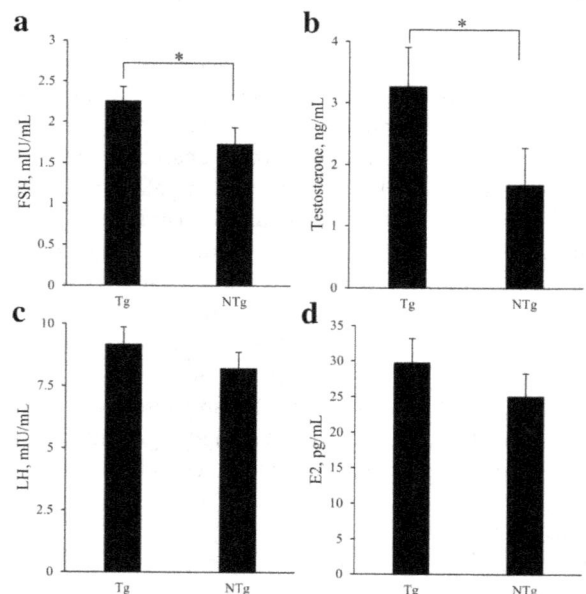

Fig. 3 Hormone assays. (**a**) Serum FSH, (**b**) testosterone, (**c**) LH, and (**d**) E2 levels in F₁ pigs. All assays were conducted in triplicate. Bars represent means ± SEM. *, $P < 0.05$

$P < 0.05$, Fig. 3a). Similarly, testosterone levels in transgenic boars were significantly higher than in non-transgenic boars (3.26 ± 0.64 ng/mL vs. 1.67 ± 0.60 ng/mL, $P < 0.05$, Fig. 3b). Although serum levels of both LH and E2 were higher in transgenic boars, the differences were not significant (LH: 9.16 ± 0.70 mIU/mL vs. 8.19 ± 0.67 mIU/mL, $P > 0.05$; E2: 29.71 ± 3.46 pg/mL vs. 25.00 ± 3.22 pg/mL, $P > 0.05$; Fig. 3c-d).

Effect of FSH overexpression on reproductive traits

Several semen quality indicators and seminal plasma components were compared between transgenic and non-transgenic boars at ~300 d of age. No significant differences were observed in any of the seven semen quality indicators ($P > 0.05$, Table 1). Transgenic and non-transgenic boars exhibited similar values for all four seminal plasma components ($P > 0.05$, Table 2).

We also compared testis and epididymis characteristics between transgenic and non-transgenic boars. As shown in Fig. 4a, the testis weight in transgenic boars was significantly higher (501.6 ± 35.6 g vs. 355.2 ± 32.8 g, $P < 0.05$). Transgenic boars exhibited higher epididymis weight but the levels were statistically indistinguishable (149.6 ± 10.6 g vs. 138.0 ± 11.0 g, $P > 0.05$, Fig. 4a). Vas deferens and seminiferous tubule diameters were also compared. Interestingly, both diameters were significantly higher in transgenic boars (vas deferens, 2216.25 ± 173.24 μm vs. 1894.72 ± 270.86 μm, $P < 0.001$; seminiferous tubules, 117.30 ± 6.65 μm vs. 107.79 ± 6.79 μm, $P < 0.001$, Fig. 4b-f). Enlargement of the vas deferens occurred mainly in the muscular layer of the wall. Finally, the number of Leydig cells in transgenic boars was significantly higher than in non-transgenic boars (337.6 ± 14.3 vs. 178.9 ± 23.4, $P < 0.01$, Fig. 4g-i).

Growth and biochemical analysis

Body weight at six growth stages (from birth to 150 d) was compared between transgenic and non-transgenic boars. There were no significant differences, although transgenic boar body weight was slightly higher from birth to 90 d, while non-transgenic boars exhibited

Table 1 Semen characteristics in transgenic and non-transgenic boars

Items	Tg	NTg	P-value
Semen volume per ejaculate, mL	218.75 ± 28.73	237.00 ± 29.54	0.079
Sperm concentration, 10^8/mL	3.58 ± 0.09	3.46 ± 0.08	0.619
Total sperm per ejaculate, 10^8	794.74 ± 28.35	832.78 ± 25.36	0.314
Sperm mobility, %	77.11 ± 2.63	73.57 ± 2.36	0.411
Teratospermia, %	8.23 ± 0.30	7.30 ± 0.38	0.764
Acrosome intactness, %	81.68 ± 0.25	81.34 ± 0.23	0.890
Normal mitochondria function, %	80.18 ± 1.52	82.22 ± 1.36	0.930

Table 2 Biochemical indicators for seminal plasma in transgenic and non-transgenic boars

Items	Tg	NTg	P-value
Seminal plasma zinc, μmol	0.76 ± 0.16	0.54 ± 0.15	0.248
Seminal plasma fructose, mIU	1.01 ± 0.05	0.95 ± 0.05	0.919
Neutral α-glucosidase, IU	0.95 ± 0.06	0.70 ± 0.05	0.324
Acid phosphatase, μmol	236.90 ± 14.17	217.89 ± 12.67	0.406

higher body weight at 150 d (Fig. 5). In addition, no significant differences in blood chemistry were observed (Table 3). We conclude that the transgenic boars in this study exhibited no detectable health defects relative to wild-type controls.

Discussion

Pig fecundity is one of the most economically important traits in pig production. Because pig reproductive traits have low heritability [20], only a few candidate genes affecting pig reproduction have been identified, such as estrogen receptor 1 (*ESR1*) and *FSHβ* [11, 21]. Transgenic mice in which porcine FSH is overexpressed exhibit significantly increased female fertility [12]. In this study, we investigated the effects of porcine FSH on reproductive traits in male transgenic pigs.

In 193 F_1 progenies, 96 transgenic pigs were identified. The transmission rate was 49.74%, consistent with ordinary Mendelian inheritance. FSH expression occurred in a pituitary-specific pattern (Fig. 2c), similar to results reported for FSHβ-overexpressing mice [12]. Because the exogenous and endogenous porcine FSHα/β are nearly identical in sequence, we could not distinguish between them using molecular methods. However, when total *FSHα/β* mRNA and serum FSH were compared in transgenic and non-transgenic pigs, transgenic animals exhibited significantly higher levels. These results suggest that pituitary-specific overexpression of FSH was successfully established in our transgenic pig model. While *FSHβ* mRNA increased approximately 10-fold in the transgenic animals, and *FSHα* mRNA increased about 3-fold, we observed only a modest increase in serum FSH levels (Figs. 2e and 3a). This is expected because serum FSH is a heterodimer, consisting of two subunits of FSHα and FSHβ, and FSH levels are probably limited by the lower level of *FSHα* mRNA expression (Fig. 2d-e) [1].

Male fertility is important in reproductive performance [22], and growing evidence suggests that FSH may be an important factor. In our study, the diameter of vas deferens and seminiferous tubules (Fig. 4b-f) increased with the increasing levels of serum FSH in transgenic boars (Fig. 3a). The enlargement of vas deferens mainly occurred in the muscular layer of the wall. In humans, the vas deferens wall is thinner after vasectomy [23].

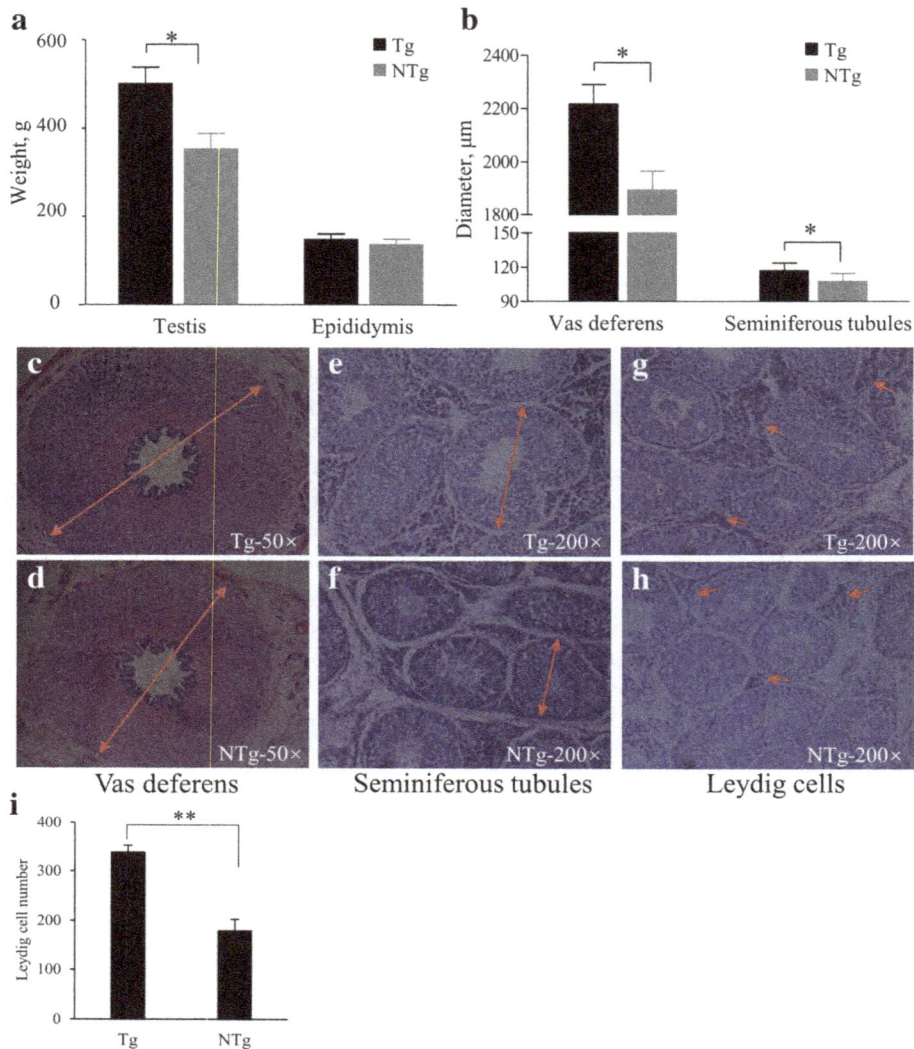

Fig. 4 Histological assessment of reproductive tissue from F_1 boars. (**a**) Comparison of testis and epididymis weight in Tg and NTg boars. (**b**) Vas deferens and seminiferous tubule diameters in Tg and NTg boars. (**c-i**) Histological sections of testis tissue. (**c-d**) Vas deferens at 50× magnification. Red arrows span the vas deferens diameter. (**e-f**) Seminiferous tubules at 200× magnification. Red arrows span tubule diameter. (**g-h**) Leydig cells at 200× magnification. Red arrows indicate Leydig cells between the seminiferous tubules. (**i**) The number of Leydig cells in Tg and NTg boars. Data are expressed as means ± SEM. *, $P < 0.05$. **, $P < 0.01$

We suggest that the thickened muscular layer of the vas deferens might affect sperm transportation and the ejaculation process, but the hypothesis has not yet been tested. Seminiferous tubule diameter correlates positively with semen quality parameters (sperm concentration, sperm motility, and total sperm per ejaculate) in rabbits [24]. In contrast, no improvement in semen quality was identified in transgenic boars in this study. In addition, semen quality in pigs does not change after treatment with FSH, although serum testosterone level increases [25]. Testosterone levels are enhanced in male mice that overexpress FSH [10]. In contrast, *FSH* and FSH receptor knockout mice have smaller testes and exhibit reduced numbers of germ and Leydig cells [5, 26]. In this study, we also observed that the serum testosterone level (Fig. 3b), testis weight (Fig. 4a) and the number of Leydig cells (Fig. 4g-i) increased in transgenic boars. The main function of Leydig cells is testosterone synthesis and secretion [27], and serum testosterone concentration is strongly related to libido in humans [28], rams [29], rats [30], and mice [31]. Testosterone also enhances libido, frequency of sexual acts, and sleep-related erections in humans [32]. If the underlying biology is similar in pigs, the increased number of Leydig cells in transgenic boars would be expected to increase testosterone levels and

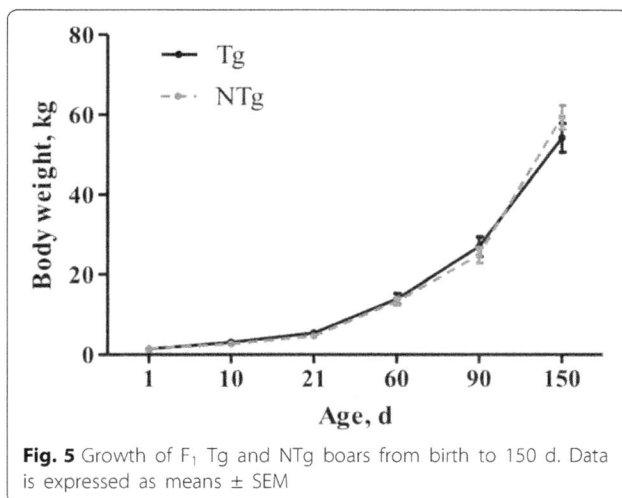

Fig. 5 Growth of F₁ Tg and NTg boars from birth to 150 d. Data is expressed as means ± SEM

thereby enhance libido, increase the frequency of sexual activity, and increase the frequency of semen collection. Because our results indicate that overexpression of FSH increases serum testosterone levels in boars, the effect is likely to be an improvement in the downstream reproductive traits.

Finally, we evaluated whether the exogenous *FSHβ* gene exerts deleterious effects on the transgenic pigs. Body weight, levels of various biochemical components in blood plasma and semen plasma, and semen quality, were similar in transgenic and non-transgenic animals.

Table 3 Blood biochemistry in transgenic and non-transgenic boars

Items	Tg	NTg	*P*-value
Glucose, mmol/L	4.58 ± 0.18	4.33 ± 0.16	0.334
Urea, mmol/L	6.87 ± 0.24	7.19 ± 0.21	0.785
Creatinine, μmol/L	118.42 ± 3.17	122.67 ± 2.83	0.0629
Blood urea nitrogen/creatinine	16.50 ± 0.52	15.80 ± 0.47	0.254
Phosphorus, mmol/L	2.28 ± 0.14	2.10 ± 0.12	0.0516
Calcium, mmol/L	2.29 ± 0.03	2.33 ± 0.03	0.580
Total protein, g/L	70.75 ± 0.36	70.27 ± 0.32	0.491
Albumin, g/L	32.50 ± 0.17	32.27 ± 0.16	0.526
Globulin, g/L	38.25 ± 0.19	38.00 ± 0.17	0.831
Albumin/ Globulin	0.84 ± 0.02	0.87 ± 0.02	0.545
Alanine aminotransferase, IU/L	59.17 ± 0.62	60.00 ± 0.56	0.809
Alkaline phosphatase, IU/L	76.75 ± 4.61	82.93 ± 4.12	0.366
γ-glutamyl transpeptidase, IU/L	32.92 ± 1.25	34.60 ± 1.22	0.802
Cholesterol, mmol/L	1.89 ± 0.07	1.80 ± 0.06	0.182
Triglyceride, mmol/L	0.59 ± 0.06	0.50 ± 0.06	0.406
Amylase, IU/L	434.03 ± 30.35	474.80 ± 27.14	0.617
Lipase, IU/L	23.50 ± 1.78	21.08 ± 1.63	0.496
Creatine kinase, IU/L	606.64 ± 18.65	591.08 ± 15.93	0.729

This suggests that FSH overexpression has no detectable adverse impact on pig health.

Conclusions

In summary, we successfully produced transgenic pigs in which exogenous *FSHα/β* genes were integrated and expressed at high levels in a pituitary-specific manner. The high level of serum FSH increases the level of serum testosterone, possibly by increasing the number Leydig cells. Higher levels of testosterone would be expected to enhance the libido and the frequency of sexual activity in transgenic boars. Nevertheless, augmented FSH levels did not improve semen quality, even though testis weight and seminiferous tubules diameter increased. Finally, the expression of exogenous *FSHα/β* genes resulted in no detectable adverse effects on growth or the overall health of transgenic boars.

Abbreviations
ACP: Acid phosphatase; ALB: Albumin; ALKP: Alkaline phosphatase; ALT: Alanine aminotransferase; AMYL: Amylase; BUN/CREA: Blood urea nitrogen/creatinine; CA: Calcium; CHOL: Cholesterol; CK: Creatine kinas; CREA: Creatinine; DIG: Digoxigenin; E2: Estradiol; ESR1: Estrogen receptor 1; FSH: Follicle-stimulating hormone; FSHα: Follicle-stimulating hormone subunit α; FSHβ: Follicle-stimulating hormone subunit β; GGT: γ-glutamyl transpeptidase; GLB: Globulin; GLU: Glucose; GnRH: Gonadotropin-releasing hormone; hCG: Human chorionic gonadotropin; LH: Luteinizing hormone; LIPA: Lipase; NAG: Neutral α-glucosidase; PHOS: Phosphorus; TP: Total protein; TRIG: Triglyceride; TSH: Thyroid-stimulating hormone

Acknowledgements
We thank for Mr. Qiao Xu, Mr. Guoying Hua, Mr. Chunzheng Fu, Mr. Yu Feng, and Mr. Linchao Wang for collecting the samples.

Funding
Funding for this study was provided by National Basic Research Program of China (973 Program, Grant 2014CB138501), the National Transgenic Animal Breeding Grand Project (2014ZX08006–005), and the Program for Changjiang Scholars and Innovation Research Teams in the University (IRT_15R62). The funding sources had no role in the study design.

Authors' contributions
KW conceived and designed the experiments. WL, YQ, MZ, MZ, YC and QL performed the experiments and collected the samples. WL and KW analyzed the data, wrote the main manuscript, and prepared the figures. All authors reviewed and approved the final manuscript.

Competing interests
The authors declare that they have no competing interests.

Author details
[1]College of Animal Science and Technology, China Agricultural University, Beijing 100193, China. [2]College of Animal Science and Veterinary Medicine, Henan Agricultural University, Zhengzhou 450002, China. [3]Institute of Zoology, Chinese Academy of Sciences, Beijing 100101, China. [4]The Department of Animal Husbandry, Rongchang Campus, Southwest University, Rongchang, Chongqing 402460, China. [5]State Key Laboratory for Agrobiotechnology, China Agricultural University, Beijing 100193, China.

References

1. Pierce JG, Parsons TF. Glycoprotein hormones: structure and function. Annu Rev Biochem. 1981;50:465–95.
2. Matzuk MM, Kumar TR, Shou W, Coerver KA, Lau AL, Behringer RR, et al. Transgenic models to study the roles of inhibins and activins in reproduction, oncogenesis, and development. In: conn PM (ed.) Recent Prog Horm Res, vol. 51; 1996: 123–157.
3. Gharib SD, Wierman ME, Shupnik MA, Chin WW. Molecular-biology of the pituitary gonadotropins. Endocr Rev. 1990;11:177–99.
4. Kato Y, Imai K, Sakai T, Inoue K. Simultaneous effect of gonadotropin-releasing hormone (GnRH) on the expression of 2 gonadotropin-beta genes by passive-immunization to GnRH. Mol Cell Endocrinol. 1989;62:135–9.
5. Kumar TR, Wang Y, Lu N, Matzuk MM. Follicle stimulating hormone is required for ovarian follicle maturation but not male fertility. Nat Genet. 1997;15:201–4.
6. McGee EA, Hsueh AJ. Initial and cyclic recruitment of ovarian follicles. Endocr Rev. 2000;21:200–14.
7. Ruwanpura SM, McLachlan RI, Meachem SJ. Hormonal regulation of male germ cell development. J Endocrinol. 2010;205:117–31.
8. Kumar TR. Mouse models for gonadotropins: a 15-year saga. Mol Cell Endocrinol. 2007;260-262:249–54.
9. Kumar TR, Fairchild-Huntress V, Low MJ. Gonadotrope-specific expression of the human follicle-stimulating hormone beta-subunit gene in pituitaries of transgenic mice. Mol Endocrinol. 1992;6:81–90.
10. Kumar TR, Low MJ. Gonadal steroid hormone regulation of human and mouse follicle stimulating hormone beta-subunit gene expression in vivo. Mol Endocrinol. 1993;7:898–906.
11. Zhao YF, Li N, Xiao L, Cao GS, Chen YZ, Zhang S, et al. FSH beta subunit gene is associated with major gene controlling litter size in commercial pig breeds. Science in China Series C-Life Sciences. 1998;41:664–8.
12. Bi M, Tong J, Chang F, Wang J, Wei H, Dai Y, et al. Pituitary-specific overexpression of porcine follicle-stimulating hormone leads to improvement of female fecundity in BAC transgenic mice. PLoS One. 2012;7:e42335.
13. Bi M. Construction of BAC transgenic mice and pigs harboring porcine follicle-stimulating hormone gene. Beijing: China Agricultural University; 2013.
14. Hu Z, Park CA, Wu X, Reecy JM. Animal QTLdb: an improved database tool for livestock animal QTL/association data dissemination in the post-genome era. Nucleic Acids Res. 2013;41:D871–9.
15. Liu W, Zhang Y, Liu Z, Guo L, Wang X, Fei J, et al. A five-fold pig bacterial artificial chromosome library: a resource for positional cloning and physical mapping. Prog Nat Sci. 2006;16:889–92.
16. Gong G, Dai Y, Fan B, Zhu H, Wang H, Wang L, et al. Production of transgenic blastocyst by nuclear transfer from different types of somatic cells in cattle. Sci China C Life Sci. 2004;47:183–9.
17. Bai Y, Zhang JB, Xue Y, Peng YL, Chen G, Fang MY. Differential expression of CYB5A in Chinese and European pig breeds due to genetic variations in the promoter region. Anim Genet. 2015;46:16–22.
18. Hu JH. Study on cryopreservation of boar semen. Yangling: Northwest A&F university; 2006.
19. Cai W, Su W, Fang Y, Chang Q, Sun S, Zhao Y, et al. Analysis of semen quality, routine blood, biochemical indexes of serum and seminal plasma in transgenic goats. Chinese. J Anim Sci. 2014;50:28–32.
20. Zhang Y. Animal breeding. Beijing: China Agricultural University Press; 2001.
21. Rothschild M, Jacobson C, Vaske D, Tuggle C, Wang L, Short T, et al. The estrogen receptor locus is associated with a major gene influencing litter size in pigs. Proc Natl Acad Sci U S A. 1996;93:201–5.
22. Berger T. Male Effects on Reproductive Performance. J. Anim. Sci. 1998; (Suppl 3):47–51.
23. Schmidt SS, Brueschke EE. Anatomical sizes of the human vas deferens after vasectomy. Fertil Steril. 1976;27:271–4.
24. Princewill Ogbuewu I, Charles Okoli I, Uwaezuoke Iloeje M. Semen quality characteristics, reaction time, testis weight and seminiferous tubule diameter of buck rabbits fed neem (Azadirachta Indica a. Juss) leaf meal based diets. International journal of reproductive. Biomedicine. 2009;7:23–8.
25. Wagner A, Claus R. The effects of postnatal FSH substitution on Sertoli cell number and the sperm production capacity of the adult boar. Anim Reprod Sci. 2009;110:269–82.
26. O'Shaughnessy PJ, Monteiro A, Abel M. Testicular development in mice lacking receptors for follicle stimulating hormone and androgen. PLoS One. 2012;7:e35136.
27. Hodgson Y, Hudson B. Leydig cell function. In: The pituitary and testis: springer; 1983. p. 107–32.
28. Travison TG, Morley JE, Araujo AB, Donnell AB O, JB MK. The relationship between libido and testosterone levels in aging men. J Clin Endocrinol Metab. 2006;91:2509–13.
29. Aguirre V, Orihuela A, Vázquez R. Effect of semen collection frequency on seasonal variation in sexual behaviour, testosterone, testicular size and semen characteristics of tropical hair rams (Ovis Aries). Trop Anim Health Prod. 2007;39:271.
30. Damassa DA, Smith ER, Tennent B, Davidson JM. The relationship between circulating testosterone levels and male sexual behavior in rats. Horm Behav. 1977;8:275–86.
31. Luttge WG, Hall NR. Differential effectiveness of testosterone and its metabolites in the induction of male sexual behavior in two strains of albino mice. Horm Behav. 1973;4:31–43.
32. Shabsigh R. The effects of testosterone on the cavernous tissue and erectile function. World J Urol. 1997;15:21–6.

17

Effects of liquid feeding of corn condensed distiller's solubles and whole stillage on growth performance, carcass characteristics, and sensory traits of pigs

Xiaojian Yang[1] (ID), Carissa Nath[2], Alan Doering[3], John Goihl[4] and Samuel Kofi Baidoo[1,5*]

Abstract

Background: The immense growth in global bioethanol production has greatly increased the supply of by-products such as whole stillage and condensed distiller's solubles, which could be potentially used for animal feeding. The objective of this study was to investigate effects of liquid feeding high levels of corn condensed distiller's solubles (CCDS) and whole stillage (CWS) on growth performance, carcass characteristics, belly firmness and meat sensory traits of pigs.

Methods: A total of 256 pigs were blocked by sex and initial BW (13.5 ± 2.5 kg), and pens of pigs (8 pigs/pen) were randomly allocated to 1 of 4 dietary treatments (8 pens/treatment): 1) corn-soybean meal based diet as control, 2) 25% CWS + 5% CCDS, 3) 19.5% CWS + 10.5% CCDS, and 4) 19.5, 26, and 32.5% CWS + 10.5, 14, and 17.5% CCDS in phases 1 (28 d), 2 (38 d), and 3 (60 d), respectively. Inclusion levels of CCDS and CWS for Treatments 1, 2, and 3 were fixed during all the three phases of the experiment. Inclusion levels of CWS and CCDS were on 88% dry matter basis. The liquid feeding system delivered feed from the mixing tank to feed troughs by high-pressure air, had sensors inside feed troughs, and recorded daily feed intake on the basis of a reference feed intake curve. The pigs were fed 5 to 10 times per day with increasing frequency during the experiment.

Results: Control pigs had greater ($P < 0.05$) average daily gain (0.91 vs. 0.84, 0.85, 0.85 kg/d) and gain to feed ratio (0.37 vs. 0.33, 0.34, 0.34) than pigs in the other three treatments during the overall period. Compared with the control, the other three groups had ($P < 0.05$) or tended to have ($P < 0.10$) lower carcass weight and backfat depth due to lighter ($P < 0.05$) slaughter body weight, but similar ($P > 0.10$) dressing percentage, loin muscle depth, and lean percentage were observed among the four treatments. Inclusion of CWS and CCDS reduced ($P < 0.05$) or tended to reduce ($P < 0.10$) belly firmness but did not influence ($P > 0.10$) the overall like, flavor, tenderness and juiciness of loin chops when compared with the control group.

Conclusion: In conclusion, our results indicate that including 30–50% of a mixture of whole stillage and condensed distiller's solubles in the growing-finishing diets may reduce growth performance, carcass weight and belly firmness, but does not affect pork sensory traits.

Keywords: Carcass, Condensed distiller's solubles, Liquid feeding, Performance, Pig, Whole stillage

* Correspondence: skbaidoo@umn.edu
[1]Southern Research and Outreach Center, University of Minnesota, Waseca, MN 56093, USA
[5]Department of Animal Science, University of Minnesota, St. Paul, MN 55108, USA
Full list of author information is available at the end of the article

Background

The U.S. ethanol production has risen tremendously from near 660 million liters in 1980 to 54 billion liters in 2014 [1]. Corn is the main raw material, which can be converted into ethanol by either the dry grind or wet milling process. The dry grind technique needs less capital investment and energy consumption and represents the method adopted in the majority of the ethanol plants in the U.S. [2]. In the dry grind process of ethanol production, the whole corn kernel is ground into flour and mixed with water to form a mash. The mash is cooked in a high-temperature cooker to decrease levels of bacteria and enzymes (e.g. alpha amylase and glucoamylase) are added to convert starch to dextrose. Then yeast is added to ferment the sugars and the resulting mixture is transferred to distillation columns where the ethanol is separated from the remaining stillage. The stillage is called whole stillage which contains about 10% dry matter and can be further centrifuged or screened to produce wet distiller's grains and thin stillage. Through the condensation process, thin stillage is concentrated by evaporation to condensed distiller's solubles containing about 30% dry matter. Recently, the interest in utilization of ethanol by-products containing relative high moisture as livestock feed ingredients has increased due to the increased availability of the co-products and high cost associated with drying. Corn condensed distiller's solubles (CCDS) and whole stillage (CWS) are high moisture ethanol by-products that may be potentially used for animal feeding. However, research on liquid feeding of swine with inclusion of high moisture ethanol by-products is limited. As reviewed by Lane [3], research in the late 1940's indicated that feeding whole stillage alone to pigs caused soft carcass, yet the carcass quality was improved if dry corn was added to the diet. Squire et al. [4] reported that liquid feeding CCDS at 15% of diet dry matter did not affect finishing pig carcass characteristics, but slightly reduced growth performance of growing pigs when compared with the corn-soybean meal control. In addition, they observed a decrease in feed intake if the level of CCDS was increased to 22.5% [4]. It has been shown that inclusion of up to 30% corn distiller's dried grains with solubles (DDGS) in growing-finishing pig diets might not adversely impact pig performance [5], but further increase of DDGS inclusion level to 45% could reduce growth performance of pigs [6]. Lee et al. [7] reported that, on a dry matter basis, CWS consisted of about 80% of distiller's grains and 20% of stillage solubles. Hence, including CWS at 20% of diet dry matter approximately contributes to 16% of distiller's grains and 4% of stillage solubles in the diet. If CCDS is also included in this diet at 5% on a dry matter basis, the total amount of stillage solubles is 9% in the diet. Similarly, on a dry matter

basis, inclusion of 25% of CWS provides 20% of distiller's grains and 5% of stillage solubles in the diet. With addition of 10% CCDS, the total amount of stillage solubles in the diet is 15%. If a diet contains, on a dry matter basis, 30% of CWS and 20% of CCDS, the total amount of distiller's grains is 24% and level of stillage solubles is 26% in the diet. We hypothesized that including CCDS and CWS with their total level at 30% of diet dry matter would not negatively influence performance of growing-finishing pigs, and that further increasing their total level to 50% would impair pig growth performance. Moreover, settling, a process by which particulates settle to the bottom of a liquid, is a concern when CWS is used in liquid feeding and including high viscosity CCDS may help reduce feed separation [8]. For example, Lu and Rosentrater [9] observed that settling happened in whole stillage after 1 h of storage. Therefore, the objective of our study was to investigate effects of liquid feeding high levels of mixture of CWS and CCDS on growth performance, carcass characteristics, belly firmness and meat sensory traits of pigs.

Methods

The experimental protocol was reviewed and approved by the University of Minnesota Institutional Animal Care and Use Committee. The experiment was conducted at the Southern Research and Outreach Center, University of Minnesota, in Waseca, MN, USA.

Nutrient composition of feed ingredients

Contents of moisture, crude protein, crude fat, neutral detergent fiber (NDF), acid detergent fiber (ADF), ash, calcium, phosphorus, sulfur, and amino acids in feed ingredients used in this study including corn, soybean meal, corn condensed distiller's solubles (CCDS) and whole stillage (CWS) were analyzed before the trial was started. The ethanol by-products CCDS and CWS were freshly obtained from a nearby ethanol plant (Guardian Energy, LLC; Janesville, MN, USA). Dry grind process is used for production of ethanol in this plant. The nutrient composition of the two by-products is presented in Table 1. During the initial phase of the feeding experiment, one batch of fresh CCDS and CWS were obtained approximately every 10 d and 5 d, respectively. Gradually the co-products were picked up more frequently. In later phase of the trial, one batch of fresh CCDS and CWS were used approximately every 5 to 7 d and 2 to 3 d, respectively. All co-products were stored in tanks that were installed indoors and the room temperature was kept at about 10 (Winter) to 20 (Fall) °C. No preservatives were added to the ethanol co-products during the experimental period. Lu and Rosentrater [9] reported that in the absence of preservatives no mold was noticed in CCDS stored up to 10 d at a temperature of 12, 22, or 35 °C, and mold appeared in CWS on the 5th and 8th

Table 1 Analyzed nutrient composition of the by-products, as-fed basis

	Corn condensed distiller's solubles (CCDS)	Corn whole stillage (CWS)
Nutrients, %		
Moisture	68.51	87.20
Crude protein	8.11	3.70
Crude fat (acid hydrolysis)	1.86	2.32
Acid detergent fiber	0.23	1.89
Neutral detergent fiber	0.32	3.37
Ash	4.17	0.71
Calcium	0.01	0.0028
Phosphorus	0.75	0.14
Sulfur	0.58	0.06
Indispensable amino acids, %		
Arg	0.44	0.17
His	0.26	0.10
Ile	0.30	0.15
Leu	0.67	0.44
Lys	0.35	0.13
Met	0.14	0.07
Phe	0.33	0.19
Thr	0.32	0.14
Trp	0.07	0.03
Val	0.42	0.19
Dispensable amino acids, %		
Ala	0.46	0.25
Asp	0.48	0.23
Cys	0.17	0.08
Glu	0.66	0.46
Gly	0.38	0.14
Pro	0.58	0.29
Ser	0.36	0.17
Tyr	0.25	0.14

day of storage at a temperature of 22 and 12 °C, respectively.

For diet formulation, metabolizable energy (ME) of CWS was assumed to be 3432 kcal/kg (88% dry matter basis), similar to the ME of corn dried distiller's grains with solubles (DDGS) with >10% oil as listed in NRC [10]. The ME of CCDS (3012 kcal/kg on 88% dry matter basis), corn (3392 kcal/kg), soybean meal (3377 kcal/kg), and choice white grease (8118 kcal/kg) were based on NRC [10, 11]. The standardized ileal digestibility (SID) of amino acids for CWS were assumed to be the same as those in NRC [10] for DDGS with >10% oil, whereas the

SID of amino acids in CCDS were based on reported values by Soares et al [12]. The SID of amino acids in corn and soybean meal were assumed to be the same as those in NRC [10]. Apparent total tract digestibility of phosphorus in corn (20%), soybean meal (31%), CCDS (60%), and CWS (60%) were assumed to be similar to the values listed in NRC [10] for the corresponding ingredients or DDGS.

Experimental design and animal management

The pigs used in the study were offspring of dam Topigs 20 (Large White × Danish Landrace) and sire Duroc (Compart's, Nicollet, MN, USA). A total of 256 crossbred pigs with an initial body weight 13.52 ± 2.54 kg were used in the present experiment. Pigs were blocked by body weight and sex and randomly assigned to 1 of 4 dietary treatments: 1) corn-soybean meal based diet (control), 2) 5% CCDS + 25% CWS, 3) 10.5% CCDS + 19.5% CWS, and 4) 10.5% CCDS + 19.5% CWS, 14% CCDS + 26% CWS, 17.5% CCDS + 32.5% CWS in phases 1 (28 d), 2 (38 d), and 3 (60 d), respectively. All percentages were on 88% dry matter basis. The ingredient and nutrient compositions of base mixes and diets are shown in Tables 2 and 3, respectively. Diets were formulated to meet or exceed estimates of requirements from NRC [10] and contain, on 88% dry matter basis, similar levels of ME and standardized ileal digestible (SID) lysine, and methionine, threonine, and tryptophan among the four treatments. Pigs were fed phases 1, 2, and 3 diets for 28, 38, and 60 d, respectively. The length of each phase was selected on the basis of pig body weight which was monitored during the experiment. At the end of each phase, the average body weight was approximately 35, 80, and 120 kg, respectively.

Each treatment had eight replicate pens with eight pigs (four barrows and four gilts) per pen and 0.95 m² per pig. Each pen had concrete slatted floors, a single liquid feeding trough (101 cm length × 28 cm width) fitted with a sensor inside the trough, a water nipple, and a water meter. Pigs were fed by a Big Dutchman liquid feeding system (Big Dutchman, Inc.; Calvesage, Germany), which was computer controlled and recorded daily feed intake that was automatically adjusted, based on a reference feed intake curve. The pigs were fed 5 to 10 times per day with frequency increasing during the trial. The mixing tank and pipelines were emptied and cleaned between feedings of each diet to avoid cross-contamination between treatments. Each time before preparation of a new batch of feed, the feed levels in all troughs were checked by the sensors inside the feed troughs. If the trough was empty, new feed would be supplied for that trough; otherwise, no new feed would be supplied for that trough for that specific feeding time. Dosing of base mixes, CCDS, CWS, and water into

Table 2 Ingredient and nutrient composition of base mixes, as-fed basis

Item	Phase 1				Phase 2				Phase 3			
Treatment[a]	1	2	3	4	1	2	3	4	1	2	3	4
Ingredient, %												
Corn	60.32	58.64	58.08	58.08	66.58	67.87	67.39	67.89	71.02	73.11	72.54	68.02
Soybean meal	35.00	34.64	34.64	34.64	29.00	25.71	25.71	23.33	25.00	21.43	21.43	22.00
Choice white grease	1.00	1.57	2.20	2.20	1.30	2.03	2.61	3.57	1.12	1.64	2.26	4.28
Limestone	0.55	2.04	2.29	2.29	0.48	1.96	2.19	2.67	0.58	2.07	2.14	3.40
Dicalcium phosphate	1.79	0.86	0.50	0.50	1.50	0.43	0.07		1.30	0.14		
Lysine HCl	0.28	0.67	0.67	0.67	0.22	0.60	0.59	0.82	0.15	0.46	0.46	0.68
DL-Methionine	0.17	0.21	0.23	0.23	0.08	0.10	0.11	0.13	0.03	0.01	0.03	
L-Threonine	0.09	0.17	0.19	0.19	0.04	0.10	0.13	0.18				
L-Tryptophan		0.06	0.06	0.06		0.06	0.06	0.08				0.02
Sodium chloride	0.30	0.43	0.43	0.43	0.30	0.43	0.43	0.50	0.30	0.43	0.43	0.60
Mineral-vitamin premix[b]	0.50	0.71	0.71	0.71	0.50	0.71	0.71	0.83	0.50	0.71	0.71	1.00
Nutrients[c], %												
Dry matter[d]	89.70	90.11	90.19	90.19	89.53	89.71	89.70	90.14	87.81	87.93	88.00	88.62
GE, kcal/kg[d]	3867	3889	3935	3935	3896	3908	3917	3947	3735	3774	3781	3853
ME, kcal/kg	3344	3344	3368	3368	3368	3392	3415	3439	3368	3368	3392	3439
Crude fat[d]	3.51	4.43	4.59	4.59	3.58	4.25	4.31	6.25	3.23	3.61	3.63	5.68
Crude protein[e]	21.38	21.58	21.57	21.57	18.88	17.89	17.87	17.07	17.20	15.96	15.93	16.07
	(19.30)	(20.65)	(21.66)	(21.66)	(18.75)	(17.28)	(17.61)	(18.56)	(18.21)	(15.59)	(16.13)	(16.56)
Calcium	0.80	1.14	1.15	1.15	0.68	0.97	0.97	1.13	0.65	0.92	0.92	1.40
Total phosphorus	0.65	0.48	0.41	0.41	0.57	0.36	0.29	0.26	0.52	0.29	0.26	0.26
ATTD phosphorus[f]	0.36	0.22	0.16	0.16	0.30	0.14	0.08	0.07	0.27	0.09	0.07	0.07
Total Lys	1.38	1.67	1.67	1.67	1.17	1.37	1.36	1.47	1.01	1.15	1.15	1.32
SID Lys[g]	1.25	1.54	1.54	1.54	1.05	1.26	1.26	1.37	0.90	1.05	1.05	1.23

[a]Treatment 1, corn-soybean meal diet; Treatment 2, 5% CCDS + 25% CWS, Treatment 3, 10.5% CCDS + 19.5% CWS; Treatment 4, graded level CCDS + graded level CWS

[b]The vitamin and trace mineral premix provided the following (per kg of final diets listed in Table 3): vitamin A, 11,000 IU; vitamin D_3, 2756 IU; vitamin E, 55 IU; vitamin B_{12}, 55 μg; riboflavin, 16,000 mg; pantothenic acid, 44.1 mg; niacin, 82.7 mg; Zn, 150 mg; Fe, 175 mg; Mn, 60 mg; Cu, 17.5 mg; I, 2 mg; and Se, 0.3 mg

[c]Calculated values, unless stated otherwise

[d]Analyzed values

[e]Analyzed values in brackets

[f]ATTD apparent total tract digestible

[g]SID standardized ileal digestible

the mixing tank which was placed on load cells was monitored by the computer. After mixing in the mixing tank for 7 min, the mixed feed was delivered to individual feed troughs by high-pressure air. The two ethanol by-products, CCDS and CWS, were stored in two separate tanks and re-circulated each time before being metered in proportion to dry base mixes with the aim to reduce settling. Dry matter content of the complete liquid feed mixture was maintained between 27 and 31% and kept similar among treatments except that, in phase 3, dry matter content in Treatment 4 was about three percentage units lower than the other treatments due to the high inclusion levels of CCDS and CWS. Animal rooms had a negative pressure ventilation system. Room temperatures

were gradually decreased from 26 °C in the beginning of the trial to 18 °C at the end of the experiment.

On d 1, 28, 84, and 126, all pigs were individually weighed and feed intake per pen recorded for calculation of average daily gain (ADG), feed intake (ADFI), and feed efficiency during the periods of d 1–28, 28–84, 84–126, and 1–126. Fresh samples of CCDS and CWS were taken roughly once a month for moisture analysis. In phase 3, diet samples were taken from each of the 32 feed troughs. ADFI was presented on 88% dry matter basis. Water disappearance from the nipple drinker in each pen was recorded by a water meter with the value of the smallest division being 0.4 L. Water consumed with feed dry matter was also

Table 3 Ingredient composition and calculated nutrient composition of diets, adjusted to 88% dry matter basis[a]

Item	Phase 1				Phase 2				Phase 3			
Dietary treatment[b]	1	2	3	4	1	2	3	4	1	2	3	4
Ingredient, %												
Base mixes	100.0	70.0	70.0	70.0	100.0	70.0	70.0	60.0	100.0	70.0	70.0	50.0
CCDS		5.0	10.5	10.5		5.0	10.5	14.0		5.0	10.5	17.5
CWS		25.0	19.5	19.5		25.0	19.5	26.0		25.0	19.5	32.5
Nutrient content[c], %												
ME, kcal/kg	3272	3296	3296	3296	3296	3320	3320	3344	3368	3368	3368	3344
Crude protein[d]	20.97 (19.83)	22.25 (21.51)	22.07 (21.90)	22.07 (22.10)	18.55 (19.13)	19.78 (20.16)	19.61 (20.25)	19.79 (19.88)	17.24 (18.19)	18.67 (18.95)	18.49 (16.90)	20.21 (17.83)
Calcium	0.78	0.78	0.79	0.79	0.67	0.67	0.67	0.66	0.65	0.65	0.65	0.70
Total phosphorus	0.64	0.67	0.69	0.69	0.56	0.59	0.61	0.70	0.52	0.55	0.59	0.81
ATTD phosphorus	0.35	0.35	0.35	0.35	0.30	0.30	0.30	0.37	0.27	0.27	0.29	0.44
Total Lys[d]	1.35 (1.30)	1.41 (1.35)	1.42 (1.28)	1.42 (1.32)	1.15 (1.21)	1.21 (1.25)	1.21 (1.25)	1.23 (1.14)	1.01 (0.95)	1.08 (1.09)	1.08 (0.98)	1.12 (1.10)
Total Met[d]	0.48 (0.46)	0.50 (0.48)	0.51 (0.49)	0.51 (0.50)	0.36 (0.37)	0.39 (0.36)	0.40 (0.38)	0.41 (0.39)	0.29 (0.32)	0.32 (0.35)	0.32 (0.32)	0.35 (0.35)
Total Met + Cys[d]	0.81 (0.79)	0.89 (0.91)	0.89 (0.89)	0.89 (0.88)	0.66 (0.65)	0.75 (0.72)	0.75 (0.70)	0.78 (0.75)	0.57 (0.64)	0.66 (0.70)	0.66 (0.64)	0.74 (0.72)
Total Thr[d]	0.88 (0.89)	0.95 (0.90)	0.95 (0.89)	0.95 (0.91)	0.73 (0.74)	0.80 (0.83)	0.82 (0.80)	0.84 (0.81)	0.63 (0.71)	0.69 (0.79)	0.68 (0.70)	0.75 (0.79)
Total Trp[d]	0.28 (0.27)	0.29 (0.28)	0.29 (0.28)	0.29 (0.27)	0.24 (0.23)	0.25 (0.24)	0.25 (0.23)	0.25 (0.24)	0.21 (0.21)	0.19 (0.22)	0.19 (0.20)	0.21 (0.23)
SID Lys	1.23	1.22	1.22	1.22	1.04	1.04	1.03	1.03	0.90	0.90	0.91	0.89
SID Met	0.45	0.45	0.45	0.45	0.34	0.34	0.34	0.34	0.27	0.27	0.27	0.27
SID Met + Cys	0.74	0.75	0.74	0.74	0.60	0.62	0.61	0.61	0.51	0.53	0.52	0.54
SID Thr	0.78	0.78	0.78	0.78	0.65	0.65	0.65	0.65	0.55	0.54	0.53	0.54
SID Trp	0.25	0.25	0.25	0.25	0.21	0.21	0.21	0.21	0.19	0.16	0.16	0.16

[a]CCDS corn condensed distiller's solubles, CWS corn whole stillage, ATTD apparent total tract digestible, SID standardized ileal digestible
[b]Treatment 1, corn-soybean meal diet; Treatment 2, 5% CCDS + 25% CWS; Treatment 3, 10.5% CCDS + 19.5% CWS; Treatment 4, graded level CCDS + graded level CWS
[c]Calculated values, unless stated otherwise
[d]Analyzed values in brackets

recorded for calculation of total daily water disappearance during the whole experimental period.

Measurement of belly firmness

In the last week of the trial, eight pigs per treatment (1 pig per pen), with similar barrow to gilt ratio and average body weight around 118 kg, were selected and sent to a local packer for evaluation of belly firmness according to the method described by Whitney et al. [13]. Briefly, belly length on a flat surface was measured and then it was placed on the sharp edge of a triangular stainless steel smoke stick. The distance between the two ends of the suspended belly was then measured for calculation of the angle, an indicator of belly firmness.

Evaluation of loin sensory quality

Loin samples were collected from the same pigs chosen for belly firmness measurement as described above. Boneless center-cut loins were removed from the right side of each carcass and processed according to the Institutional Meat Purchasing Specifications [14]. These boneless center-cut loins were then transported to the meat lab of Minnesota's Agricultural Utilization Research Institute for subsequent sensory evaluation. Consumer taste panels were conducted according to guidelines set forth by the American Meat Science Association [15]. Panelists were recruited from Marshall, Minnesota, USA. All consumers were 18 years of age or older and consumed pork on a regular basis. Seventy-one consumers participated in the study over eight different panel sessions. Prior to the taste panel sessions, chops were thawed for 48 h at 4 °C. On the appropriate day and time, chops were cooked on clamshell grills to a target internal temperature of 71 °C. Chops were cut into 1.27 cm × 1.27 cm × 2.54 cm cubes and placed in glass bowls. These glass bowls were covered in aluminum foil and then placed in a 44 °C warming oven until served. Panels were conducted in booths preventing panelist interaction. Each sample was coded with a random 3-digit number. All samples were served under red lights to limit differences in visual color. Panelists were instructed to rate pork chops for overall liking and liking for flavor, tenderness and juiciness on a 9-point hedonic scale (1 = dislike extremely, 9 = like extremely). Each panelist evaluated samples of all four treatments.

Evaluation of carcass characteristics

To evaluate carcass characteristics including hot carcass weight, carcass yield, fat depth, loin depth, and percentage of lean at the end of the trial, pigs were transported to a commercial pork packing plant (Tyson Foods; Waterloo, IA, USA), where they were slaughtered and carcass measured with the Animal Ultrasound System (Animal Ultrasound Services, Ithaca, NY, USA) by placing the ultrasound parallel to the midline of the carcass for measurement of the average fat and loin depth spanning the last rib to the tenth rib and for prediction of lean with hot carcass weight, fat and loin depth.

Chemical analysis

AOAC methods [16] were used for analysis of moisture (method 934.01), crude protein (method 984.13), crude fat by acid hydrolysis (method 954.02), acid detergent fiber (ADF; method 973.18), ash (method 942.05), calcium (method 975.03B), phosphorus (method 966.01), and sulfur (method 956.01) in feed ingredients, base mixes or diets sampled from feed troughs. A heat-stable alpha-amylase was utilized for analysis of neutral detergent fiber (NDF) using the Ankom^{2000} Fiber Analyzer (Ankom Technology, Macedon, NY, USA) according to the method of Van Soest et al. [17]. Amino acid composition in feed ingredients was determined according to the AOAC method 982.30 E (a, b, c) [16] by the University of Missouri Agricultural Experiment Station Chemical Laboratories in Columbia, MO, USA. Gross energy (GE) was measured using an adiabatic bomb calorimeter (IKA Works, Inc., NC, USA) with benzoic acid as a calibration standard.

Statistical analysis

All data were analyzed using the MIXED procedure of SAS. Pen was the experimental unit for all responses. Growth performance data for the three phases were analyzed with repeated measures in time and the first order autoregressive covariance structure. Treatment, time, and interaction between treatment and time were included as fixed effects, and pen was considered as a random effect. The PLM procedure and the slice option of the MIXED procedure were used for multiple comparisons of means within different phases [18]. Overall (d 1–126) growth performance (ADG, ADFI, and GF), water disappearance, and carcass characteristics data were analyzed by analysis of variance, with treatment as a fixed effect. Belly firmness data were subjected to analysis of covariance with belly thickness as a covariate. Sensory quality was analyzed by mixed model with dietary treatment as a fixed effect and panelist as a random effect. Tukey test was employed for multiple comparisons in all analyses. Statistical significance level was set at 0.05 and probabilities between 0.05 and 0.10 were considered as a tendency. The least squares mean and standard error of the mean (SEM) were presented unless otherwise indicated.

Results

Growth performance

Crude protein was used as an indicator to evaluate how uniformly the feed was mixed and delivered to the feed troughs in phase 3. The coefficient of variation of crude protein for troughs within a treatment was 15.5, 7.5, 10.0, and 13.7% for Treatments 1, 2, 3, and 4, respectively. The average coefficient of variation for the four treatments was 11.7%.

Two pigs (1 pig from Treatments 1 and 2, respectively) were removed during the experiment for reasons unrelated to the dietary treatments. Growth performance is shown in Table 4. No difference ($P > 0.05$) was observed for body weight across treatments on d 28. But on d 84 and 126, pigs fed diets containing CCDS and CWS (Treatments 2, 3, and 4) were lighter ($P < 0.05$) than pigs fed the control diet. Body gain followed a similar pattern to that of body weight. No difference ($P > 0.05$) in ADG was noticed during the period of d 1–28. Pigs in the control group gained more ($P < 0.05$) than the other three groups from d 28 to 84. However, from d 84 to 126, only pigs fed the diet containing 5% CCDS + 25% CWS (Treatment 2) had lower ($P < 0.05$) ADG than pigs fed the control diet. Overall, inclusion of CCDS and CWS (Treatments 2, 3, and 4) led to a reduction ($P < 0.05$) of ADG in comparison with the corn-soybean meal based control diet.

During the period of d 1–28, ADFI of pigs fed diets containing CCDS and CWS was greater ($P < 0.05$) or tended to be higher ($P < 0.10$) compared with the control group. No difference ($P > 0.05$) in ADFI was noticed across the treatments after d 28. Overall, Treatment 2 had greater ($P < 0.05$) ADFI than the control group and tended to have higher ($P < 0.10$) ADFI than Treatments 3 and 4.

Treatment 1 had higher ($P < 0.05$) feed efficiency (ADG:ADFI) than the other three treatments from d 1 to 84. During the period of d 84–126, the feed efficiency did not differ ($P > 0.10$) between treatments. Overall,

Table 4 Effects of liquid feeding ethanol byproducts on growth performance and water consumption of pigs[1, 2]

Item	Dietary treatment				SEM[3]	P value
	Treatment 1 (corn-soybean meal)	Treatment 2 (5% CCDS + 25% CWS)	Treatment 3 (10.5% CCDS + 19.5% CWS)	Treatment 4 (graded level CCDS + graded level CWS)		
Body weight, kg						
d 1	13.4	13.6	13.6	13.5	0.7	1.00
d 28	34.8	34.7	34.8	34.2	1.1	0.984
d 84	84.8 [A]	80.0 [B]	79.9 [B]	78.8 [B]	1.5	0.010
d 126	127.7 [A]	119.9 [B]	120.8 [B]	120.5 [B]	1.6	0.0002
Average daily gain, kg/d						
d 1–28	0.76	0.75	0.76	0.74	0.02	0.791
d 28–84	0.89 [A]	0.81 [B]	0.81 [B]	0.80 [B]	0.02	0.004
d 84–126	1.02 [A]	0.95 [B]	0.97 [AB]	0.99 [AB]	0.01	0.02
d 1–126	0.91 [A]	0.84 [B]	0.85 [B]	0.85 [B]	0.01	0.003
Average daily feed intake (kg/d, 88% dry matter basis)						
d 1–28	1.15 [Bb]	1.28 [A]	1.27 [A]	1.25 [ABa]	0.04	0.008
d 28–84	2.23	2.32	2.25	2.26	0.03	0.176
d 84–126	3.70	3.74	3.65	3.66	0.001	0.113
d 1–126	2.48 [B]	2.56 [Aa]	2.50 [ABb]	2.50 [ABb]	0.02	0.014
Feed conversion efficiency (weight gain : feed intake)						
d 1–28	0.67 [A]	0.59 [B]	0.60 [B]	0.59 [B]	0.01	<0.0001
d 28–84	0.40 [A]	0.35 [B]	0.36 [B]	0.35 [B]	0.01	<0.0001
d 84–126	0.28	0.25	0.27	0.27	0.004	0.225
d 1–126	0.37 [A]	0.33 [B]	0.34 [B]	0.34 [B]	0.004	< 0.0001
Water disappearance during d 1–126 (l/pig·d)						
Water from feed	5.6 [C]	5.8 [B]	5.6 [C]	6.3 A	0.03	< 0.0001
Water from drinkers	2.0	1.3	1.5	1.8	0.3	0.391
Total water	7.6 [ab]	7.1 [b]	7.1 [b]	8.1 [a]	0.3	0.056

[1]*CCDS* corn condensed distiller's solubles, *CWS* corn whole stillage
[2]Means within a row without common upper case letters differ ($P < 0.05$). Means within a row without common lower case letters tend to differ ($0.05 < P < 0.10$)
[3]SEM listed in the table were calculated for each time point or period without consideration of repeated measures. When a repeated measures design was taken into consideration, SEM were 1.3, 0.02, 0.03, and 0.01 for body weight, average daily gain, average daily feed intake, and gain to feed ratio, respectively

inclusion of CCDS and CWS (Treatments 2, 3, and 4) resulted in a decrease ($P < 0.05$) of feed efficiency in comparison with the corn-soybean meal based control diet.

Daily total water disappearance including water consumed with feed dry matter and nipple drinkers was 7.6, 7.1, 7.1, and 8.1 L per pig for Treatments 1 to 4, respectively, during the whole experimental period, with a tendency ($P < 0.10$) for Treatment 4 being higher than Treatments 2 and 3 (Table 4). This is mainly due to difference ($P < 0.05$) in amount of water from feed between treatments.

Carcass characteristics

Data on carcass characteristics are presented in Table 5. Compared with the control, the other three groups had lower carcass weight ($P < 0.05$ for Treatments 2 and 3, $P < 0.10$ for Treatment 4) and backfat depth ($P < 0.10$) due to lower slaughter body weight ($P < 0.05$) in the three treatments supplemented with CCDS and CWS. Nevertheless, no differences ($P > 0.10$) in dressing percentage, muscle depth, and lean percentage were observed among the treatments.

Belly firmness

Data on belly firmness are shown in Table 5. A higher score indicates a firmer belly. As expected, inclusion of CCDS and CWS reduced ($P < 0.05$) or tended to reduce ($P < 0.10$) firmness of belly. There was no difference ($P > 0.05$) in belly firmness among the three groups fed diets containing CCDS and CWS.

Sensory quality

Inclusion of CWS and CCDS did not influence ($P > 0.10$) the overall liking, flavor, tenderness and juiciness of loin chops (Table 6).

Discussion

This study was designed to evaluate CCDS and CWS for feeding growing-finishing pigs. Dry matter content in CCDS (315 g/kg) and CWS (128 g/kg) used in our study was within the range of previously reported values [8, 9, 19]. However, the nutrient profile of our CCDS sample was, on a dry matter basis, quite different from the values for corn distillers solubles listed in NRC [10], with our sample having much lower ADF (7 vs. 85 g/kg), NDF (10 vs. 282 g/kg), ether extract (59 vs. 137 g/kg), calcium (0.2 vs. 3.3 g/kg), and sulfur (18 vs. 42 g/kg), but higher crude protein (258 vs. 213 g/kg), ash (132 vs. 99 g/kg), and phosphorus (24 vs. 14 g/kg), likely due to differences in ethanol production processes. During a period of 4 months, we monitored the temporal changes in nutrient composition of CCDS and CWS sampled once a month from the same ethanol plant (Guardian Energy, LLC; Janesville, MN, USA) and found that the coefficient of variation for dry matter, gross energy, crude protein, ether extract, NDF, ADF, and amino acids was 6.0, 0.8, 6.6, 44.6, 10.1, 15.4, and 10.8% (range 2.0–24.7%) for CCDS and 12.4, 0.8, 4.9, 7.9, 6.8, 17.2, and 6.0% (range 0.0–12.5%) for CWS, respectively. The temporal changes of nutrients in CCDS and CWS used in this trial might be partially related to variations of corn grain composition and/or stability of the production process over time. For example, different batches of corn might be used; the inconsistence in water inclusion rate and efficiency of distillation and centrifugation systems might also contribute to the variation. Tanghe et al. [19] analyzed five samples of condensed distiller's solubles derived from fermentation of wheat or mixture of grains and found large variation of nutrient compositions. For example, there was 2 to 4-fold difference in contents of NDF, ADF, crude ash, and phosphorus among the five samples. The crude fat content in the five samples of condensed distiller's solubles varied from 61 to 95 g/kg dry matter, with mean as 71 g/kg dry matter which is

Table 5 Effects of liquid feeding ethanol byproducts on carcass characteristics and belly firmness of pigs[1,2]

| Item | Dietary treatment | | | | SEM | P value |
	Treatment 1 (corn-soybean meal)	Treatment 2 (5% CCDS + 25% CWS)	Treatment 3 (10.5% CCDS + 19.5% CWS)	Treatment 4 (graded level CCDS + graded level CWS)		
Carcass weight, kg	96.6 [A a]	89.8 [B]	90.3 [B]	91.3 [AB b]	1.5	0.012
Dressing, %	75.6	75.0	74.8	75.8	0.35	0.171
Fat depth, mm	21.9 [a]	19.8 [b]	19.6 [b]	19.6 [b]	0.6	0.030
Muscle depth, mm	67.8	65.4	65.9	65.9	0.8	0.143
Lean, %	54.2	54.3	54.4	54.4	0.2	0.849
Belly thickness, mm	34.1	32.2	33.3	31.2	2.0	0.752
Belly firmness score, degree	56.3 [A a]	27.3 [AB b]	24.6 [B]	20.6 [B]	8.5	0.021

[1]CCDS corn condensed distiller's solubles, CWS corn whole stillage
[2]Means within a row without common upper case letters differ ($P < 0.05$). Means within a row without common lower case letters tend to differ ($0.05 < P < 0.10$)

Table 6 Effects of liquid feeding ethanol byproducts on sensory characteristics of loins[a]

Item	Dietary treatment				SEM	P value
	Treatment 1 (corn-soybean meal)	Treatment 2 (5% CCDS + 25% CWS)	Treatment 3 (10.5% CCDS + 19.5% CWS)	Treatment 4 (graded level CCDS + graded level CWS)		
Overall liking score	5.72	5.80	5.72	5.56	0.17	0.531
Flavor score	5.50	5.59	5.47	5.38	0.18	0.671
Tenderness score	5.86	5.97	5.75	5.63	0.19	0.250
Juiciness score	5.40	5.54	5.25	5.15	0.19	0.153

[a]CCDS corn condensed distiller's solubles, CWS corn whole stillage

similar to the fat content in our CCDS sample but much lower than the fat level (190 g/kg dry matter) in CCDS reported previously [8]. Nutrient composition of our CWS sample was within the range of values reported by Han and Liu [20], who also indicated variation of amino acids contents in CWS (coefficients of variation from 1.1 to 17.5%) and distiller solubles (coefficients of variation from 8.4 to 27.3%) sampled from different plants. Compared with the CCDS sample, CWS used in this study had, on a dry matter basis, lower contents of ash, phosphorus, sulfur, but greater levels of ADF, NDF, and ether extract, and similar concentration of crude protein.

Total water disappearance (7–8 L/pig·d) for growing-finishing pigs in the current study was similar to the findings of Vermeer et al. [21] for pigs raised on liquid feeding systems from 22 to 115 kg. Water to feed ratio, or dry matter content in liquid diet, may affect pig growth performance [22–24] and nutrient digestibility [25, 26]. For example, Hurst et al. [24] found that feed efficiency of growing-finishing pigs was improved by 3.6% and feed intake was reduced by about 3% as the ratio of water to feed was increased from 1.5:1 to 3:1 when pigs were fed at the level of 90 to 95% of ad libitum feed intake. In the present study, dry matter content in feed was kept similar, with an attempt to remove the potential confounding effect of dry matter content.

In the current study, feeding a mixture of CCDS and CWS at a level of 30 to 50% of the dietary dry matter reduced daily gain and feed efficiency of growing-finishing pigs when compared with the corn-soybean meal-based control diet. Assuming that CWS consists of 30% of distiller's grains and 70% of thin stillage (as-fed basis), or 80% of distiller's grains and 20% of solubles (dry matter basis) [7], there was approximately 10, 14.5, and 14.5–24% of solubles and 20, 15.5, and 15.5–26% of distiller's grains for Treatments 2, 3, and 4, respectively, on a dry matter basis in the current study. Research has shown that performance of growing and finishing pigs is generally not adversely impacted when including up to 30% corn DDGS [5, 27], but might be reduced in case that more than 30% corn DDGS is included [28]. However, Stein and Shurson [5] also indicated that inferior growth performance might be observed in pigs fed diets

containing less than 30% DDGS due to poor quality of DDGS and/or the associated increase in dietary crude protein concentration. The crude protein level was about 1 to 3 percentage units higher in the diets containing bioethanol co-products compared with the corn-soybean meal based control diet in the current experiment. Mycotoxins in the ethanol co-products are another concern although their levels were not analyzed in the current study. Squire et al. [4] reported that feeding growing pigs 15% CCDS on a dry matter basis reduced daily gain by 6% and feed intake by 8%, and feed intake was further reduced when CCDS inclusion level was increased to 22.5% compared with the corn-soybean meal control diet.

With limited information on nutrient digestibility in CCDS and CWS, attempts were made to formulate diets containing similar levels of nutrients (ME; SID lysine, methionine plus cysteine, threonine, and tryptophan) among treatments. We measured apparent ileal and fecal digestibility of nutrients in phase 3 diets (unpublished data). Ileal digestibility of dry matter and gross energy was similar among the 4 treatments. Nevertheless, fecal digestibility of dry matter, nitrogen and energy in Treatment 4 was about 6 percentage units lower than that in the other 3 treatments, likely due to higher inclusion levels of CCDS and CWS in phase 3 of Treatment 4. Nutrient digestibility in CCDS and/or CWS might be overestimated in diet formulation of this study. We speculated that if the diets were formulated on the basis of net energy, performance of pigs fed diets containing CCDS and CWS might be improved. Nevertheless, there is a scarcity of data on net energy concentration of ethanol byproducts used in our study. Tanghe et al. [19] used the difference method to measure digestible energy (DE) in five samples of condensed distiller's solubles derived from fermentation of wheat or mixture of grains, which varied from 3559 to 4371 kcal/kg dry matter, with average as 3989 kcal/kg dry matter. Assuming the ratio of ME to DE is 0.96, average of the ME of the five samples would be 3822 kcal/kg dry matter, which is greater than the ME value of CCDS (3415 kcal/kg dry matter) used for diet formulation in our study. The predicted net energy of the five samples ranged from 2293 to 2580 kcal/kg and averaged at 2388 kcal/kg on a dry matter basis

[19], whereas the net energy of CCDS is 2436 and 2627 kcal/kg dry matter documented by NRC [11] and NRC [10], respectively. It was reported that ileal digestibility of nitrogen and most amino acids was lower in condensed distiller's solubles than in DDGS and increasing percentage of solubles in conventional DDGS reduced ileal digestibility of amino acids in pigs [12, 29]. Furthermore, Tanghe et al. [19] found large variation of apparent fecal and ileal digestibility of energy and nutrients among five samples of condensed distiller's solubles derived from fermentation of wheat or mixture of grains. For example, among the five condensed distiller's solubles samples, a wide range for apparent total tract digestibility of gross energy (76–89%) and phosphorus (46–78%), and for apparent ileal digestibility of lysine (61 to 94%), methionine (75 to 86%), cysteine (50 to 75%), threonine (58 to 85%), and tryptophan (60 to 88%) was observed, indicating challenges to accurate formulation of swine diets containing ethanol co-products by using mean nutrient values of the by-products.

Dietary treatments did not affect carcass traits dramatically in this experiment when pigs were slaughtered at the same age. It was reported that including 15% of, on a dry matter basis, CCDS did not impact carcass traits (dressing, fat depth, muscle depth, lean) and loin meat quality when pigs were fed diets containing similar digestible energy and digestible amino acids and slaughtered at similar body weight [4, 8]. A recent study indicated that feeding diets containing 0, 15, 30, or 45% of DDGS and similar levels of ME and SID lysine and tryptophan to growing-finishing pigs did not have much effect on carcass leanness [6]. For carcass traits evaluation, slaughter body weight might be taken into consideration, because an increase of slaughter weight may decrease carcass leanness [30]. In the current study, slaughter body weight was about 8 kg heavier in the control group compared to the other three treatments and dressing percentage dropped numerically by 0.6 percentage unit when including 30% of ethanol by-products. Cisneros et al. [31] reported that an increase of 10 kg in slaughter live weight was associated with an increase of hot carcass weight and dressing percentage by 8.1 kg and 0.32 percentage unit, respectively. It has been suggested that high fiber diets may increase gut fill and hence intestinal mass, thus reducing dressing percentage. However, reduction of dressing percentage by the inclusion of corn DDGS is only observed in some experiments but not in other studies [5], probably partly due to differences in inclusion levels. For example, Cromwell et al. [6] reported that, at similar slaughter body weight, inclusion of up to 30% DDGS did not influence dressing percentage, but including 45% DDGS reduced dressing percentage by 0.5 percentage

unit. Thirty percent DDGS may provide about 10% NDF in diet. In the current study, the CCDS and CWS samples contained, on a dry matter basis, approximately 1% and 26% NDF, respectively. If CCDS and CWS are included at levels of 20 and 30% of diet dry matter, respectively, together the two ingredients would provide 8% NDF in the diet, which is lower than the level contributed by 30% DDGS.

Reduction in belly firmness of pigs fed diets containing CCDS and CWS in the current study was most likely due to the amount and profile of fatty acids in these two ingredients. The authors are not aware of any published data regarding effects of CCDS and CWS on pork sensory characteristics. It was demonstrated that including up to 45% of corn DDGS in pig diets resulted in softer belly and an increase of iodine values, but did not affect sensory traits of loin chops [32–34]. Nevertheless, please be aware that sensory traits were evaluated with limited number of pigs in the current trial.

Conclusion

Results from this study indicate that growth performance, carcass weight, and belly firmness were reduced in growing-finishing pigs fed diets containing high level (30–50% on a dry matter basis) of a mixture of corn condensed distiller's solubles and whole stillage. Nevertheless, the high inclusion level of the two co-products did not adversely affect sensory traits of loin chops.

Acknowledgements
We would like to thank A. Fossen, A. Tekeste, J. Kim, D. Pangeni, P. Ren, H. Manu, and M. Woldeab for their assistance with the animal trial and lab analysis.

Funding
Financial support from Minnesota Pork Board, Agricultural Utilization and Research Institute, and Minnesota Corn Growers Association is greatly appreciated. The funding agencies were not involved in the design of the study and collection, analysis, and interpretation of data and in writing the manuscript.

Authors' contributions
XY and SKB designed the study. XY, SKB, CN, AD and JG were involved in performing the experiment and data interpretation. XY drafted the manuscript and SKB revised the manuscript. All authors read and approved the final version of the manuscript.

Competing interests
The authors declare that they have no competing interests.

Author details
[1]Southern Research and Outreach Center, University of Minnesota, Waseca, MN 56093, USA. [2]Agricultural Utilization and Research Institute, Marshall, MN 56258, USA. [3]Agricultural Utilization and Research Institute, Waseca, MN 56093, USA. [4]Agri-Nutrition Services, Inc., Shakopee, MN 55379, USA. [5]Department of Animal Science, University of Minnesota, St. Paul, MN 55108, USA.

References

1. Renewable Fuels Association. 2015. http://www.ethanolrfa.org/resources/industry/statistics/. Accessed 18 Jan 2016.
2. Bothast RJ, Schlicher MA. Biotechnological processes for conversion of corn into ethanol. Appl Microbiol Biotechnol. 2005;67:19–25.
3. Lane AM. Distillers by-products as cattle and swine feeds. 1980. http://arizona.openrepository.com/arizona/bitstream/10150/199713/1/370051-099-103.pdf. Accessed 9 Jan 2017.
4. Squire JM, Zhu CL, Jeaurond EA, de Lange CFM. Condensed corn distillers' solubles in swine liquid feeding: growth performance and carcass quality. J Anim Sci. 2005;83 Suppl 1:165. Abstr.
5. Stein HH, Shurson GC. Board-invited review: the use and application of distillers dried grains with solubles in swine diets. J Anim Sci. 2009;87:1292–303.
6. Cromwell GL, Azain MJ, Adeola O, Baidoo SK, Carter SD, Crenshaw TD, et al. Corn distillers dried grains with solubles in diets for growing-finishing pigs: a cooperative study. J Anim Sci. 2011;89:2801–11.
7. Lee WJ, Sosulski FW, Sokhansanj S. Yield and composition of soluble and insoluble fractions from corn and wheat stillages. Cereal Chem. 1991;68:559–62.
8. de Lange CFM, Zhu CH. Liquid feeding corn-based diets to growing pigs: practical considerations and use of co-products. In: Patience JF, editor. Feed efficiency in swine. Wageningen: Wageningen Academic Publishers; 2012. p. 63–80.
9. Lu Y, Rosentrater KA. Physical and chemical properties of whole stillage, thin stillage and syrup. 2015. http://lib.dr.iastate.edu/cgi/viewcontent.cgi?article=1462&context=abe_eng_conf. Accessed 18 Jan 2016.
10. NRC. Nutrient requirements of swine. 11th ed. Washington: Nat Acad Press; 2012.
11. NRC. Nutrient requirements of swine. 10th ed. Washington: Nat Acad Press; 1998.
12. Soares JA, Stein HH, Singh V, Shurson GC, Pettigrew JE. Amino acid digestibility of corn distillers dried grains with solubles, liquid condensed solubles, pulse dried thin stillage, and syrup balls fed to growing pigs. J Anim Sci. 2012;90:1255–61.
13. Whitney MH, Shurson GC, Johnston LJ, Wulf DM, Shanks BC. Growth performance and carcass characteristics of grower-finisher pigs fed high-quality corn distillers dried grain with solubles originating from a modern Midwestern ethanol plant. J Anim Sci. 2006;84:3356–63.
14. NAMP. (North American Meat Processors Association). The Meat Buyers Guide. 1st ed. Reston: Wiley; 1997.
15. AMSA (American Meat Science Association). Research guidelines for cookery, sensory evaluation and instrumental tenderness measurements of fresh meat. Chicago: American Meat Science Association in cooperation with National Live Stock and Meat Board; 1995.
16. AOAC (Association of Official Analytical Chemists). Official methods of analysis. 18th ed. Arlington: AOAC Int.; 2006.
17. Van Soest PV, Robertson J, Lewis B. Methods for dietary fiber, neutral detergent fiber, and nonstarch polysaccharides in relation to animal nutrition. J Dairy Sci. 1991;74:3583–97.
18. Tao J, Kiernan K, Gibbs P. Advanced techniques for fitting mixed models using SAS/STAT® software. 2015. https://support.sas.com/resources/papers/proceedings15/SAS1919-2015.pdf. Accessed 1 May 2016.
19. Tanghe S, De Boever J, Ampe B, De Brabander D, De Campeneere S, Millet S. Nutrient composition, digestibility and energy value of distillers dried grains with solubles and condensed distillers solubles fed to growing pigs and evaluation of prediction methods. Anim Feed Sci Technol. 2015;210:263–75.
20. Han J, Liu K. Changes in composition and amino acid profile during dry grind ethanol processing from corn and estimation of yeast contribution toward DDGS proteins. J Agric Food Chem. 2010;58:3430–7.
21. Vermeer HM, Kuijken N, Spoolder HAM. Motivation for additional water use of growing-finishing pigs. Livest Sci. 2009;124:112–8.
22. Gill BP, Brooks PH, Carpenter JL. Voluntary water use by growing pigs offered a liquid feed of differing water to meal ratios. In: Smith AT, Lawrence TLJ, editors. Pig housing and the environment: occasional publication No. 11. Edinburgh: British Society of Animal Production; 1987. p. 131–3.
23. Geary TM, Brooks PH, Morgan DT, Campbell A, Russell PJ. Performance of weaner pigs fed ad libitum with liquid feed at different dry matter concentrations. J Sci Food Agric. 1996;72:17–24.
24. Hurst D, Clarke L, Lean IJ. Effect of liquid feeding at different water-to-feed ratios on the growth performance of growing-finishing pigs. Animal. 2008;2:1297–302.
25. Barber J, Brooks PH, Carpenter JL. The effects of water to food ratio on the digestibility, digestible energy and nitrogen retention of a grower ration. Anim Prod. 1991;52:601. Abstr.
26. Pedersen C, Stein HH. Effects of liquid and fermented liquid feeding on energy, dry matter, protein and phosphorus digestibility by growing pigs. Livest Sci. 2010;134:59–61.
27. Graham AB, Goodband RD, Tokach MD, Dritz SS, DeRouchey JM, Nitikanchana S, et al. The effects of low-, medium-, and high-oil distillers dried grains with solubles on growth performance, nutrient digestibility, and fat quality in finishing pigs. J Anim Sci. 2014;92:3610–23.
28. Stender D, Honeyman MS. Feeding pelleted DDGS-based diets to finishing pigs in deep-bedded hoop barns. J Anim Sci. 2008;86 Suppl 2:50. Abstr.
29. Li P, Xu X, Zhang Q, Liu JD, Li QY, Zhang S, et al. Effect of different inclusion level of condensed distillers solubles ratios and oil content on amino acid digestibility of corn distillers dried grains with solubles in growing pigs. Asian-Australas J Anim Sci. 2015;28:102–10.
30. Latorre MA, Lazaro R, Valencia DG, Medel P, Mateos GG. The effects of gender and slaughter weight on the growth performance, carcass traits, and meat quality characteristics of heavy pigs. J Anim Sci. 2004;82:526–33.
31. Cisneros F, Ellis M, McKeith FK, McCaw J, Fernando RL. Influence of slaughter weight on growth and carcass characteristics, commercial cutting and curing yields, and meat quality of barrows and gilts from two genotypes. J Anim Sci. 1996;74:925–33.
32. Xu G, Baidoo SK, Johnston LJ, Bibus D, Cannon JE, Shurson GC. Effects of feeding diets containing increasing content of corn distillers dried grains with solubles to grower-finisher pigs on growth performance, carcass composition, and pork fat quality. J Anim Sci. 2010;88:1398–410.
33. McClelland KM, Rentfrow G, Cromwell GL, Lindemann MD, Azain MJ. Effects of corn distillers dried grains with solubles on quality traits of pork. J Anim Sci. 2012;90:4148–56.
34. Widmer MR, McGinnis LM, Wulf DM, Stein HH. Effects of feeding distillers dried grains with solubles, high-protein distillers dried grains, and corn germ to growing-finishing pigs on pig performance, carcass quality, and the palatability of pork. J Anim Sci. 2008;86:1819–31.

18

Intermittent suckling with or without co-mingling of non-littermate piglets before weaning improves piglet performance in the immediate post-weaning period when compared with conventional weaning

Diana L. Turpin[1*], Pieter Langendijk[2], Kate Plush[3,4] and John R. Pluske[1]

Abstract

Background: In this experiment, intermittent suckling (IS) with or without the co-mingling (CoM) of piglets was studied as a method to stimulate solid feed intake and reduce post-weaning stress.

Methods: Three weaning regimes using 30 multiparous sows were compared: (1) conventional weaning (CW) ($n = 10$ litters), where piglets had continuous access to the sow until weaning (d 0, farrowing = d −25 relative to weaning); (2) intermittent suckling (IS) ($n = 10$ litters), where piglets were separated from the sow for 8 h/d starting at d −7 (relative to weaning); and (3) intermittent suckling with co-mingling (ISCo) ($n = 10$ litters) where IS started at d −7 and two litters were housed together during separation and then returned to their original sow. Ad libitum creep feed was available from d −17. At weaning pigs were housed in pens of 11 pigs, 27 pens in total. The ISCo treatment was divided in half to examine effects of different mixing strategies after weaning. Half of the ISCo litters were kept in familiar groups (ISCoF, familiar, $n = 4$) and the other half were mixed within treatment resulting in groups of unfamiliar pigs (ISCoNF, not familiar, $n = 5$), the same as IS ($n = 9$) and CW ($n = 9$) treatments.

Results: The ISCo piglets ate more creep feed in the week before weaning ($P < 0.01$), but also showed more aggressive and manipulative behaviour on first day of CoM compared with CW piglets ($P < 0.05$). IS with or without CoM increased exploratory and play behaviour on the first day of treatment intervention ($P < 0.001$) and increased sleeping behaviour on the last day of treatment intervention compared with CW ($P < 0.001$). Mixing strategy at weaning had an effect on performance data with the highest growth and feed intake seen in ISCoF pigs 2 to 8 d after weaning ($P < 0.001$). IS and ISCoNF pigs also grew faster and ate more than CW pigs 2 to 8 d after weaning ($P < 0.001$). Post-weaning injury scores suggested reduced aggression in ISCo as evidenced by reduced redness (skin irritation) ($P < 0.05$), and a tendency for ISCo to have less scratches than CW ($P < 0.1$). The IS pigs slept the most and displayed less manipulative behaviours on the day of weaning and plasma haptoglobin levels remained low in IS pigs after weaning ($P \leq 0.01$).

Conclusions: Both intermittent suckling techniques improved production indices in the immediate post-weaning period. However, the addition of co-mingling before weaning in combination with grouping familiar pigs together after weaning improved performance in an additive manner.

Keywords: Behaviour, Co-mingling, Intermittent suckling, Piglet, Weaning

* Correspondence: d.turpin@murdoch.edu.au
[1]School of Veterinary and Life Sciences, Murdoch University, Murdoch, WA 6150, Australia
Full list of author information is available at the end of the article

Background

Under current commercial pork production conditions, weaning is an abrupt transition to independency that takes place at 21 to 28 d of age. Despite efforts to familiarise piglets with creep feed during lactation, there is often large within and between litter variation in intake [1]. A lack of familiarity with feed, in combination with other stressors such as a change in housing, maternal separation and mixing with unfamiliar piglets, generally causes a period of reduced nutrient intake immediately after weaning [2, 3]. To avoid a consequential reduction in growth and gastrointestinal tract (GIT) inflammation and dysfunction that generally results from underfeeding in combination with other weaning-associated stressors [4, 5], there has been a large focus on finding methods that stimulate feeding behaviour in this period.

Intermittent suckling (IS), a gradual weaning regime that also mimics the increasing time a sow would spend away from her piglets under natural conditions, has shown an improvement in post-weaning feed intake and growth in litters compared with conventional weaning [6, 7]. This improvement in post-weaning performance is likely mediated through increase familiarisation with creep feed as piglets are forced to explore sources of nutrition other than milk. Alternatively, habituation with maternal separation may also prevent or attenuate the weaning-associated stress response [8], reducing the development of altered behaviour patterns such as aggression, manipulation and a lack of play behaviour in recently weaned piglets [9]. With a worldwide shift towards housing systems that reduce sow confinement, IS has also received renewed interest from a reproductive point of view as a potential way to mate sows during lactation rather than after weaning [10].

Piglets will also generally have a higher solid feed intake after weaning if they are kept in alternative housing that resembles natural conditions, such as multi-suckling and get-away systems with a communal piglet area [11–13]. In a study that used a get-away system with a communal piglets area, sows spent 14 h/d away from the piglets by the end of lactation (d 27) and piglets spent 40% of the observation time in the pens of other litters [11]. By allowing non-littermates to co-mingle (CoM) in these systems, piglets are more likely to have positive play experiences [14] and improved social skills, which results in better acceptance of unfamiliar pigs and reduced aggression after weaning [15–17].

In this study, we developed a novel IS system to include the opportunity for pre-weaning socialisation (CoM), where non-litter mates could interact with each other before weaning to potentially improve social development and reduce post-weaning aggression. We hypothesised that piglets subjected to IS (8 h separation per day for 7 d) would show better growth, higher feed intake, reduced negative behaviour patterns and reduced blood markers of inflammation, stress and lipid mobilisation supporting an increase in more feed directed behaviours before and after weaning. Intermittent suckling was applied in two regimes, with and without CoM (the mixing of piglets from two litters during separation) and compared with conventional weaning. It was expected that combining IS with CoM would improve post-weaning performance in an additive manner and further reduce aggressive behaviour in the immediate post-weaning period.

Methods

This study was conducted at a commercial piggery in Western Australia. At the conclusion of the study, the pigs continued on a grower/finisher facility at another location.

Animals, housing and diet

A total of 30 primiparous and multiparous sows (Large White x Landrace) and their offspring were selected based on their farrowing date from the farm and used in a single replicate. One week before farrowing, pregnant sows were housed individually across three adjacent rooms consisting of 5, 16 and 9 sows, respectively, with parities ranging from one to seven.

Individual housing consisted of farrowing crates (0.6 × 2.4 m) within farrowing pens (1.8 × 2.4 m) with slatted flooring. The pens consisted of a covered, heated creep area with artificial lighting provided between 0700 and 1700 h, a single feeder, and two nipple waterers. Water was provided on an ad libitum basis. An average litter size of 10.7 ± 0.48 (mean ± SD) was achieved by cross-fostering within the first 3 d of farrowing within each pre-allocated treatment group. Within 1 week of farrowing, piglets were made individually identifiable with numbered ear tags, their tails were docked, males were castrated and all piglets received a 1 ml intramuscular (IM) iron injection (PigDex100, Aventis Animal Nutrition, Carole Park, QLD, 4300) and a 2 mL IM injection of Respisure One® (*Mycoplasma hypopneumoniae* vaccine; Pfizer, West Ryde, NSW, 2114). The beginning of the experiment, d −25 (25 d before weaning), was designated as the day on which most litters were born. Litters were born from 1 d before to 1 d after d −25.

Sows were fed a standard lactation diet ad libitum from entry into the farrowing house (approximately 1 wk before farrowing) until weaning (14.5 MJ/kg digestible energy (DE); crude protein (CP), 19.1%). Piglets were offered creep feed (15 MJ/kg DE; CP, 23%) *ad libitum* from a single rotary hopper feeder (27 cm diameter) from 17 d before weaning (d −17). Immediately preceding weaning, piglets received a second 2 mL IM dose of Respisure One® (*Mycoplasma hypopneumoniae* vaccine;

Pfizer, West Ryde, NSW, 2114) as well as a 2 mL IM injection of Relsure® PCV (Porcine circovirus Type 1 vaccine, Zoetis, Florham Park, NJ, 07932). At weaning, piglets were moved to a separate building and placed into weaner pens with slatted flooring (1.13 m × 2.5 m) in a temperature controlled building. Feed was provided on an ad libitum basis (14.6 MJ/kg DE; CP, 20.8%) via a 5-hole weaner hopper. Piglets remained in the weaner pens until the end of the experiment. Pigs had unlimited access to drinking water provided by a single water nipple in each pen.

Experiment design

Once farrowing was completed, selected sows and their litters were randomly allocated to one of three treatments: conventional weaning (CW) (n = 10 litters), IS (intermittent suckling) (n = 10 litters) or ISCo (intermittent suckling with CoM) (n = 10 litters). Litters receiving different treatments were spread evenly across the three adjacent rooms and parity remained similar across the three treatments (3.6 ± 2.1 for CW, 3.5 ± 2.0 for IS and 3.6 ± 2.0 for ISCo; mean ± SD). The piglets in the CW treatment remained with the sow continuously through lactation until weaning. The piglets in the IS and ISCo treatments were separated from their sow and housed in an empty farrowing crate in another room for 8 h/d (0700 to 1500 h) for 7 d before weaning. Separation was achieved by transporting piglets by litter in a trolley from one room to another. The empty farrowing pens in which separated piglets were housed were identical to those that housed the sows, but no sows were housed in the separation room. For the ISCo treatment, two litters were housed in one farrowing pen during the separation time to allow for pre-weaning socialisation. The space allowance was 0.42 m² and 0.21 m² for IS and ISCo piglets respectively. In the ISCo treatment, the same two litters were socialised each day and the original litter was returned to the sow at the end of separation. When piglets were separated from the sow, the rotary feeder with creep feed was moved with the litter into the separation pen. The separation pens housing the ISCo piglets therefore had two rotatory feeders per pen during separation.

Weaning age across treatments was equal and averaged 25.3 ± 0.7 (mean ± SD). Piglets were mixed and housed in pens of 10 to 13 with an average of 11.1 ± 0.82

(mean ± SD) per pen. Mixing for CW and IS was done within treatment by randomly allocating 2–3 piglets per litter into each pen. The grouping of ISCo piglets after weaning was achieved in one of two ways: (i) four litters were only grouped with piglets they had previously had contact with before weaning (called ISCoF; familiar), and (ii) six litters were mixed with unfamiliar piglets (similar to CW and IS) (called ISCoNF; not familiar). This arrangement was necessary to examine the effect of familiarity on post-weaning behaviours since reduced aggression in pigs socialised before weaning has been partly attributed to a higher number of familiar pigs in the group rather than previous social experience [18].

Measurements

Body weights, feed intake and injury scores

All piglets were individually weighed on d −17, −7 and −4 before weaning, at weaning, and on d 2 and 8 after weaning. Creep and weaner feed residuals were measured simultaneously with body weights. Minimal wastage was observed due to twice daily checking of the feeders by staff to ensure the pan was not too full. Therefore, disappeared creep feed was considered eaten.

During the weighing procedure on the day of weaning (d 0) and 2 d after weaning, an injury score was recorded from all piglets as an indicator of aggression. The injury scoring system was adapted from Widowski et al. [19] and consisted of a four-point scale for scratches and redness around the head, ears and flank (Table 1).

Behavioural measurements

At the beginning of the experiment, five focal piglets were randomly selected from each litter (a total of 150 piglets, 50 piglets per treatment group) and made individually identifiable with marker spray. Instantaneous scan sampling by one observer was then used to record the main activity of the individual selected piglets on d −10, −7, −1, 0, 1 and 7. All focus piglets were observed every 30 min for two, 2-h periods (morning 0900 and afternoon 1400), excluding d 0 and d 1 during which only one, 2- h period was used to record behaviour. The different types of behaviour recorded during scan sampling were adapted from behaviour categories previously used by Pluske and Williams [20] and Bolhuis et al. [21] and are presented in Table 2.

Table 1 Injury scoring system using scratches and redness adapted from Widowski et al. [19]

Score	0	1 (Mild)	2 (Moderate)	3 (Severe)
Scratches	No scratches were evident on the head, ears or flank	1 to 3 small (≤2 cm) scratches or areas of abraded skin on head, ears or flank	1 to 3 large (>2 cm) scratches or areas of abraded skin on head, ears or flank	More than 3 scratches or larger areas of superficial skin loss on head, ears or flank
Redness	No redness or swelling on the head, ears or flank	Redness or swelling barely detectable on head, ears or flank	Redness or swelling were obvious on head, ears or flank	Irritation easily observed as darker reddening and/or moderate to severe swelling on head, ears or flank

Table 2 Ethogram used during instantaneous scan sampling observations adapted from Pluske and Williams [20] and Bolhuis et al. [21]

Behaviour	Description
Sleeping behaviour	Lying on the side or belly with eyes closed, not performing any other described behaviour
Lying/sitting behaviour	Lying on the side or belly with eyes open or passive sitting, not performing any other described behaviour
Standing behaviour	Standing without performing any other described behaviour
Aggressive behaviour	Head knocking, biting or fighting with another pen or littermate
Exploring/play behaviour	Standing up and investigating the surroundings such as nosing the floor, scrapping the floor with one of the forelegs, nosing or nibbling on fixtures. Running across the pen and pivoting with or without the gentle nudging of a pen or littermate
Manipulative behaviour	Belly nosing Mounting Oral manipulation of other pen or littermates Laying down and biting the metal frame work of the weaner pens (after weaning only)
Ingestive-related behaviour	Eating (chewing feed) Drinking from water nipple Eliminating (defecating or urinating)
Sow directed behaviour	Suckling or massaging the sow Manipulation of the sow during the pre-weaning period

Plasma analysis

Blood samples were taken from two randomly selected piglets per litter or pen on d −6 and −4 before weaning and d 2 and 4 after weaning (piglets were not bled more than two times within a 7 d period). Piglets were held in dorsal recumbancy and blood samples were collected via jugular venipuncture with the procedure lasting no more than 90 s. Nine millilitres of blood was collected into lithium heparin coated tubes. Blood samples were taken at the same time of day (noon) to minimise the effects of diurnal variation. Plasma was separated by centrifugation (20 min at 2,800 × g, 4 °C) and then stored as 0.5 mL aliquots at −20 °C until analysis.

Plasma cortisol levels were determined using a commercial ELISA kit (Enzo Life Sciences, Cortisol ELISA kit, AD-901-071, Farmingdale, NY) in accordance with the manufacturers' instructions with the exception of optical density, which was read at 415 nm instead of the recommended 405 nm. Intra-assay CV was 10.5% (low standard), 6.6% (medium standard) and 7.3% (high standard). Plasma was analysed at Animal Health Laboratories (Department of Agriculture and Food Western Australia) for the determination of (i) haptoglobin (Hp), using an enzymatic colorimetric in-house assay

based on modified methods of Eckersall et al. [22], and (ii) glycerol, using a Randox Reagent Kit (GY105, Crumlin, United Kingdom) and Olympus AU400 analyser (Tokyo, Japan).

Statistical analysis

Statistical analyses were performed using SPSS (v.21; IBM). Residuals were tested for normality and data were transformed or non-parametric tests were used if needed. Analysis of post-weaning data included four treatment groups after weaning (CW, IS, ISCoF and ISCoNF; as described above). Differences between ISCoF and ISCoNF were present for post-weaning ADG (average daily gain), ADFI (average daily feed intake) and FCR (feed conversion ratio) ($P < 0.05$). Therefore, post-weaning performance data are presented with four treatment groups (CW, IS, ISCoF and ISCoNF). However, no differences between ISCoF and ISCoNF were present for behaviour, injury score or blood parameter data ($P > 0.05$; with the exception of sleeping behaviour on d 7 after weaning during which ISCoNF slept more than ISCoF, $P < 0.01$). Therefore, analysis was repeated with three treatment groups for post-weaning data for these parameters (CW, IS and ISCo).

Data for pre-weaning piglet mortality, body weight (BW), ADG, ADFI and FCR were analysed on a per litter or pen basis using a general linear model with treatment as the fixed factor. Since two feeders were included in each of the ISCo separation pens, ADFI data for the ISCo treatment group before weaning were analysed on a per socialised group basis (i.e. $n = 5$). To normalise the distribution of pre-weaning ADFI data, feed disappearance at −7 to −4 d before weaning and −4 to 0 d before weaning were pooled together and presented as feed disappearance during the last week of lactation. Unexpectedly, feed disappearance from d −17 to d −7 for the IS treatment groups tended to be higher than that of the controls ($9 ± 1.7$, $8 ± 2.3$ and $3 ± 1.6$ g/piglet/day for IS, ISCo and CW respectively, $P < 0.1$), therefore, ADFI from d −17 to d −7 was included as a covariate in the analysis of the pre-weaning ADFI data. Feed conversation ratio was calculated by dividing ADFI (d 2 to 8) by ADG (d 2 to 8).

Measurements for injury scores, plasma cortisol, Hp and glycerol were averaged per crate or pen and then analysed using a general linear model with treatment as a fixed factor. Redness scores before and after weaning, as well as plasma Hp concentration were not normally distributed. Data were transformed using a square root transformation to force normality. The mean values and confidence intervals were then back-transformed and expressed as least square means.

Behaviour data obtained from instantaneous scan sampling are expressed as a percentage of total observations for each behaviour on a specific day. The distribution of all behaviours observed was not normal and transformation of the data did not correct this. The percentage of scan samples for a specific behaviour was therefore compared between treatments on different days using a Kruskal-Wallis test with post-hoc analysis to determine which groups were different from one another. A non-parametric Friedman test was used to compare differences in proportion of total observations spent on a specific behaviour within treatment. If this test detected an overall treatment effect, data were subsequently tested pairwise using a Wilcoxon test.

All post-hoc analyses included a Bonferroni correction for pairwise comparisons and correlations were performed using a Pearson correlation test. Statistical significance was accepted at $P \leq 0.05$ and a trend was considered at $P > 0.05$ and $P \leq 0.1$. Data are presented as raw means $\pm$ SEM, except when n is different between treatments, in which case data are presented as raw means $\pm$ SE unless otherwise stated.

Results
Piglet mortality
Pre-weaning piglet mortality was similar between all treatments ($P > 0.05$). Mortality mostly occurred before treatment intervention (CW: 0.4 ± 0.70, IS: 0.3 ± 0.67, ISCo: 0.1 ± 0.32 piglets per litter, $P > 0.05$). Litter sizes on the first day of treatment intervention did not differ (CW: 10.4 ± 0.70, IS: 10.3 ± 0.95, ISCo: 10.5 ± 0.53, $P > 0.05$) and mortality after the start of IS with or without CoM was negligible, with only 1 piglet in the IS treatment group dying due to crushing on d −4 of the experiment. Therefore, litter sizes at weaning were 10.4 ± 0.70 for CW, 10.2 ± 1.23 for IS and 10.5 ± 0.53 for ISCo ($P > 0.05$). None of the treatment groups experienced post-weaning mortalities, however, 9 piglets were not weaned due to having a BW less than 4.6 kg (5 piglets from CW, 1 piglet from IS and 3 piglets from ISCo) and were returned to the herd with a nurse sow as per regular farm protocol.

Production data
Piglet BW and ADG were similar between all treatments between d −17 to −7 (Table 3, $P > 0.05$). During the last week of lactation (d −7 to 0), exposure to IS or IS with CoM did not affect BW or growth compared with CW litters (Table 3, $P > 0.05$). Therefore, BW did not differ between the three treatment regimes at weaning (Table 3, $P > 0.05$). However, ISCo litters ate more creep feed than IS and CW litters between d −7 and 0 (Table 3, $P < 0.01$).

After weaning, all treatment groups suffered a reduction in weight gain from d 0 to 2. Between d 2 and 8, ISCoF pens grew the fastest, CW pens grew the slowest and IS and ISCoNF pens were intermediate (Table 4, $P < 0.001$).

Table 3 Mean values for piglet BW, ADG and ADFI before weaning for three different weaning treatments

Day[1]	Treatment[2]			
	CW	IS	ISCo	SEM
BW, kg				
D −17	2.8	2.9	3.0	0.14
D −7	5.4	5.7	5.6	0.22
D −4	6.3	6.4	6.4	0.22
D 0	7.3	7.4	7.5	0.23
ADG, g				
D −17 to −7	265	278	272	8.76
D −7 to −4	283	229	270	19.44
D −4 to 0	267	253	251	12.84
ADFI[3],g				
D −7 to 0	$7^a \pm 2.1$	$15^a \pm 2.1$	$22^b \pm 3.0$	

[1]Day = day in relation to weaning with weaning = d 0, d −25 is the day on which most the litters were born
[2]CW = conventional weaning ($n = 10$), IS = intermittent suckling ($n = 10$), ISCo = intermittent suckling with co-mingling ($n = 10$)
[3]D −17 to −7 ADFI included as covariate in analysis
[a,b]Values within a row not having the same superscript are significantly different

At the end of the experiment (d 8), BW was similar across all four treatment groups (Table 4, $P > 0.05$).

Weaning markedly increased feed intake in all treatment groups relative to intake in lactation, however ADFI did not differ between treatments during the first 2 d after weaning. Between d 2 to 8, all IS treatment pens were eating more than CW pens (Table 4,

Table 4 Mean values for pig BW, ADG, ADFI and FCR after weaning for three different weaning treatments

Day[1]	Treatment[2]			
	CW	IS	ISCoF	ISCoNF
BW, kg				
D 2	7.5 ± 0.23	7.3 ± 0.23	7.3 ± 0.34	7.6 ± 0.31
D 8	8.4 ± 0.24	8.7 ± 0.24	9.0 ± 0.37	9.0 ± 0.33
ADG, g				
D 0 to 2	$−25 \pm 13.9$	$−56 \pm 13.7$	$−78 \pm 21.6$	$−48 \pm 17.7$
D 2 to 8	165 ± 8.6^a	234 ± 8.5^b	294 ± 13.4^c	242 ± 11.1^b
ADFI, g				
D 0 to 2	80 ± 5.3	73 ± 5.3	67 ± 7.9	72 ± 7.1
D 2 to 8	192 ± 4.5^a	256 ± 4.5^b	291 ± 6.7^{cy}	269 ± 6.0^{bcx}
FCR, g/g				
D 2 to 8	1.2 ± 0.03^a	1.1 ± 0.03^{ab}	1.0 ± 0.05^b	1.1 ± 0.04^{ab}

[1]Day = day in relation to weaning with weaning = d 0, d −25 is the day on which most the litters were born
[2]CW = conventional weaning ($n = 9$), IS = intermittent suckling ($n = 9$), ISCoF = intermittent suckling with co-mingling, familiar pigs ($n = 4$), ISCoNF = intermittent suckling with co-mingled, not familiar pigs ($n = 5$)
[a,b]Values within a row not having the same superscript are significantly different
[x,y]Values within a row not having the same superscript are a trend

$P < 0.001$), however, ISCoF were eating more than IS pens ($P < 0.001$) and there was a tendency for ISCoF to eat more than ISCoNF pens (P < 0.1). Feed conversion ratio was lower for ISCoF than CW between 2 and 8 days after weaning, while IS and ISCoNF were intermediate (Table 4, $P < 0.05$).

Behavioural measurements

Pre-weaning behaviours

Before the start of IS intervention (d −10), there was no difference ($P > 0.05$) between treatments in the proportion of total observations spent per behaviour category (Table 5). On d −7, treatment intervention caused an increase in lying and sitting behaviour for both IS and ISCo piglets compared with d −10 behaviour observations for each treatment (Table 5, $P < 0.001$). Furthermore, exploratory and play behaviour also increased ($P < 0.001$) in IS piglets between d −10 to −7 whereas aggressive behaviour increased in ISCo piglets ($P < 0.001$). Compared with CW piglets, ISCo piglets showed a greater proportion of standing, manipulative and aggressive behaviours on d −7 while IS piglets were intermediate (Table 5, $P < 0.05$ for all behaviours). Only piglets subjected to ISCo slept more, had higher levels of other inactive behaviour (lying and sitting) and showed a greater proportion of total observations on ingestive-related behaviours compared with IS and CW piglets on d −7 (Table 5, $P \leq 0.001$ for all behaviours).

Over the course of treatment intervention (d −7 to −1), sleeping increased in both IS and ISCo piglets ($P < 0.001$) with piglets from both IS treatments spending over 90% of total observations during separation sleeping in their crates, which was higher than the level of sleeping for CW piglets ($P < 0.001$). Consequently, the expression of all other behaviour categories reduced over time (d −7 to −1) for ISCo ($P < 0.001$) and lying/sitting, standing and exploratory and play behaviour reduced

over time (d −7 to −1) for IS piglets ($P < 0001$). As a result, observations for lying/sitting ($P < 0.001$), standing ($P < 0.001$), exploration/play ($P < 0.001$) and manipulation ($P < 0.01$) were lower in IS and ISCo piglets on day −1 compared with CW piglets (Table 5). There was no difference in the expression of ingestive-related behaviour or aggressive behaviour between treatments on day −1 (Table 5, $P < 0.05$).

Post-weaning behaviours

On the day of weaning (approximately 3 h after the event), a considerable proportion of observations was spent on sleeping behaviour in IS pigs with IS pigs sleeping more than CW and ISCo pigs ($P < 0.001$) and CW and ISCo pigs lying or sitting more than IS pigs (Table 6, $P < 0.001$). More manipulative and exploratory/play behaviour was also observed for both CW and ISCo pigs compared with IS pigs ($P < 0.001$) on the day of weaning and CW pigs displayed the highest level of standing behaviour compared with both IS treatments (Table 6, $P < 0.001$).

Over the initial 24 h after weaning (d 0 to d 1), sleeping and lying/sitting behaviour varied differently over time for each of the different treatments (see Table 6) resulting in higher levels of sleeping for ISCo than CW and IS pigs and higher levels of lying or sitting behaviour for CW and ISCo pigs than IS pigs 24 h after weaning (Table 6, $P < 0.001$). While the proportion of total observations spent on manipulative behaviour did not change for CW and IS pigs, there was a reduction in observations for ISCo pigs over the initial 24 h after weaning (Table 6, $P < 0.001$). Therefore, on d 1 after weaning, ISCo pigs expressed less manipulative behaviour than CW pigs, and IS pigs were intermediate ($P < 0.05$). Alternatively, exploratory and play behaviour increased over the initial 24 h after weaning for IS pigs while the proportion of total observations for exploratory and play

Table 5 Pre-weaning behaviour (proportion of total observations, %) of selected CW, IS and ISCo piglets[1]

Behaviour[2], % of total observations	D[3] −10				D −7				D −1			
	CW	IS	ISCo	SEM	CW	IS	ISCo	SEM	CW	IS	ISCo	SEM
Sleeping	50.5	44.0[x]	39.0[x]	3.9	54.8[a]	55.0[a,y]	24.5[b,y]	2.7	47.8[a]	91.2[b,z]	95.0[b,z]	2.3
Inactive	2.0[x]	5.5[x]	3.0[x]	1.3	4.5[a,xy]	13.5[b,y]	32.5[c,y]	1.6	7.0[a,y]	3.0[b,x]	0.5[b,z]	1.6
Standing	6.0	7.5[x]	5.5[x]	1.8	4.3[a]	6.3[ab,x]	7.8[b,x]	1.1	2.0[a]	0.3[b,y]	0.0[b,y]	0.4
Aggressive behaviour	3.0	2.0	0.5[x]	1.1	0.5[a]	1.0[ab]	3.0[b,y]	0.6	0.0	0.0	0.3[x]	0.1
Exploratory and play behaviour	6.0	10.0[x]	12.0[x]	2.1	7.8[a]	17.0[b,y]	16.8[b,x]	1.9	7.3[a]	1.5[b,z]	0.5[b,y]	0.9
Manipulative behaviour	4.5	6.0	5.5[x]	1.7	3.0[a]	4.5[ab]	7.5[b,x]	1.2	3.0[a]	0.8[b]	0.5[b,y]	0.6
Ingestive-related behaviour	4.0	5.0	5.5	1.5	2.3[a]	1.0[a]	5.8[b]	0.9	3.8	3.2	3.0	1.0
Sow-directed behaviour	24.0	20.0	29	3.1	22.3	-	-	1.2	28.5	-	-	1.3

[1]CW = conventional weaning ($n = 50$), IS = intermittent suckling ($n = 50$), ISCo = intermittent suckling with co-mingling ($n = 50$)

[2]Results obtained by scan sampling observations. Data were tested nonparametrically

[3]Day = day in relation to weaning with weaning = d 0, d −25 is the day on which most the litters were born

[a-c]For each day, means with different superscripts are different ($P < 0.05$), indicating differences between treatments per day

[x-z]For each day treatment, means with different superscripts ($P < 0.05$), indicating differences between days within each treatment

Table 6 Post-weaning behaviour (proportion of total observations, %) of CW, IS and ISCo pigs[1]

Behaviour[2], % of total observations	D[3] 0				D 1				D 7			
	CW	IS	ISCo	SEM	CW	IS	ISCo	SEM	CW	IS	ISCo	SEM
Sleeping	20.4[a,x]	91.3[b,x]	22.4[a,x]	2.8	53.1[a,y]	58.2[a,y]	78.0[b,y]	4.1	39.8[a,z]	58.9[b,y]	54.6[b,z]	3.1
Inactive	32.1[a,x]	0.5[b,x]	18.4[c,x]	2.7	19.9[a,y]	15.8[a,y]	8.1[b,y]	2.3	15.1[a,y]	6.6[b,z]	11.2[ab,y]	1.4
Standing	13.3[a,x]	2.6[b]	4.1[b]	1.9	4.6[y]	6.6	5.1	1.9	8.7[x]	4.8	4.8	1.3
Aggressive behaviour	2.0	0.0	1.5	0.9	0.0	0.0	0.0	0.0	0.0	0.8	0.0	0.0
Exploratory and play behaviour	15.8[a,x]	0.0[b,x]	31.6[ax]	2.5	5.6[y]	5.1[y]	2.6[y]	1.5	8.2[y]	6.6[y]	3.1[y]	1.5
Manipulative behaviour	10.7[a,xy]	1.0[b,x]	12.8[a]	2.1	7.7[ax]	4.6[ab,x]	3.1[b]	1.5	13.0[y]	10.7[y]	15.3	1.8
Ingestive-related behaviour	2.0[a,x]	1.5[a,x]	8.7[b,x]	1.3	6.6[y]	9.2[y]	3.1[y]	2.1	13.0[z]	9.9[y]	11.0[x]	1.5

[1]CW = conventional weaning (*n* = 50), IS = intermittent suckling (*n* = 50), ISCo = intermittent suckling with co-mingling (*n* = 50)
[2]Results obtained by scan sampling observations. Data were tested nonparametrically
[3]Day = day in relation to weaning with weaning = d 0, d −25 is the day on which most the litters were born
[a-c]For each day, means with different superscripts are different (*P* < 0.05), indicating differences between treatments per day
[x-z]For each day treatment, means with different superscripts (*P* <0.05), indicating differences between days within each treatment

behaviour decreased over the same timeframe for ISCo and CW pigs (*P* < 0.001 for IS and ISCo and *P* =0.001 for CW).

Apart from the day of weaning during which ISCo pigs had the highest level of eating behaviour (Table 6, *P* < 0.001), no difference in eating patterns were observed between treatments after weaning (*P* > 0.05).

One week after weaning, the proportion of total observations spent on sleeping in both IS treatments was more than the CW pigs (*P* < 0.001), but lying and sitting was less in both IS treatments compared with CW pigs (*P* = 0.001). In contrast, the proportion of total observations spent on the other behaviour categories was similar for all treatments 7 d after weaning (Table 6, *P* > 0.05), with the exception of aggressive behaviour where IS pigs fought more than CW and ISCo pigs (*P* > 0.05). However, this type of behaviour was still rarely observed.

Injury scores

On the day of weaning, there was no difference in scratch scores between treatments (Table 7, *P* > 0.05), however CW piglets had higher redness scores than piglets in either IS treatment (Table 7, *P* < 0.05). Two days after weaning, there was a tendency for pigs in CW pens to have more scratches than pigs in ISCo pens (Table 7, *P* < 0.1). At the same point in time, higher redness scores were seen in CW and IS pigs compared with ISCo pigs (Table 7, *P* < 0.001).

Before weaning, scratch scores did not correlate with redness scores (*r* = 0.11, P > 0.05). However, scratch and redness scores were positively correlated after weaning (*r* = 0.67, *P* < 0.001). There was a weak positive correlation across treatments between aggressive behaviour the day before weaning (d −1) and scratch scores measured on the day of weaning (just prior to the event of weaning) (*r* = 0.22, *P* < 0.01).

Blood measures

Pre-weaning treatment did not affect plasma cortisol concentration before weaning and concentrations were still similar between treatments 2 d after weaning (Table 8, *P* > 0.05). Four days after weaning, cortisol levels for IS pigs were higher than that of ISCo pigs (Table 8, *P* = 0.001) and tended to be higher than that of CW pigs (Table 8, *P* < 0.1).

Plasma Hp concentrations were higher in IS piglets compared to CW piglets on d −6 before weaning (the second day of IS intervention), and ISCo Hp levels were intermediate (Fig. 1, *P* < 0.05). Two days later, there was no difference in Hp concentration between treatments (Fig. 1, *P* > 0.05). After weaning, ISCo pigs had the highest Hp concentration on d 2 (*P* < 0.01), while on d 4, Hp levels were higher in CW pigs than IS pigs and ISCo pigs were intermediate (Fig. 1, *P* = 0.01).

The concentration of glycerol in plasma was highest in CW piglets on d −6 and −4 before weaning (Table 9, *P* < 0.05 and < 0.01 respectively for d −6 and −4). Two days after weaning, there was no difference in glycerol concentration between treatments (Table 9, P > 0.05) and on the final blood sampling day (d 4), there was a tendency for ISCo pigs to have a lower glycerol concentration than CW pigs (Table 9, *P* < 0.1).

Discussion

Intermittent suckling systems mimic (semi) natural housing conditions where piglets do not have continuous access to the sow through lactation. From a production perspective, the potential benefits of IS are twofold: (i) periods of separation force piglets to explore nutrient options other than milk, stimulating an interest in creep feed [6, 7], and (ii) with enforced separation, sows may show oestrus meaning they can be mated while still

---

Table 7 Effects of three different weaning treatments on injury score on the day of weaning (just prior to the process of weaning) and at 2 d after weaning

	Treatment[1]			SEM	P-value
	CW	IS	ISCo		
Scratch injury score					
Weaning[2]	0.48	0.30	0.51	0.09	0.237
D 2 after weaning	1.03[x]	0.97[xy]	0.55[y]	0.15	0.074
Redness injury score[3]					
Weaning[2]	0.21[a](0.11-0.35)	0.05[b](0.01-0.13)	0.04[b](0.01-0.12)		0.016
D 2 after weaning	0.48[a](0.32-0.67)	0.35[a](0.21-0.52)	0.02[b](0.00-0.08)		<0.001

[1]CW = conventional weaning ($n = 10$ pre-weaning, $n = 9$ post-weaning), IS = intermittent suckling ($n = 10$ pre-weaning, $n = 9$ post-weaning), ISCo = intermittent suckling with co-mingling ($n = 10$ pre-weaning, $n = 9$ post-weaning)
[2]Injury scores were assessed just prior to weaning
[3]Data were square-root transformed and then back transformed and expressed as least square means with 95% confidence intervals
[ab]Values within a row not having the same superscript are significantly different
[xy]Values within a row not having the same superscript are a trend

lactating, which allows piglets to be weaned later [10] and gives them even more time to become familiar with solid food [8, 23]. The focus of the present study was to use two different IS regimes to stimulate feed intake and reduce post-weaning stress in piglets in order to improve post-weaning performance without an extended lactation. Co-mingling was used in addition to IS as a technique to reduce post-weaning aggression.

Performance before and after weaning

In contrast to previous IS studies, IS for 8 h/d, 7 d before weaning in the current study did not increase creep feed intake during lactation compared with piglets continuously kept with the sow [6, 7, 23, 24]. Despite this result however, ADG and ADFI at 2 to 8 d after weaning were higher in IS pigs than in their CW counterparts. A similar outcome was reported by Berkeveld et al. [8]

Table 8 Plasma cortisol (ng/ml) concentrations during three different weaning treatments before and after weaning

Day[1,2]	Treatment[3]			P-value
	CW	IS	ISCo	
D −6	19 (11.0-29.6)	19 (10.6-28.8)	23 (14.1-34.6)	0.724
D −4	20 (12.1-29.6)	27 (18.0-38.5)	22 (13.4-31.7)	0.467
D 2	28 (14.9-45.5)	20 (9.2-35.0)	24 (12.2-40.7)	0.672
D 4	25[abx](17.9-32.6)	32[axy](24.4-41.1)	14[by](8.9-19.8)	0.001

[1]Day in relation to weaning with 0 representing weaning (e.g. 2 is 2 d after weaning)
[2]Data has been transformed via square root and then back transformed and expressed as least square means with 95% confidence intervals
[3]CW = conventional weaning ($n = 10$ pre-weaning, $n = 9$ post-weaning), IS = intermittent suckling ($n = 10$ pre-weaning, $n = 9$ post-weaning), ISCo = intermittent suckling with co-mingling ($n = 10$ pre-weaning, $n = 9$ post-weaning)
[ab]Values within a row not having the same superscript are significantly different
[xy]Values within a row with different superscripts are trends

where piglets subjected to IS for 10 h/d, 7 d before weaning (26 d of age) had a comparable creep feed intake to CW piglets before weaning, but grew faster and ate more between 2 and 7 d after weaning. Also, Kuller et al. [6] reported that IS litters (12 h/d, 11 d before weaning) that had little or no creep feed intake during lactation still tended to have a higher weight gain after weaning than CW litters with a similar creep feed intake during lactation. Both authors hypothesised that IS litters were experiencing weaning as a less stressful event because there were already habituated with separation from the sow. The current study attempted to study this theory further by measuring behaviour and injury scores as well as plasma cortisol and Hp, two blood markers

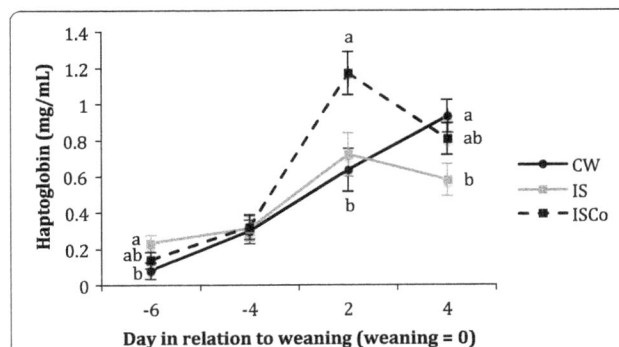

Fig. 1 Plasma haptoglobin (mg/mL) concentrations in the three different weaning treatments before and after weaning. Values are presented as actual means ± SE. Analysis involved transformation of the data using a square root calculation; CW = conventional weaning ($n = 10$ pre-weaning, $n = 9$ post-weaning), IS = intermittent suckling ($n = 10$ pre-weaning, $n = 9$ post-weaning), ISCo = intermittent suckling with co-mingling ($n = 10$ pre-weaning, $n = 9$ post-weaning). [a,b] On each experimental day, values not having the same superscript are significantly different.

Table 9 Plasma glycerol concentrations in three different weaning regimens before and after weaning

Day[1,2]	Treatment[3]			P-value
	CW	IS	ISCo	
D −6	81[a](68.0-95.7))	57[b](46.1-69.3)	53[b](42.8-65.2)	0.004
D −4	78[a](61.7-95.3)	44[b](32.4-57.8)	45[b](33.2-59.0)	0.002
D 2	20 (12.4-29.5)	27 (16.4-40.1)	12 (6.7-19.9)	0.613
D 4	20[x](13.9-28.1)	12[xy](6.7-19.9)	10[y](5.7-15.7)	0.053

[1]Day in relation to weaning with 0 representing weaning (e.g. -6 is 6 d before weaning). [2]Data has been transformed via square root and then back transformed and expressed as least square means with 95% confidence intervals
[3]CW = conventional weaning (n = 10 pre-weaning, n = 9 post-weaning), IS = intermittent suckling (n = 10 pre-weaning, n = 9 post-weaning), ISCo = intermittent suckling with co-mingling (n = 10 pre-weaning, n = 9 post-weaning)
[ab]Values within a row not having the same superscript are significantly different
[xy]Values within a row with different superscripts are trends

that are often increased at the time of weaning [25–29] (to be discussed in more detail below).

Socialising piglets in larger groups before weaning likely implements the social learning of eating behaviours with inexperienced eaters learning from experienced eaters [30]. In the current study, ISCo pigs ate the most creep feed during lactation, which was supported by the observations of ingestive-related behaviour on d −7. However, feed intake before weaning from d −17 to d −7 for both IS treatment groups tended to be higher than that of the controls. Therefore, in the case of ISCo, creep feed was already stimulated before the start of treatment intervention. This, in combination with the fact that a control litter in combination with CoM could not be included in the study, means that some caution must be used when interpreting the positive effects of the ISCo treatment.

Feed intake during lactation stimulates feed intake after weaning [31], and the higher creep feed intake for ISCo piglets during lactation was the likely cause of better growth and better feed intake in ISCoNF pigs 2 to 8 d after weaning compared with CW pigs. The fact that ISCoF pigs grew faster and tended to eat more solid feed than ISCoNF highlights the impact familiarity can have on piglet performance. Differences in how the ISCo piglets were grouped after weaning (ISCoF versus ISCoNF) only influenced the performance data and had no effects on behavioural data, injury scores or blood parameters measured in this study, which could have been due to the small sample size. Therefore, the mechanism by which familiarity improves post-weaning performance is not immediately obvious from the results of the current study. Nevertheless, a study by Kanitz et al. [32], where piglets were subjected to 4 h of maternal and littermate deprivation at different ages reported that the presence

of an age-matched conspecific has a direct calming effect as measured by hypothalamic-pituitary- adrenal axis activity and behavioural responses in test situations. This was further influenced by the degree of familiarity between the piglets, with a familiar conspecific providing the most social support, allowing pigs to cope with stress better. Alternatively, other studies have shown that aggression between familiar dyads is lower than that between unfamiliar dyads [17, 33], with the exception of aggression around the feeder [34].

Behavioural effects of IS and ISCoM before and after weaning

In the current study, ISCo piglets displayed the greatest proportion of total observations on aggressive behaviour on the first day of separation and mixing (d −7). This corresponds to results of other investigations [35–37]. In the current study, however and similar to results from Kanaan et al. [36], aggressive behaviour appeared to be transient with no difference between treatments for aggression or scratch scores closer to the time of weaning. The difference in available floor space for IS and ISCo piglets during times of separation from the sow (0.42 m^2 for IS versus 0.21 m^2 for ISCo) may have also been a confounding factor contributing to increased aggression [38]. After weaning, aggression was rarely observed across all treatments, which may have been partly due to the use of instantaneous behaviour sampling since events like the delivery of aggression are generally short in duration and continuous observations may be a more reliable sampling technique to capture the behaviour [18]. Injury scores including scratches and redness were also included in the current study as another way to measure aggression. Before weaning, injury scores were weakly correlated with aggressive behaviour, but after weaning, there was no correlation. This is most likely due to the low number of observations for aggression given other studies have shown skin lesions scores to be a reliable method to assess aggression [18, 39]. A reduced scratch and redness score in ISCo pigs 2 d after weaning combined with no difference in injury scores between ISCoF and ISCoNF may suggest that pre-weaning socialisation can teach piglets how to avoid fights with foreign pigs, be more tolerant of unfamiliar pigs, and (or) allow them to establish dominance hierarchies quicker through learnt social skills [16, 17, 36]. The reason CW piglets had a higher redness score than both IS treatments just before the event of weaning is not known. However, since there was no correlation between scratch and redness scores before weaning, the possibility of CW piglets rubbing up against each other in competition for the teat may have caused more redness at the time the injury scores were recorded.

Piglets subjected to IS with or without CoM during lactation showed a more than two-fold increase in percentage of scan samples in exploratory and play behaviour on the first day of separation compared with piglets that remained with their sow continuously. Similarly, Berkeveld et al. [40] reported a greater total activity at the start of IS when piglets were separated from the sow for 12 h per day. It was hypothesised that the increase in activity may have been due to restlessness associated with sudden and previous unexperienced separation from the sow because, similar to the present study, the behaviour reduced over time (d –7 to –1) and sleeping increased as IS and ISCo piglets became familiar with the process.

Manipulative behaviours such as belly nosing and manipulating pen or littermates can be signs of piglet distress [9]. To the authors' knowledge, few studies have reported on manipulative behaviour in relation to CoM techniques. Alternatively, in an IS study by Berkeveld et al. [40], manipulative behaviour was considerably less in IS piglets compared with CW pigs on d 37 of the experiment, but this comparison was made at a time when CW pigs had been fully weaned, but the IS litters were still intermittently in contact with the sow. In the current study, IS piglets expressed the lowest level of manipulative behaviour compared with the other treatment groups before and after weaning suggesting that IS is not associated with the development of behaviour patterns indicative of piglet distress. In contrast, manipulative behaviour did not appear to dominate in the CW treatment more than the ISCo treatment before or after weaning, making it difficult to draw firm conclusions. On the day of weaning, while ISCo and CW piglets engaged in active behaviours such as manipulation, exploration/play, and ingestive-related behaviours (ISCo only), IS pigs spent virtually all their time sleeping in their pen. Other inactive behaviours including lying and sitting, which can be considered as symptoms of stress [41], were measured separately to sleeping in the current study. Therefore, the high level of sleeping observed in IS piglets on the day of weaning (2 h after the event) likely indicates that IS pigs were comfortable with their new surroundings [40, 42]. This, in combination with the reduced level of manipulative behaviour on the day of weaning, may have contributed to the improvement in post-weaning performance observed 2 to 8 d after weaning. In contrast, CW pigs had the highest level of sitting and lying behaviour on the day of weaning and d 7 after weaning, which could suggest that these pigs were experiencing more distress [9].

Effects of IS and ISCo on bloods parameters before and after weaning

The possibility that CW pigs experienced more distress after weaning as suggested by the behavioural data, is not supported by the plasma cortisol results with no difference between treatments detected 2 d after weaning. This lack of difference in cortisol between treatments 2 d after weaning might be related to the timing of blood sampling. At weaning, cortisol levels have been shown to reduce over a 24 h period [43]. Therefore, taking a blood sample 2 d after the initial mixing period might have missed the maximum value. Pluske and Williams [20] and Turpin et al. [24] had similar outcomes sampling at two and one day after weaning, respectively. The reason why IS pigs had a higher cortisol concentration 4 d after weaning is unknown. However, the mean concentration of cortisol at d 4 after weaning for IS pigs (32 ng/mL) was not as high as levels associated with weaning associated stress in other studies [27, 44] (62 ng/mL and 41 ng/mL respectively). In contrast to the cortisol results, IS pigs had the lowest Hp concentrations on both measurement days after weaning. Haptoglobin is a major acute phase protein in the pig that serves as an integral part of the innate immune response, making it a suitable non-specific marker to monitor animal health and well-being [45]. While previous results from the research group have not reported an influence of IS on Hp values [24], post-weaning Hp results in the current study suggest that IS (without CoM) had an influence on the general wellness of pigs [46]. In saying this however, Hp values for all treatments remained below the acute range of 3,000 µg/mL to 8,000 µg/mL [47] throughout the experiment, suggesting that acute infection or inflammation was not likely present. Furthermore, the increase in Hp in IS pigs on d –6 may reflect the transient increase in stress associated with the beginning of IS as previously reported by Turpin et al. [24]. The reason why there was no increase in plasma Hp for ISCo piglets at the same time might have been related to the increase in creep feed intake for these piglets around this time.

Plasma glycerol was measured in this experiment to give an indication of lipid mobilisation under the three different weaning regimes. The CW piglets had the highest pre-weaning values, most likely reflecting a greater overall lipid metabolism as a result of lipid intake from milk. However, at 4 d after weaning, the tendency for glycerol levels to be higher in CW piglets compared with ISCo piglets most likely reflects a greater rate of lipolysis consistent with the differences in feed intake.

Conclusion

These results show that an IS regime (without CoM) involving an 8 h/d separation for 7 d before weaning improved ADG and ADFI in the immediate post-weaning period compared with conventional weaning. However, this improvement did not seem to occur through increased familiarisation with creep feed, but rather through the prevention or attenuation of the weaning-associated stress response as evidenced by increased sleeping behaviour and reduced manipulative behaviour immediately after weaning as well as reduced post-weaning Hp levels. The addition of CoM to the IS regime also improved post-weaning performance between 2 and 8 d after weaning most likely due to social learning facilitating more eating before weaning and reduced aggression after weaning also reducing stress, as evidenced by a reduction post-weaning injury scores. Furthermore, grouping familiar pigs together after weaning additively improved ADG, ADFI and FCR between 2 and 8 d after weaning. Overall, these results suggest that mimicking certain aspects of weaning under natural conditions, such as gradual maternal separation and the opportunity to mix with non-litter mates in lactation, can positively affect post-weaning performance, and highlights opportunities for potential housing systems that enhance piglet welfare whilst also potentially facilitating mating in lactation.

Abbreviations
ADFI: Average daily feed intake; ADG: Average daily gain; BW: Body weight; CoM: Co-mingling; FCR: Feed conversion ratio; GIT: Gastrointestinal tract; Hp: Haptoglobin; IM: Intramuscular; IS: Intermittent suckling.

Acknowledgments
The technical assistance of Josie Mansfield and Ingunn Stensland is gratefully acknowledged. Appreciation is extended to the Cooperative Research Centre for High Integrity Australian Pork (Pork CRC) for funding and provision of a postgraduate scholarship to the first author.

Funding
Financial support by Australian Cooperative Research Centre for High Integrity Australian Pork.

Author's contributions
DT participated in the design and coordination of the study, carried out the on farm data collection and the ELISAs, performed the statistical analysis, interpreted the data and drafted the manuscript. PL participated in the design of the study and helped with the statistical analysis and the drafting of the manuscript. KP participated in the design of the study and helped with the statistical analysis and the drafting of the manuscript. JP participated in the design and coordination of the study, helped with on farm data collection and helped with the drafting of the manuscript. All authors read and approved the final manuscript.

Competing interests
The authors declare that they have no competing interests.

Author details
[1]School of Veterinary and Life Sciences, Murdoch University, Murdoch, WA 6150, Australia. [2]Trouw Nutrition, Veerstraat 38, Boxmeer 5831 JN, The Netherlands. [3]South Australian Research and Development Institute, Roseworthy Campus, JS Davis Building, Roseworthy, SA 5371, Australia. [4]Present address: SunPork Farms, 563 Coleman Road, Pinkerton Plains, SA 5400, Australia.

References
1. Pajor EA, Fraser D, Kramer DL. Consumption of solid food by suckling pigs: individual variation and relation to weight gain. Appl Anim Behav Sci. 1991;32:139–55.
2. Leibbrandt VD, Ewan RC, Speer VC, Zimmerman DR. Effect of weaning and age at weaning on baby pig performance. J Anim Sci. 1975;40:1077–80.
3. Brooks P, Tsourgiannis C. Factors affecting the voluntary feed intake of the weaned pig. In: Pluske JR, Le Dividich J, Verstegen MWA, editors. Weaning the pig: concepts and consequences. Wageningen: Wageningen Academic Publishers; 2003. p. 81–116.
4. Pluske JR, Hampson DJ, Williams IH. Factors influencing the structure and function of the small intestine in the weaned pig: a review. Livest Prod Sci. 1997;51:215–36.
5. Pié S, Lalles J, Blazy F, Laffitte J, Sève B, Oswald I. Weaning is associated with an upregulation of expression of inflammatory cytokines in the intestine of piglets. J Nutr. 2004;134:641–7.
6. Kuller W, Soede N, van Beers-Schreurs H, Langendijk P, Taverne M, Verheijden J, et al. Intermittent suckling: effects on piglet and sow performance before and after weaning. J Anim Sci. 2004;82:405–13.
7. Kuller W, Soede N, van Beers-Schreurs H, Langendijk P, Taverne M, Kemp B, et al. Effects of intermittent suckling and creep feed intake on pig performance from birth to slaughter. J Anim Sci. 2007;85:1295–301.
8. Berkeveld M, Langendijk P, Soede NM, Kemp B, Taverne MA, Verheijden JH, et al. Improving adaptation to weaning: effect of intermittent suckling regimens on piglet feed intake, growth, and gut characteristics. J Anim Sci. 2009;87:3156–66.
9. Dybkjær L. The identification of behavioural indicators of 'stress' in early weaned piglets. Appl Anim Behav Sci. 1992;35:135–47.
10. Kemp B, Soede N. Should weaning be the start of the reproductive cycle in hyper-prolific sows? a physiological view. Reprod Domest Anim. 2012;47:320–6.
11. Weary DM, Pajor EA, Bonenfant M, Fraser D, Kramer DL. Alternative housing for sows and litters: part 4. Effects of sow-controlled housing combined with a communal piglet area on pre-and post-weaning behaviour and performance. Appl Anim Behav Sci. 2002;76:279–90.
12. van Nieuwamerongen S, Soede N, van der Peet-Schwering C, Kemp B, Bolhuis J. Development of piglets raised in a new multi-litter housing system vs. Conventional single-litter housing until 9 weeks of age. J Anim Sci. 2015;93:5442–54.
13. Pajor EA, Weary DM, Fraser D, Kramer DL. Alternative housing for sows and litters: part 1. Effects of sow-controlled housing on responses to weaning. Appl Anim Behav Sci. 1999;65:105–21.
14. Petersen H, Vestergaard K, Jensen P. Integration of piglets into social groups of free-ranging domestic pigs. Appl Anim Behav Sci. 1989;23:223–36.
15. Kutzer T, Bünger B, Kjaer JB, Schrader L. Effects of early contact between non-littermate piglets and of the complexity of farrowing conditions on social behaviour and weight gain. Appl Anim Behav Sci. 2009;121:16–24.
16. D'Eath RB. Socialising piglets before weaning improves social hierarchy formation when pigs are mixed post-weaning. Appl Anim Behav Sci. 2005;93:199–211.
17. Li Y, Wang L. Effects of previous housing system on agonistic behaviors of growing pigs at mixing. Appl Anim Behav Sci. 2011;132:20–6.
18. Verdon M, Morrison RS, Hemsworth PH. Rearing piglets in multi-litter group lactation systems: effects on piglet aggression and injuries post-weaning. Appl Anim Behav Sci. 2016;183:35–41.
19. Widowski T, Cottrell T, Dewey C, Friendship R. Observations of piglet-directed behavior patterns and skin lesions in eleven commercial swine herds. J Swine Health Prod. 2003;11:181–5.
20. Pluske J, Williams I. Reducing stress in piglets as a means of increasing production after weaning: administration of amperozide or co-mingling of piglets during lactation? J Anim Sci. 1996;62:121–30.
21. Bolhuis JE, Schouten WGP, Schrama JW, Wiegant VM. Behavioural development of pigs with different coping characteristics in barren and substrate-enriched housing conditions. Appl Anim Behav Sci. 2005;93:213–28.
22. Eckersall P, Duthie S, Safi S, Moffatt D, Horadagoda N, Doyle S, et al. An automated biochemical assay for haptoglobin: prevention of interference from albumin. Comp Haematol Int. 1999;9:117–24.
23. Berkeveld M, Langendijk P, van Beers-Schreurs HM, Koets AP, Taverne MA, Verheijden JH. Postweaning growth check in pigs is markedly reduced by intermittent suckling and extended lactation. J Anim Sci. 2007;85:258–66.

24. Turpin DL, Langendijk P, Chen T-Y, Lines D, Pluske JR. Intermittent suckling causes a transient increase in cortisol that does not appear to compromise selected measures of piglet welfare and stress. Animals. 2016;6:24.

25. van der Meulen J, Koopmans S, Dekker R, Hoogendoorn A. Increasing weaning age of piglets from 4 to 7 weeks reduces stress, increases post-weaning feed intake but does not improve intestinal functionality. Animal. 2010;4:1653–61.

26. Worsaae H, Schmidt M. Plasma cortisol and behaviour in early weaned piglets. Acta Vet Scand. 1980;21:640–57.

27. Moeser AJ, Vander Klok C, Ryan KA, Wooten JG, Little D, Cook VL, et al. Stress signaling pathways activated by weaning mediate intestinal dysfunction in the pig. Am. J. Of physiol. Gastrointest. Liver Physiol. 2007;292:G173–81.

28. Sauerwein H, Schmitz S, Hiss S. The acute phase protein haptoglobin and its relation to oxidative status in piglets undergoing weaning-induced stress. Redox Rep. 2005;10:295–302.

29. Capozzalo MM, Kim JC, Htoo JK, de Lange CF, Mullan BP, Hansen CF, et al. Effect of increasing the dietary tryptophan to lysine ratio on plasma levels of tryptophan, kynurenine and urea and on production traits in weaner pigs experimentally infected with an enterotoxigenic strain of escherichia coli. Arch Anim Nutr. 2015;69:17–29.

30. Morgan C, Lawrence A, Chirnside J, Deans L. Can information about solid food be transmitted from one piglet to another? J Anim Sci. 2001;73:471–8.

31. Bruininx E, Binnendijk G, Van der Peet-Schwering C, Schrama J, Den Hartog L, Everts H, et al. Effect of creep feed consumption on individual feed intake characteristics and performance of group-housed weanling pigs. J Anim Sci. 2002;80:1413–8.

32. Kanitz E, Hameister T, Tuchscherer M, Tuchscherer A, Puppe B. Social support attenuates the adverse consequences of social deprivation stress in domestic piglets. Horm Behav. 2014;65:203–10.

33. Stookey JM, Gonyou HW. Recognition in swine: recognition through familiarity or genetic relatedness? Appl Anim Behav Sci. 1998;55:291–305.

34. Puppe B. Effects of familiarity and relatedness on agonistic pair relationships in newly mixed domestic pigs. Appl Anim Behav Sci. 1998;58:233–9.

35. Parratt CA, Chapman KJ, Turner C, Jones PH, Mendl MT, Miller BG. The fighting behaviour of piglets mixed before and after weaning in the presence or absence of a sow. Appl Anim Behav Sci. 2006;101:54–67.

36. Kanaan VT, Pajor EA, Lay DC, Richert BT, Garner JP. A note on the effects of co-mingling piglet litters on pre-weaning growth, injuries and responses to behavioural tests. Appl Anim Behav Sci. 2008;110:386–91.

37. Wattanakul W, Stewart A, Edwards S, English P. Effects of grouping piglets and changing sow location on suckling behaviour and performance. Appl Anim Behav Sci. 1997;55:21–35.

38. Hvozdik A, Kottferova J, Da Silva Alberto J. Ethological study of social behaviour of pigs from the point of view of housing restriction. Archiv Fur Tierzucht. 2002;45:557–64.

39. Turner SP, Farnworth MJ, White IM, Brotherstone S, Mendl M, Knap P, et al. The accumulation of skin lesions and their use as a predictor of individual aggressiveness in pigs. Appl Anim Behav Sci. 2006;96:245–59.

40. Berkeveld M, Langendijk P, Bolhuis JE, Koets AP, Verheijden JH, Taverne MA. Intermittent suckling during an extended lactation period: effects on piglet behavior. J Anim Sci. 2007;85:3415–24.

41. Colson V, Orgeur P, Foury A, Mormède P. Consequences of weaning piglets at 21 and 28 days on growth, behaviour and hormonal responses. Appl Anim Behav Sci. 2006;98:70–88.

42. Morgan T, Pluske J, Miller D, Collins T, Barnes AL, Wemelsfelder F, et al. Socialising piglets in lactation positively affects their post-weaning behaviour. Appl Anim Behav Sci. 2014;158:23–33.

43. Rault J-L, Dunshea FR, Pluske JR. Effects of oxytocin administration on the response of piglets to weaning. Animals. 2015;5:545–60.

44. Moeser AJ, Ryan KA, Nighot PK, Blikslager AT. Gastrointestinal dysfunction induced by early weaning is attenuated by delayed weaning and mast cell blockade in pigs. Am J Physiol Gastrointest Liver Physiol. 2007;293:G413–21.

45. Cray C, Zaias J, Altman NH. Acute phase response in animals: a review. Comp Med. 2009;59:517–26.

46. Knura S, Lipperheide C, Petersen B, Wendt M. Impact of hygienic environment on haptoglobin concentration in pigs. In: Tielen MJM, Voets MT, editors. Proceedings of the 10th international congress on animal hygiene. Maastricht: Animal Health Service; 2000. p. 537–41.

47. Sales N, Collins D, Collins A, McKenna T, Bauer M, Parke C, et al. Porcine haptoglobin levels measured at 7–14 days after weaning were independent of age, weight or gender. Anim Prod Sci. 2015;55:1457.

19

Transfer of β-hydroxy-β-methylbutyrate from sows to their offspring and its impact on muscle fiber type transformation and performance in pigs

Haifeng Wan[†], Jiatao Zhu[†], Caimei Wu[†], Pan Zhou, Yong Shen, Yan Lin, Shengyu Xu, Lianqiang Che, Bin Feng, Jian Li, Zhengfeng Fang and De Wu[*]

Abstract

Background: Previous studies suggested that supplementation of lactating sows with β-hydroxy-β-methylbutyrate (HMB) could improve the performance of weaning pigs, but there were little information in the muscle fiber type transformation of the offspring and the subsequent performance in pigs from weaning through finishing in response to maternal HMB consumption. The purpose of this study was to determine the effect of supplementing lactating sows with HMB on skeletal muscle fiber type transformation and growth of the offspring during d 28 and 180 after birth. A total of 20 sows according to their body weight were divided into the control (CON, $n = 10$) or HMB groups (HMB, $n = 10$). Sows in the HMB group were supplemented with β-hydroxy-β-methylbutyrate calcium (HMB-Ca) 2 g /kg feed during d 1 to 27 of lactation. After weaning, 48 mixed sex piglets were blocked by sow treatment and fed standard diets for post-weaning, growing, finishing periods. Growth performance was recorded during d 28 to 180 after birth. Pigs were slaughtered on d 28 ($n = 6$/treatment) and 180 ($n = 6$/treatment) postnatal, and the longissimus dorsi (LD) was collected, respectively.

Results: The HMB-fed sows during lactation showed increased HMB concentration ($P < 0.05$) in milk and LD of weaning piglets ($P < 0.05$). In addition, offsprings in HMB group had a higher finishing BW and lean percentage than did pigs in CON group ($P < 0.05$), meanwhile, compared with pigs from sows fed the CON diet, pigs from sows fed HMB diet showed higher type II muscle fiber cross-sectional area (CSA), elevated myosin heavy chain (MyHC) *IIb* and *Sox6* mRNA, and fast-MyHC protein levels in LD ($P < 0.05$).

Conclusions: HMB supplemented to sow diets throughout lactation increases the levels of HMB in maternal milk and skeletal muscle of pigs during d 28 after birth and promotes subsequent performance of pigs between d 28 and 180 of age by enhancing glycolytic muscle fiber transformation.

Keywords: β-hydroxy-β-methylbutyrate, Lactating sow, Muscle fiber, Offspring, Performance

* Correspondence: sow_nutrition@sina.com
[†]Equal contributors
Institute of Animal Nutrition, Sichuan Agricultural University, No. 211, Huimin Road, Wenjiang District, Chengdu 611130, Sichuan, People's Republic of China

Background

It has been suggested that establishing sufficient skeletal muscle is essential for lasting metabolic health [1]. Growing evidence has further indicated that maternal nutrition during lactation plays an important role in postnatal skeletal muscle development and growth of the offspring [2]. Skeletal muscle satellite cells provide myonuclei with muscle fibers and fuse with them, leading to muscle fiber hypertrophy. In addition, the transitions in myosin heavy chain isoforms are accelerated at postnatal 2 to 3 wk and muscle fiber type is quite vulnerable to various stimuli, which might affect subsequent growth and health [3]. Research has suggested that a high-protein/ low-carbohydrate diet fed to mice during lactation impeded skeletal muscle growth in offspring and caused a transient shift towards oxidative versus glycolytic muscle metabolism [4]. Meanwhile, Lefaucheur et al. [5] reported that early postnatal under-nutrition of piglets reduced the glycolytic capacity of skeletal muscle, which was reflected a delay in muscle maturation. In addition, studies have shown that insulin resistance correlates with skeletal muscle fiber distribution and a decreased percentage of slow oxidative type I fibers [6, 7]. Conde-Aguilera et al. also found that feeding a diet with 30% deficient in total sulfur amino acid to piglets during 10 d after weaning decreased fast-twitch muscle growth and glycolytic metabolism [8]. As a result, it is of great interest that maternal nutrition interventions during lactation are being developed to increase or maintain skeletal muscle mass, which might have a lasting impact on future performance.

Studies have suggested that leucine could activate the mammalian target of rapamycin (mTOR) to promote skeletal muscle protein synthesis in pigs [9]. Moreover, β-hydroxy-β-methylbutyrate (HMB) is a metabolite of leucine, which has also similar effects with leucine in promoting skeletal muscle growth [10]. Moore et al. [11] showed that when the early post-hatch poult was fed with 0.1% β-hydroxy-β-methylbutyrate calcium (HMB-Ca), HMB may improve muscle development via an increase in myogenic satellite cell mitotic activity. Meanwhile, previous studies have also shown that HMB could stimulate satellite cell proliferation and differentiation [12, 13], and dietary HMB supplementation to piglets during early postnatal period promoted skeletal muscle fiber development [14]. However, limited research existed evaluating the impact of maternal HMB consumption during lactation on skeletal muscle development of the offspring. Therefore, the study was to assess the effect of feeding HMB to sows during lactation on muscle fiber type transformation and performance of the offspring during d 28 and 180 after birth.

Methods

All procedure including use and treatment of animals was in accordance with guidelines set by the Animal Care and Use Committee of Sichuan Agricultural University. The calcium salt (monohydrate) of β-hydroxy-β-methylbutyrate (purity 93%), was purchased from Jiangyin Sanyi Chemical Co., Ltd., (Jiangsu, China).

Animal, diets and treatments

A total of 20 Landrace × Yorkshire (the third parity) pregnant sows with similar body weight (252.9 ± 3.6 kg) were used in the experiment. At d 110 of gestation, sows were moved to farrowing room and were also individually penned (2.06 m × 1.50 m), and all sows were randomly assigned to control (CON, $n = 10$) or experimental groups (HMB, $n = 10$) according to their body weights. All sows were fed 2.60 kg/d gestation diet from d 110 of gestation to parturition. After farrowing, the CON sows were fed a basal lactation diet (Table 1), and the sows in the HMB group were fed basal lactation diet supplemented with 2.0 g/kg of HMB-Ca. The feed level was progressively

Table 1 Composition and nutrient levels of the basal diets for sows during lactation (as-fed basis)[a]

Ingredient	Percent
Corn	63.50
Soybean meal	23.00
Fish meal	3.00
Wheat bran	5.00
Soybean oil	2.50
L-Lysine HCl (98%)	0.08
Limestone	0.87
Dicalcium phosphate	1.00
Choline chloride (50%)	0.15
Salt	0.40
Vitamin and mineral premix[b]	0.50
Calculated nutrient levels[c]	
CP, %[d]	17.90
ME, MJ /kg[d]	13.65
Ca, %[d]	0.78
TP, %[d]	0.63
AP, %[d]	0.41
Total Lys, %[e]	1.02
Total Met + Cys, %[d]	0.58
Total Leu, %[e]	1.51

[a]The HMB group diet was formed by supplementing 2.0 g/kg HMB-Ca to the basal diets
[b]Provided per kg of diet: copper, 20 mg; iron, 80 mg; zinc, 100 mg; manganese, 25 mg; selenium, 0.15 mg; iodine, 0.14 mg; vitamin A, 4000 IU; vitamin D3, 800 IU; vitamin E, 441 IU; menadione, 0.5 mg; thiamine, 1.0 mg; riboflavin, 3.75 mg; vitamin B6, 1.0 mg; vitamin B12, 15 μg; niacin, 10 mg; D-pantothenic acid, 12 mg; folic acid, 1.3 mg; D-biotin, 200 μg for gestation and lactation diet
[c]The nutrient composition of diets were calculated using nutrition value for the ingredients obtained from China Feed Information Database 2013 (24)
[d]Calculated values
[e]Analyzed values

increased during the first 5 d of lactation, and was allowed free access to feed during the d 6 of lactation to weaning, and sow feed intake was record daily during lactation. In addition, each pig from each litter was individually weighted at birth. In addition, the litter size was standardized to 10 piglets per litter by cross-fostering within the treatment group beyond 24 h after birth, and piglets had free access to water and creep feed was not provided to piglets at any point during the lactation period. Male piglets were castrated on d 7 after birth and all piglets were weaned at 28 d of age.

At weaning (d 28 of lactation), due to housing restriction, a total of 48 mixed sex pigs (24 barrows and 24 gilts, $n = 24$/treatment) from ten randomly litters per treatment approaching an average weight were selected and blocked according to maternal treatment, and were moved to 12 nursery pen (1.46 m× 1.30 m per pen, 2 barrows and 2 gilts per pen) until d 66 of age. At the d

67 of age, these pigs were moved to 12 grow-finish pens (2.03 m× 1.90 m per pen, 2 barrows and 2 gilts per pen) until the d 180 of age. In the whole experimental period, all the pigs had free access to water and were fed ad libitum, and they were phased-fed standard nursery and growing-finishing diets twice daily until d 180 after birth (Table 2).

Performance measurement and sample collection
For the growth performance study, pigs were individually weighted at birth and postnatal d 28, 66, 108, 150 and 180, respectively. The average daily gain (ADG) was calculated as (weight change/number of days) ×1000. Average daily feed intake was obtained from all pens in the nursery period and growing–finishing period, and the Feed/Gain (F/G) was calculated as the average daily feed intake (ADFI) /ADG. Plasma samples (10 mL) were also collected from the jugular vein of six piglets closest

Table 2 Composition and nutrient levels of the basal diets for the offspring (as-fed basis)

	Growth stage			
	D 28 to 66	D 67 to 108	D 109 to 150	D 151 to 180
Ingredient, %				
Corn	59.6	64.3	70.4	75.3
Soybean meal	18.0	24.0	22.0	16.0
Fish meal	3.0	2.0	–	–
Full fat soybean	5.0	–	–	–
Wheat bran	–	5.0	4.0	5.0
Whey powder	8.0	–	–	–
Sucrose	2.0	–	–	–
L-Lysine HCl (98%)	–	0.26	0.28	0.27
D,L-Methionine (99%)	–	0.08	0.06	0.06
L-Threonine (98.5%)	–	0.06	0.06	0.06
Limestone	0.50	0.60	0.80	0.85
Dicalcium phosphate	1.00	1.30	1.00	1.06
Choline chloride (50%)	0.10	0.10	0.10	0.10
Salt	0.30	0.30	0.30	0.30
Soybean oil	1.00	1.00	–	–
Vitamin and mineral premix[a]	1.50	1.00	1.00	1.00
Calculated nutrient levels[b]				
ME, MJ/kg	13.39	13.35	13.19	13.15
CP, %	18.0	17.7	15.9	13.8
Ca, %	0.72	0.71	0.63	0.63
TP, %	0.60	0.67	0.55	0.54
Total Lys, %	1.35	1.12	0.97	0.84

[a]Provided per kg of diet for piglets from d 28 to 66 of age: iron, 100 mg; copper, 6 mg; zinc, 100 mg; manganese, 4 mg; selenium, 0.30 mg; and iodine, 0.14 mg; vitamin A, 2,200 IU; vitamin D_3, 220 IU; vitamin E, 16; menadione, 0.5 mg; thiamine, 1.5 mg; riboflavin, 4.0 mg; vitamin B_6, 7.0 mg; vitamin B_{12}, 20 µg; niacin, 30 mg; D-pantothenic acid, 12 mg; folic acid, 0.3 mg; and D-biotin, 80 µg Provided per kg of diet for pigs from d 67 to 180 of age: iron, 60 mg; copper, 6 mg; zinc, 60 mg; manganese, 4 mg; selenium, 0.30 mg; and iodine, 0.14 mg; vitamin A, 1,300 IU; vitamin D_3, 150 IU; vitamin E, 11 IU; menadione, 0.5 mg; thiamine, 1.0 mg; riboflavin, 3.0 mg; vitamin B_6, 1.0 mg; vitamin B_{12}, 10 µg; niacin, 30 mg; D-pantothenic acid, 8 mg; folic acid, 0.3 mg; D-biotin, 50 µg
[b]The nutrient composition of diets were calculated using nutrition value for the ingredients obtained from China Feed Information Database 2013 (24)

to the average body weight of each treatment on d 28 of lactation. Samples were collected in heparinized tubes kept on the ice and centrifuged at 2,550 × g for 10 min at 4 °C. Milk samples (30 mL) were collected from all functional glands of sows at d 28 of lactation by injecting 1 mg oxytocin into the ear vein. The plasma supernatant and milk samples were refrigerated at −20 °C immediately prior to subsequent analysis. At d 28 and 180 after birth, a total of 24 pigs (one castrated male pig with mean body weight per pen) were slaughtered after overnight fasting by electrical stunning followed by exsanguination, according to the method of Refeldt et al. [4]. After slaughter, the abdomen was opened, and the entire intestine was rapidly removed. Then, all samples for mRNA, histological and biochemical analyses were collected as quickly as possible. Longissimus dorsi (LD) samples from the left half of the carcass were dissected at the level of the $12^{th}/13^{th}$ ribs for pigs during d 28 and 180 after birth, as previously described by Cerisuelo et al. [15], and were rapidly frozen in liquid nitrogen and stored at −80 °C for subsequent RNA and biochemical analyses. The LD at the level of the $11^{th}/12^{th}$ ribs was excised and stored in liquid nitrogen for subsequent muscle morphological assessment. The right half of the carcass was dissected into primary cuts, such as the loin, neck and ham, which were further manually separated into lean meat, subcutaneous fat, bones, and skin for pigs at d 180 after birth, as described by Rehfeldt et al. [4].

Milk and muscle HMB determination

Milk and muscle tissue samples were measured for HMB content by a modification of the plasma and milk analysis method previously described by Deshpande et al. [16] and Ehling et al. [17]. Briefly, acidified muscle tissue homogenates were extracted with methyl-t-butyl ether for 2 h, and a-hydroxy-a-methylbutyric acid (Sigma H40009; Sigma-Aldrich, Saint Louis, MO, USA) was used as an internal standard. The extracting solution was centrifuged at 12,000 × g and 4 °C for 5 min. Then, the supernatant was transferred to a clean test tube and evaporated to dryness by nitrogen flushing at 40 °C. After drying, 1 and 4 mL of 0.10 mol/L HCl in 90/10 (v/v) water-acetonitrile were added the same test tube for milk and muscle sample, respectively. After a brief vortex mixing, the mixture was passed through an OASIS® MCX cartridge (60 mg) and the eluate was collected for the muscle sample, and this step was omitted for milk sample. The HMB content was analysed via high-performance liquid chromatography with tandem mass spectrometry (LC-MS/MS, Agilent, USA). The recovery intra-day and inter-day are all > 85 and 70% for sow milk and piglet skeletal muscle, respectively, and the detectable minimum amount of HMB in the sow milk and piglet skeletal muscle were 0.02 μmol/L or 0.02 nmol/g.

Biochemical analyses and amino acid content analysis

Plasma insulin concentration was analyzed using a porcine insulin RIA kit (Tianjin Jiuding Medical Biological Engineering Co., LTD, Tianjin, China), and the detection limit of plasma insulin was 2 μU/mL. Plasma glucose and urea levels were conducted by using specific assay kits (Nanjing Jiancheng Bioengineering Institute, Nanjing, China) according to the manufacturer's instructions, moreover, the detection range were 0–28 mmol/L and 0.03-19 mmol/L for plasma glucose and urea, respectively. Plasma free amino acids content were determined according to the method as described by Li et al. [18], and briefly, 300 μL of plasma sample and 900 μL of 10% sulfosalicylic acid were mixed well and centrifuged at 12, 000 × g and 4 °C for 15 min. Then the supernatant fluid was filtered through a 0.22-μm-pore-size PTFE syringe filter (Millpore) into a 2-mL auto-sampler vial and determined for amino acid level by an L8800 high-speed analyzer (Hitachi, Tokyo, Japan) for the amino acid. Amino acid standard solutions type B and AN-II (Wako Pure Chemical Industries, Ltd., Osaka, Japan) were used for calibration.

Cross-sectional area and total number of muscle fiber determination

For histological analyses of muscle fiber, frozen muscle samples were equilibrated to −25 °C. Then transverse serial section (10 μm) was cut in a cryostat at −20 °C and subsequently mounted on glass microscopic slides. These sections were then stained with acid-preincubated ATPase treatment at pH 4.20 and subsequent alkaline (pH = 9.40) treatment to evaluate muscle fiber morphology using a modification of the method of Guth et al. [19]. All sections were photographed using a digital microscope (Nikon), and muscle fibers were counted over 5 randomly selected fields of known size (1.01 mm^2, 200 to 300 fibers). The total number of fibers was calculated by multiplication of the number of fibers per $centimeter^2$ with the loin meat area (LMA), and the mean muscle fiber cross-sectional area (CSA) in the united area was measured by Image-Pro Plus 6.0 software (Media Cybernetics, Bethesda, MD).

RNA isolation, cDNA synthesis and real-time PCR

Approximately 50 mg of muscle samples were crushed and total RNA was extracted using RNAiso Plus reagent (TaKaRa). The integrity of RNA was estimated by electrophoresis on a 1% agarose gel stained with ethidium bromide and the purity of total RNA was verified using a Nanovue^{TM} Plus Spectrophotometer (GE Healthcare, UK) at 260 and 280 nm. The OD_{260}/OD_{280} ratios of the RNA samples were all between 1.8 and 2.0. Subsequently, reverse transcription was performed from 1 μg of total RNA using the PrimeScript^{TM} RT Reagent Kit

(TaKaRa) according to the manufacturer's recommendations, and the reverse transcription products (cDNA) were stored at −20 °C for relative quantification by PCR. Primers were designed by Primer Express 3.0 (Applied Biosystems, Foster, CA, USA) based on known sequence deposited in GenBank, and the detected genes of muscle samples included myogenic genes (*Pax7*, *MRF4* and *MSTN*), protein synthesis and degradation factors (*IGF-I*, *mTOR*, *MAFbx* and *MuRF1*), transcription factor (*Sox6* and *FoxO1*), myosin heavy chain (*MyHC*) isoform (*slow/I*, *IIa*, *IIx* and *IIb*), and the information of primers was presented in Table 3. Real-time-qPCR analysis was carried out using the SYBR Green method and the target genes were quantified using CFX manager V1.1 software (Bio-Rad Laboratories). The mixture (10 µL) contained 5 µL SYBR® Premix Ex Taq™ II (TaKaRa), 1 µL cDNA,

0.5 µL of each gene-specific primer, and 3 µL ddH$_2$O. All measurements were performed in triplicate. The thermal cycling conditions were follows: denaturation at 95 °C for 15 s, followed by 40 cycles of denaturation at 95 °C for 5 s and annealing at 61.5 °C for 30 s, and collection of fluorescent signals. The relative mRNA abundances of the target genes were calculated using the $2^{-\Delta\Delta CT}$ method, and glyceraldehyde-3-phosphate dehydrogenase (*GAPDH*) was used as a reference gene in this study. The mRNA level of each target gene for Con group was set to 1.0.

Tissue protein extraction and Western blot analysis
Frozen LD samples were lysed, and the protein concentration was measured using a protein assay kit (Nanjing Jiancheng Bioengineering Institute, Nanjing, China) according

Table 3 The primer sequences of the target genes

Target gene	Primer sequence (5′to 3′)	Product size, bp	GenBank No.
MyHC1	F: GTTTGCCAACTATGCTGGGG	95	NM_213855.1
	R: TGTGCAGAGCTGACACAGTC		
MyHCIIa	F: CTCTGAGTTCAGCAGCCATGA	127	NM_214136.1
	R: GATGTCTTGGCATCAAAGGGC		
MyHCIIx	F: TTGACTGGGCTGCCATCAAT	111	NM_001104951.1
	R: GCCTCAATGCGCTCCTTTTC		
MyHCIIb	F: GAGGTACATCTAGTGCCCT	83	NM_001123141.1
	R: GCAGCCTCCCCAAAAATAGC		
Pax7	F: TGAAGGTCGGAGTGAACGGAT	74	NM_001206359.1
	R: CACTTTGCCAGAGTTAAAAGCA		
IGF-1	F: CACAGACGGGCATCGTGGAT	90	NM_214256.1
	R: ACTTGGCAGGCTTGAGGGGT		
mTOR	F: CATTGGAGATGGTTTGGTGA	160	XM_003127584.5
	R: ATGGGATGTGGCTTGTTTGA		
MRF4	F: CCTTCGGTGCCTTTCTTCCAT	88	NM_001244672.1
	R: GAGTTATTTCTCCCCCACTTCC		
MSTN	F: TTTACCTGTTTATGCTGATTGTTG	194	NM_214435.2
	R: TTTGCTAATGTTAGGAGCTGTTTC		
FoxO1	F: TCAAGGATAAGGGCGACAGC	95	NM_214014.2
	R: AATGTCATTATGGGGAGGAGAGT		
Sox6	F: CGGATTGGGGAGTATAAGCA	159	XM_01399454.1
	R: CATCTGAGGTGATGGTGTGG		
MAFbx	F: CCCTCTCATTCTGTCACCTTG	104	NM_001044588
	R: ATGTGCTCTCCCACCATAGC		
MuRF1	F: GCTGGATTGGAAGAAGATGTAT	144	NW_001184756
	R: AGGAAAGAATGTGGCAGTGTCT		
GAPDH	F: TGAAGGTCGGAGTGAACGGAT	74	NM_001206359.1
	R: CACTTTGCCAGAGTTAAAAGCA		

MyHC myosin heavy chain, *Pax7* paired box 7, *IGF-I* insulin-like growth factor-I, *mTOR* mammalian target of rapamycin, *MRF4* muscle regulator factor 4, *MSTN* myostatin, *FoxO1* forkhead box transcription factor O1, *MAFbx* muscle atrophy F-box, *MuRF1* muscle Ring finger 1, GAPDH glyceraldehyde-3-phosphate dehydrogenase to the manufacturer's instructions. Western blot analysis of fast-MyHC protein (anti-fast myosin skeletal heavy chain antibody, Abcam, ab91506, diluted 1: 5,000) was conducted according to a previous publication [20]. The relative expression of the target protein was normalized to GAPDH (bio-rich042, BMSX, diluted 1: 20,000) as an internal control. The band density of fast-MyHC was normalized to that of GAPDH and fast-MyHC protein content was presented as the fold change relative to the CON group.

Statistical analysis
For performance of fattening pigs, individual pen served as the experimental unit, and other data including skeletal muscle characteristics, plasma amino acid and metabolites, milk and muscle HMB levels, muscle fiber histological analyses, gene and protein expressions were analyzed with the individual pig as the experimental unit. For means separation, differences among the least square means of the CON vs. HMB treatments were compared using t-tests analysis. The results for mRNA level and fast-MyHC protein expression are presented as the least square mean and their SEM, and other results in tables are shown as the mean and pooled SEM. A difference was considered statistically significant when $P < 0.05$, if the P-values were between 0.05 and 0.10, the difference was considered a tendency.

Results
Performance and skeletal muscle composition
The results of sow feed intake during lactation was provided in Fig. 1. Sows in the HMB group had lower feed intake than those in the CON group ($P < 0.05$). The fattening

pig performance body composition are presented in Tables 4 and 5. The average feed intake (ADFI) of pigs during d 28 to 180 was not different between treatments. However, pigs from sows fed HMB diet during lactation exhibited higher final weight (+6.0%, $P < 0.05$), ADG (+6.4%, $P < 0.05$), and had a decreased tendency in F/G (−3.5%, $P = 0.069$) than did pigs from sows fed the CON diet during postnatal d 28 to 180. In addition, fattening pigs from sows fed HMB diets also showed superior carcass weight ($P < 0.05$) and lean meat percentage ($P = 0.07$) than did pigs from sows fed the CON diet.

Milk and muscle HMB, plasma amino acid and metabolites concentrations
As shown in Table 6, sows supplemented with HMB diet had an increased HMB concentration in milk at d 28 of lactation ($P < 0.01$) than sows fed with CON diet, with piglets from HMB-supplemented sows having an elevated HMB concentration ($P < 0.01$) of LD than those of their CON counterparts at d 28 of age. In addition, in comparison with CON piglets, maternal HMB treatment increased plasma leucine ($P < 0.05$) and essential amino acid (EAA) contents ($P < 0.05$) of the piglets at d 28 of age. Similarly, plasma glucose content exhibited an increased level ($P < 0.05$) for piglets from HMB than CON groups, Furthermore, there was an increased tendency in the plasma insulin content for piglets from sows supplemented HMB diet than those from sows supplemented CON diet ($P = 0.078$).

Muscle fiber characteristics
Data on the measures of CSA and numbers of muscle fiber in LD of fattening pigs are shown in Table 7. An

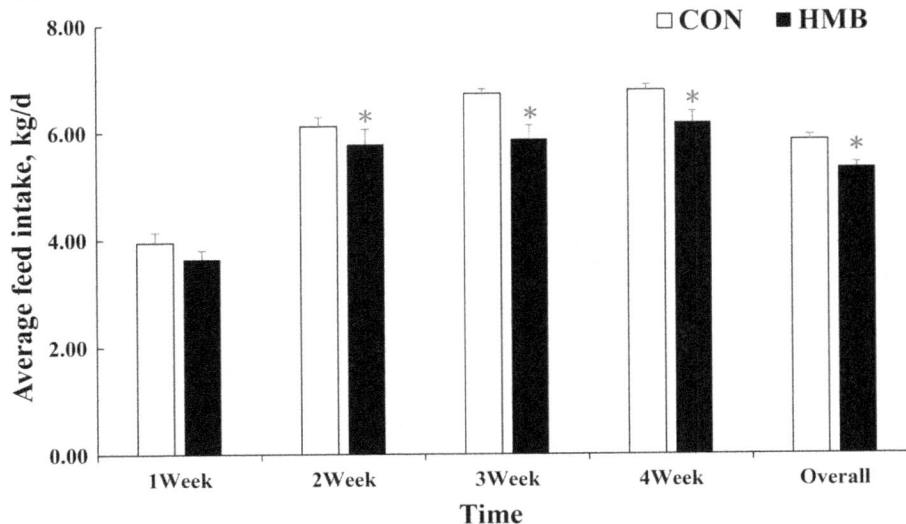

Fig. 1 Effects of dietary supplementation of HMB on average feed intake of sows during lactation. Values are means, with their standard errors represented by vertical bars ($n = 10$). *Represents mean values between the two groups differ significantly at $P < 0.05$. CON control; HMB β-hydroxy-β-methylbutyrate

Table 4 Effects of feeding HMB to sows during lactation on performance of the offspring

	CON	HMB	SEM	P value
Number of pigs on trial (n)	24	24		
Pig BW, kg				
At birth	1.41	1.42	0.07	0.887
D 28	7.65	7.81	0.32	0.670
D 66	22.6	22.9	0.5	0.677
D 108	47.6	49.4	1.1	0.293
D 150	79.8	83.9	1.4	0.080
D 180	105.7	112.0	1.8	0.041
Pig ADFI, g/d				
D 28 to 66	586	579	24	0.836
D 67 to 108	1,087	1,158	29	0.122
D 109 to 150	2,187	2,250	64	0.505
D 151 to 180	3,048	3,071	83	0.845
Overall	1,652	1,693	30	0.368
Pig ADG, g/d				
D 28 to 66	394	398	13	0.828
D 67 to 108	588	631	19	0.153
D 109 to 150	773	820	27	0.247
D 151 to 180	864	939	40	0.226
Overall	645	686	11	0.031
F/G				
D 28 to 66	1.49	1.45	0.03	0.457
D 67 to 108	1.85	1.84	0.03	0.720
D 109 to 150	2.83	2.74	0.05	0.200
D 151 to 180	3.56	3.28	0.15	0.232
Overall	2.56	2.47	0.03	0.069

ADG average daily gain, *BW* body weight, *ADFI* average daily feed intake, *CON* control, *HMB* β-hydroxy-β-methylbutyrate

increased tendency in the loin meat area (LMA) of LD was observed in fattening pigs from sows fed HMB diet compared with fattening pigs from sows fed the CON diet ($P = 0.069$). Similarly, higher type II muscle fiber CSA ($P < 0.05$) was discovered in offspring from sows fed HMB diet during lactation than in those from sows fed the CON diet, and pigs the HMB group had also an elevated tendency in the total mean myofiber CSA than those in the CON group ($P = 0.065$).

Gene and protein expression

According to Figs. 2, 3, 4 and 5, maternal HMB treatment during lactation induced increased *mTOR* and *Sox6* mRNA levels in LD of the offspring at d 28 of age ($P < 0.05$). Meanwhile, no significant effects were observed in of mRNA levels of *MAFbx* and *MuRF1* in skeletal muscle of weaning piglets between treatments ($P > 0.05$). However, higher mRNA levels of

Table 5 Effects of feeding HMB to sows during lactation on skeletal muscle composition of the offspring

	CON	HMB	SEM[a]	P value
Weaning pigs				
Body weight, kg	7.83	7.98	0.27	0.705
LD weight, g	127.2	130.7	9.0	0.792
Finishing pigs				
Body weight, kg	106.8	114.0	1.2	0.002
Carcass weight, kg	77.8	83.8	0.9	0.001
LD weight, kg	3.02	3.27	0.09	0.065
Lean meat, %	63.05	66.25	1.09	0.067
Bone, %	10.67	10.66	0.45	0.983
SCAT, %	20.01	16.21	1.19	0.051
Skin, %	6.27	6.88	0.23	0.100

LD longissimus dorsi, *SCAT* subcutaneous adipose tissue, *CON* control, *HMB* β-hydroxy-β-methylbutyrate
[a]Pooled standard error of the mean ($n = 6$/treatment)

MyHC-IIb were observed in LD of offspring in HMB group than those in the CON group during weaning and finishing stages ($P < 0.05$). In addition, weaning pigs in HMB group had an increased tendency in mRNA rate of *MyHC-IIb* /*MyHC-I* of LD than those in the CON group ($P = 0.073$). Similarly, compared with those in pigs in the CON group, pigs in HMB group exhibited higher the fast-MyHC protein levels in LD of the pigs at weaning and finishing stages ($P < 0.05$).

Discussion

The first aim of the present study was to investigate the effects of maternal HMB consumption during lactation

Table 6 Effects of feeding HMB to lactating sows on HMB and amino acid metabolism in piglets during d 28 of lactation

	CON	HMB	SEM[a]	P value
HMB level				
Maternal milk, μmol/L	0.42	10.50	0.01	<0.001
Piglet' LD, nmol/g tissue	6.04	7.51	0.33	0.025
Plasma metabolite				
Insulin, μU/mL	20.98	23.98	1.05	0.078
Glucose, mmol/L	7.71	9.77	0.46	0.029
Urea, mmol/L	4.79	4.40	0.34	0.860
Leucine, nmol/mL	155.6	227.7	20.8	0.044
BCAA, nmol/mL	516.8	561.2	37.8	0.437
EAA, nmol/mL	1,449	2,521	92	0.002
NEAA, nmol/mL	2,389	2,690	182	0.319

LD longissimus dorsi, *BCAA* branched-chain amino acids, *EAA* essential amino acid, *NEAA* nonessential amino acids, *CON* control, *HMB* β-hydroxy-β-methylbutyrate
[a]Pooled standard error of the mean ($n = 10$/treatment for maternal milk; $n = 6$/treatment for piglet's LD and plasma metabolite)

Table 7 Effects of feeding HMB to lactating sows on muscle fiber characteristics of fattening pigs

	CON	HMB	SEM[a]	P value
LMA, cm²	75.30	82.15	2.38	0.069
Total number of muscle fiber, ×10³	1013	1018	42	0.935
Mean myofiber CSA, μm²				
Total	3,615	4,268	183	0.065
Type I	2,732	3,203	213	0.169
Type II	3,719	4,774	252	0.025

LMA loin meat area, *CSA* cross sectional area, *CON* control, *HMB* β-hydroxy-β-methylbutyrate
[a]Pooled standard error of the mean (*n* = 6/treatment)

on performance and skeletal muscle characteristics of the offspring at weaning (d 28 after birth). It has established that maternal nutrition during lactation has previously been reported as one of the important factors affecting the growth of weaning pigs [21]. Given that pig performance was, to a great extent, influenced by birth BW, therefore, piglets with similar birth BW were selected and used in this experiment for CON and HMB groups, respectively. However, in the current trial, we found that maternal HMB treatment did not affect the BW of pigs at weaning, which disagreed with the findings of Nissen et al. [22], who reported that piglets from sows fed total HMB-Ca level of 2.0 g/d from d 108 of gestation to d 21 of lactation had higher weaning BW compared with those from sows fed the CON diet, and the discrepancy are likely related to the higher dosage (10.6 g/d) by multiplying dietary HMB level (2 g/kg) by average feed intake of sows (5.3 kg /d) during d 1 to 27 of lactation in this study compared with the used dosage (2 g/d) in previous study by Nissen et al. [22]. In addition, we found that dietary supplementation of higher level of HMB significantly reduced feed intake of

sows during lactation (Fig. 1), which might be the important proof leading to the non-difference in BW of piglets at weaning. But till now it has not been demonstrated in the reduced feed intake of sows fed HMB, and research in the rat has shown that dietary HMB supplementation could impair insulin resistance and increase plasma non-essential fatty acid level [23], which could lead to the decreased feed intake during lactation, and the underlying mechanism of which remains to be explained further. However, studies have suggested that neonatal pigs have great growth potential in the immediate neonatal period, especially in terms of skeletal muscle growth, which is dependent on maternal milk intake and quality [24] and the increased milk production is associated with an increased lactation feeding level [25, 26], therefore, lower feed intake of sows during lactation might affect to some extent piglet growth in this study.

However, we found that feeding HMB to sows during lactation increased HMB level in maternal milk and LD of piglets in this study, which is consistent with the result of Nissen et al. [22] and Michelle et al. [27]. Meanwhile, higher plasma leucine, glucose, and insulin levels were observed in offspring in the HMB group, which is accordance with the results of Tatara et al. [28] and Wan et al. [14], and HMB could increase plasma amino acid and glucose levels of pigs [14, 28]. In this respect, insulin, glucose and amino acid were the important anabolic promoters affecting skeletal muscle protein synthesis [24], moreover, no difference were observed in mRNA expressions of *MuRF1* and *MAFbx*, which also provided the further evidence of the similar protein degradation in skeletal muscle of pigs between treatments. On the contrary, the increased mTOR mRNA expression was observed in the skeletal muscle of piglets in the HMB group. Therefore, these results suggested that

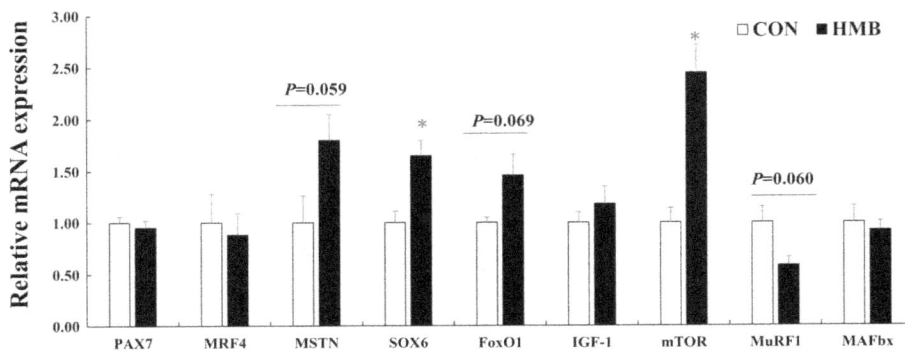

Fig. 2 Effects of feeding HMB to lactating sows on myogenic, protein synthetic and proteolytic gene expressions in skeletal muscle of weaning piglets. Values are means, with their standard errors represented by vertical bars (*n* = 6). * Represents mean values between the two groups differ significantly at *P* < 0.05. Data was normalized against *GAPDH*, with results expressed relative to the CON sample using the ΔΔCt method (where Ct is cycle threshold) with efficiency correction. *Pax7* paired box 7; *IGF-I* insulin-like growth factor-I; *mTOR* mammalian target of rapamycin; *MRF4* muscle regulator factor 4; *MSTN* myostatin; *FoxO1* forkhead box transcription factor O1; *MAFbx*, muscle atrophy F-box; *MuRF1*, muscle Ring finger 1; CON control; HMB β-hydroxy-β-methylbutyrate

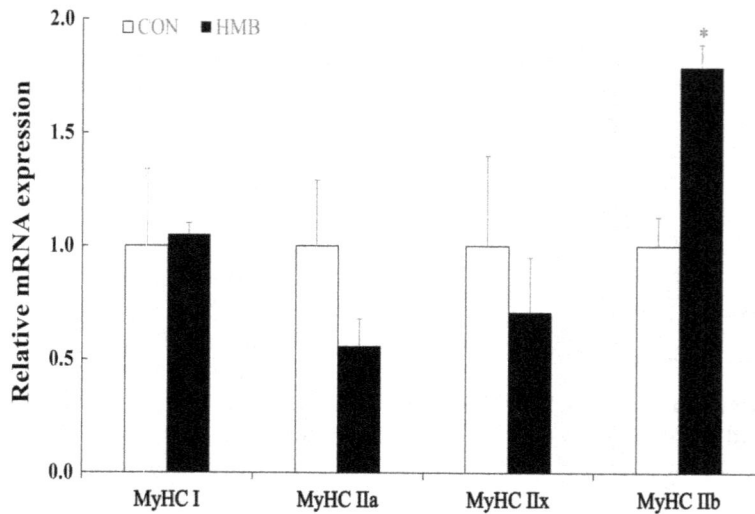

Fig. 3 Effects of supplementing lactating sows with HMB on MyHC isoforms mRNA levels in skeletal muscle of the offspring at d 28 of age. Values are means, with their standard errors represented by vertical bars ($n = 6$). * Represents mean values between the two groups differ significantly at $P < 0.05$. Data was normalized against *GAPDH*, with results expressed relative to the CON sample using the $\Delta\Delta Ct$ method (where Ct is cycle threshold) with efficiency correction. *MyHC* myosin heavy chain; CON control; HMB β-hydroxy-β-methylbutyrate

HMB could not lead to the protein degradation of skeletal muscle in piglets in this study. In addition, It is appeared that the rate of *MyHC-II* mRNA level is positive to skeletal muscle maturity of piglets [5], in this experiment, maternal HMB treatment significantly increased *MyHC-IIb* mRNA level and fast-MyHC protein expression in LD of the offspring at weaning. Therefore, supplementation of sows with HMB during lactation might promote skeletal muscle transformation by transfer of

HMB from maternal milk to skeletal muscle of the offspring during early postnatal period.

The secondary aim of this study was to explore the subsequent performance and skeletal muscle growth of pigs during finishing stage in response to maternal HMB consumption during lactation. Interestingly, an important observation was that pigs from sows in HMB group had greatly higher finishing BW compared with pigs from sows in the CON group, which was similar with

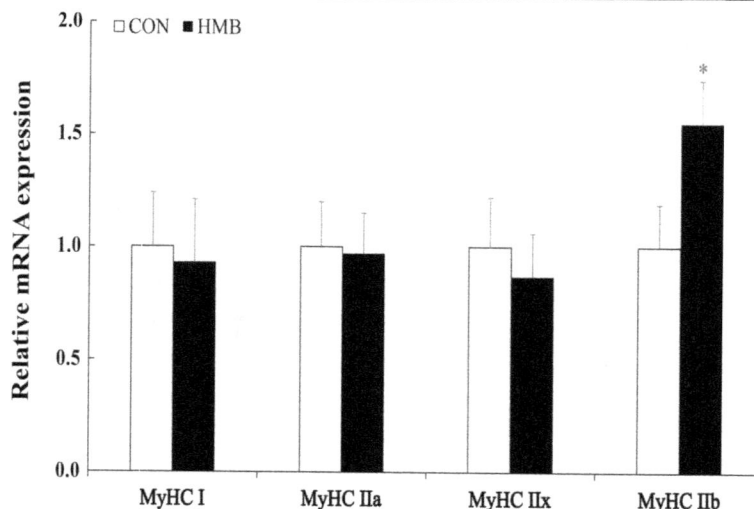

Fig. 4 Effects of feeding HMB to lactating sows on MyHC isoforms mRNA level in skeletal muscle of the offspring at d 180 of age. Values are means, with their standard errors represented by vertical bars ($n = 6$). * Represents mean values between the two groups differ significantly at $P < 0.05$. Data was normalized against *GAPDH*, with results expressed relative to the control sample using the $\Delta\Delta Ct$ method (where Ct is cycle threshold) with efficiency correction. *MyHC* myosin heavy chain; CON control; HMB β-hydroxy-β-methylbutyrate

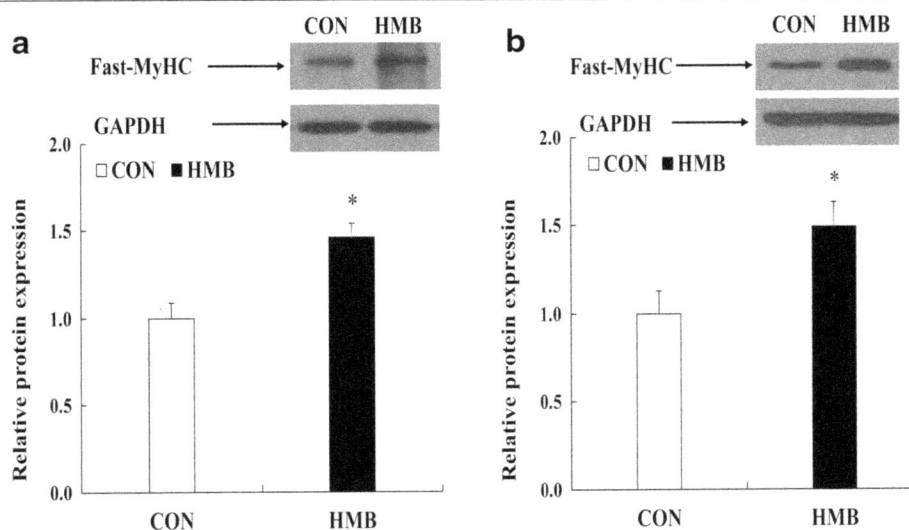

Fig. 5 Effects of feeding HMB to lactating sows on fast-MyHC protein expression in skeletal muscle of the offspring during d 28 and 180 after birth. Values are means, with their standard errors represented by vertical bars ($n = 6$). *Represents mean values between the two groups differ significantly at $P < 0.05$. **a** (the offspring during d 28 after birth); **b** (the offspring during d 180 after birth); MyHC myosin heavy chain; GAPDH glyceraldehyde-3-phosphate dehydrogenase. CON control; HMB β-hydroxy-β-methylbutyrate

the results of Tatara et al. [28, 29], who showed that when sow supplemented daily with 0.05 g/kg BW of HMB-Ca during the last two weeks of gestation, fattening pigs exhibited higher final BW at d 180 of age. Meanwhile, we found no statistical difference in ADFI between treatments from the weaning to d 180 after birth. Moreover, we also found that maternal HMB treatment contributed to an elevated carcass weight and lean meat percentage of pigs at d 180 of age. In this respect, a previous study found that feeding HMB to sows during lactation improved the performance of weaning pigs and increased HMB concentration in milk during d 20 of lactation, which are responsible for the enhanced skeletal muscle development of the offspring [22]. In addition, some studies also showed that pigs with heavier body weights exhibited greater muscle fiber cross-sectional areas and /or a greater number of muscle fibers than did pigs with lighter body weights at the same age [30]. A higher mean cross-sectional area of type II fiber was observed in LD of the offspring from sows fed HMB diet during lactation in our study, which also provided the important evidence for the improvement of performance by changing muscle fiber size of pigs. Therefore, it is possible that the increase in muscle fiber size in LD of pigs directly contributes to skeletal muscle growth and performance at d 180 after birth.

In addition, research has suggested that MyHC-IIb is the determining fiber contributing to the differentiation of large and small muscle area in the pig, and high muscularity is positively correlated with a high abundance of MyHC-IIb transcript [31], and more oxidative fibers might convert to glycolytic fibers with increasing age or weight, and that the early developmental stage might be a key stage for this conversion [32]. In this experiment, we found that maternal HMB treatment during lactation greatly increased MyHC-IIb mRNA and fast-MyHC protein levels in LD of the offspring at weaning and finishing stages (Figs. 3, 4, 5), which led to the improved subsequent performance of fattening pigs. Similarly, Rehfeldt et al. [33] reported that high live weight pig had more proportion of fast-twitch glycolytic fiber in semitendinosus muscle compared with low live weight pig by maternal daidzein feeding. Moreover, in this study, higher level of Sox6 mRNA expression was observed in LD of weaning pigs from sows fed HMB diet during lactation. At present, there are few related studies in this respect. But, Wen et al. and Quiat et al. reported that overexpression of Sox6 down-regulated MyHC-I expression and up-regulated MyHC-IIb expression in pig and mice [34, 35]. Based on the above results and the previous reports, it is possible that the increased HMB level of skeletal muscle accelerated MyHC-IIb isoform transformation of piglets during lactation, which might contribute to skeletal muscle growth during post-weaning stages. Further studies should be conducted in relation to the specific mechanisms of fast-twitch fiber development that are mediated by HMB in piglets.

Conclusion

In conclusion, the present study showed that HMB supplemented to sows during lactation increases the levels of HMB in maternal milk and skeletal muscle of the weaning pigs, which contributes to the *MyHC-IIb* isoform transformation in muscle fiber of piglets at weaning, and subsequent performance of the offspring between d 28 and 180 of age.

Abbreviations

CSA: Cross-sectional area; FoxO1: Forkhead box transcription factor O1; GAPDH: Glyceraldehyde-3-phosphate dehydrogenase; HMB: β-hydroxy-β-methylbutyrate; HMB-Ca: β-hydroxy-β-methylbutyrate calcium; IGF-I: Insulin-like growth factor-I; LC-MS/MS: Liquid chromatography with tandem mass spectrometry; LD: Longissimus dorsi; LMA: Loin meat area; *MAFbx*: Muscle atrophy F-box; MRF4: Muscle regulator factor 4; MSTN: Myostatin; mTOR: Mammalian target of rapamycin; *MuRF1*: Muscle Ring finger 1; MyHC: Myosin heavy chain; Pax7: Paired box 7

Acknowledgements

The authors thank their laboratory colleagues for their assistance, and would like to thank John A. Rathmacher and John C. Fuller, Jr for their immeasurable help in determining the purity of HMB-Ca.

Funding

This work was supported by the National Special Research Fund for Non-Profit Sector (Agriculture) (No.201203015), Academy of Kechuang Feed Industry in Sichuan (2013NZ0056), Research Team of Youth Scientific and Technical Innovation of Sichuan (13CXTD0004), and the Program for Changjiang Scholars and Innovative Research Team in University (IRT13083).

Authors' contributions

DW designed the study, HW, JZ, CW, RZ, PZ, YS, carried out the study, and HW, SX, BF analyzed the data; HW wrote the paper; DW, LC, ZF, YL, made some modification in the manuscript. All authors read and approved the final manuscript.

Competing interests

The authors declare that they have no competing interests.

References

1. Brown LD. Endocrine regulation of fetal skeletal muscle growth: impact on future metabolic health. J Endocrinol. 2014;221:R13–29.
2. Bayol SA, Bruce CR, Wadley GD. Growing healthy muscles to optimise metabolic health into adult life. J Dev Org Health Dis. 2014;5:420–34.
3. Lefaucheur L, Gerrard D. Muscle fiber plasticity in farm mammals. J Anim Sci. 2000;77:1–19.
4. Rehfeldt C, Stabenow B, Pfuhl R, Block J, Nurnberg G, Otten W, et al. Effects of limited and excess protein intakes of pregnant gilts on carcass quality and cellular properties of skeletal muscle and subcutaneous adipose tissue in fattening pigs. J Anim Sci. 2012;90:184–96.
5. Lefaucheur L, Ecolan P, Barzic YM, Marion J, Le Dividich J. Early postnatal food intake alters myofiber maturation in pig skeletal muscle. J Nutr. 2003; 133:140–7.
6. Oberbach A, Bossenz Y, Lehmann S, Niebauer J, Adams V, Paschke R, et al. Altered fiber distribution and fiber-specific glycolytic and oxidative enzyme activity in skeletal muscle of patients with type 2 diabetes. Diabetes Care. 2006;29:895–900.
7. Gaster M, Staehr P, Beck-Nielsen H, Schrøder HD, Handberg A. GLUT4 is reduced in slow muscle fibers of type 2 diabetic patients is insulin resistance in type 2 diabetes a slow, type 1 fiber disease? Diabetes. 2001;50:1324–9.
8. Conde-Aguilera JA, Lefaucheur L, Tesseraud S, Mercier Y, Le Floc'h N, van Milgen J. Skeletal muscles respond differently when piglets are offered a diet 30% deficient in total sulfur amino acid for 10 days. Eur J Nutr. 2015;1–10.
9. Boutry C, El-Kadi SW, Suryawan A, Wheatley SM, Orellana RA, Kimball SR, et al. Leucine pulses enhance skeletal muscle protein synthesis during continuous feeding in neonatal pigs. Am J Physiol Endocrinol Metab. 2013; 305:E620–31.

10. Wilkinson DJ, Hossain T, Hill DS, Phillips BE, Crossland H, Williams J, et al. Effects of leucine and its metabolite beta-hydroxy-beta-methylbutyrate on human skeletal muscle protein metabolism. J Physiol. 2013;591:2911–23.
11. Moore DT, Ferket PR, Mozdziak PE. The effect of early nutrition on satellite cell dynamics in the young turkey. Poult Sci. 2005;84:748–56.
12. Kornasio R, Riederer I, Butler-Browne G, Mouly V, Uni Z, Halevy O. Beta-hydroxy-beta-methylbutyrate (HMB) stimulates myogenic cell proliferation, differentiation and survival via the MAPK/ERK and PI3K/Akt pathways. BBA-Mol Cell Res. 2009;1793:755–63.
13. Alway SE, Pereira SL, Edens NK, Hao Y, Bennett BT. Beta-Hydroxy-beta-methylbutyrate (HMB) enhances the proliferation of satellite cells in fast muscles of aged rats during recovery from disuse atrophy. Exp Gerontol. 2013;48:973–84.
14. Wan H, Zhu J, Su G, Liu Y, Hua L, Hu L, et al. Dietary supplementation with β-hydroxy-β-methylbutyrate calcium during the early postnatal period accelerates skeletal muscle fibre growth and maturity in intra-uterine growth-retarded and normal-birth-weight piglets. Br J Nutr. 2016;1–10.
15. Cerisuelo A, Baucells MD, Gasa J, Coma J, Carrion D, Chapinal N, et al. Increased sow nutrition during midgestation affects muscle fiber development and meat quality, with no consequences on growth performance. J Anim Sci. 2009;87:729–39.
16. Deshpande P, Jie Z, Subbarayan R, Mamidi VK, Chunduri RH, Das T, et al. Development and validation of LC-MS/MS method for the estimation of beta-hydroxy-beta-methylbutyrate in rat plasma and its application to pharmacokinetic studies. Biomed Chromatogr. 2013;27:142–7.
17. Ehling S, Reddy TM. Investigation of the presence of β-hydroxy-β-methylbutyric Acid and α-hydroxyisocaproic acid in bovine whole milk and fermented dairy products by a validated liquid chromatography–mass spectrometry method. J Agric Food Chem. 2014;62:1506–11.
18. Li H, Wan H, Mercier Y, Zhang X, Wu C, Wu X, et al. Changes in plasma amino acid profiles, growth performance and intestinal antioxidant capacity of piglets following increased consumption of methionine as its hydroxy analogue. Br J Nutr. 2014;112:855–67.
19. Guth L, Samaha FJ. Procedure for the histochemical demonstration of actomyosin ATPase. Exp Neurol. 1970;28:365–7.
20. Wang J, Li X, Yang X, Sun Q, Huang R, Xing J, et al. Maternal dietary protein induces opposite myofiber type transition in Meishan pigs at weaning and finishing stages. Meat Sci. 2011;89:221–7.
21. Ramanau A, Kluge H, Spilke J, Eder K. Supplementation of sows with L-carnitine during pregnancy and lactation improves growth of the piglets during the suckling period through increased milk production. J Nutr. 2004;134:86–92.
22. Nissen S, Faidley TD, Zimmerman DR, Izard R, Fisher CT. Colostral milk fat percentage and pig performance are enhanced by feeding the leucine metabolite beta-hydroxy-beta-methyl butyrate to sows. J Anim Sci. 1994;72:2331–7.
23. Yonamine C, Teixeira S, Campello R, Gerlinger-Romero F, Rodrigues C, Guimarães-Ferreira L, et al. Beta hydroxy beta methylbutyrate supplementation impairs peripheral insulin sensitivity in healthy sedentary Wistar rats. Acta Physiol. 2014;212:62–74.
24. Davis TA, Fiorotto ML. Regulation of muscle growth in neonates. Curr Opin Clin Nutr. 2009;12:78.
25. Verstegen MW, Mesu J, van Kempen GJ, Geerse C. Energy balances of lactating sows in relation to feeding level and stage of lactation. J Anim Sci. 1985;60:731–40.
26. van den Brand H, Heetkamp MJ, Soede NM, Schrama JW, Kemp B. Energy balance of lactating primiparous sows as affected by feeding level and dietary energy source. J Anim Sci. 2000;78:1520–8.
27. Kao M, Columbus DA, Suryawan A, Steinhoff-Wagner J, Hernandez-Garcia A, Nguyen HV, et al. Enteral β-hydroxy-β-methylbutyrate supplementation increases protein synthesis in skeletal muscle of neonatal pigs. Am J Physiol-Endoc M. 2016;310(11): E1072–84.
28. Tatara MR, Krupski W, Tymczyna B, Studzinski T. Effects of combined maternal administration with alpha-ketoglutarate (AKG) and beta-hydroxy-beta-methylbutyrate (HMB) on prenatal programming of skeletal properties in the offspring. Nutr Metab. 2012;9:39.
29. Tatara MR, Sliwa E, Krupski W. Prenatal programming of skeletal development in the offspring: effects of maternal treatment with beta-hydroxy-beta-methylbutyrate (HMB) on femur properties in pigs at slaughter age. Bone. 2007;40:1615–22.
30. Rehfeldt C, Tuchscherer A, Hartung M, Kuhn G. A second look at the

Transfer of β-hydroxy-β-methylbutyrate from sows to their offspring and its impact on muscle fiber...

197

influence of birth weight on carcass and meat quality in pigs. Meat Sci. 2008;78:170–5.

31. Wimmers K, Ngu N, Jennen D, Tesfaye D, Murani E, Schellander K, et al. Relationship between myosin heavy chain isoform expression and muscling in several diverse pig breeds. J Anim Sci. 2008;86:795–803.

32. Men XM, Deng B, Xu ZW, Tao X, Qi KK. Age-related changes and nutritional regulation of myosin heavy-chain composition in longissimus dorsi of commercial pigs. Animal. 2013;7:1486–92.

33. Rehfeldt C, Adamovic I, Kuhn G. Effects of dietary daidzein supplementation of pregnant sows on carcass and meat quality and skeletal muscle cellularity of the progeny. Meat Sci. 2007;75:103–11.

34. Wen W, Chen X, Chen D, Yu B, Luo J, Huang Z. Cloning and functional characterization of porcine Sox6. Turk J Biol. 2016;40:160–5.

35. Quiat D, Voelker KA, Pei J, Grishin NV, Grange RW, Bassel-Duby R, et al. Concerted regulation of myofiber-specific gene expression and muscle performance by the transcriptional repressor Sox6. Proc Natl Acad Sci U S A. 2011;108:10196–201.

PERMISSIONS

LIST OF CONTRIBUTORS

Dagmar Jezierny, Rainer Mosenthin and Pia Rosenfelder-Kuon
University of Hohenheim, Institute of Animal Science, Emil-Wolff-Strasse 10, 70599 Stuttgart, Germany

Nadja Sauer
Agricultural Analytic and Research Institute Speyer, Obere Langgasse 40, 67346 Speyer, Germany

Klaus Schwadorf
University of Hohenheim, Core Facility Hohenheim, Emil-Wolff-Strasse 12, 70599 Stuttgart, Germany

Julián Valencia, Germán Gómez, Walter López, Henry Mesa and Francisco Javier Henao
Universidad de Caldas, Faculty of Agricultural Sciences, A.A. 275, Manizales, Caldas, Colombia

Isabel Ortiz, Jesús Dorado and Manuel Hidalgo
Veterinary Reproduction Group, Department of Animal Medicine and Surgery, Faculty of Veterinary Medicine, University of Cordoba, 14071 Cordoba, Spain

Jane Morrell
Department of Clinical Sciences, Division of Reproduction, Swedish University of Agricultural Sciences, Uppsala, Sweden

Jaime Gosálvez
Department of Biology, Universidad Autónoma de Madrid, 28049 Madrid, Spain

Francisco Crespo
Department of Reproduction, Centro Militar de Cría Caballar (FESCCR-Ministry of Defense), 05005 Ávila, Spain

Juan M. Jiménez
Department of Chemistry and Physics of Materials, University of Salzburg, Hellbrunner Straße 34/III, A-5020 Salzburg, Austria

Hans Vergauwen, Sara Prims, Christophe Casteleyn, Steven Van Cruchten and Chris Van Ginneken
Laboratory of Applied Veterinary Morphology, Department of Veterinary Sciences, Faculty of Biomedical, Pharmaceutical and Veterinary Sciences, University of Antwerp, Campus Drie Eiken, Universiteitsplein 1, D.U.015, 2610 Wilrijk, Belgium

Jeroen Degroote and Joris Michiels
Department of Applied Biosciences, Faculty of Bioscience Engineering, Ghent University, Ghent, Belgium

Wei Wang
Department of Applied Biosciences, Faculty of Bioscience Engineering, Ghent University, Ghent, Belgium
Laboratory for Animal Nutrition and Animal Product Quality (LANUPRO), Department of Animal Production, Faculty of Bioscience Engineering, Ghent University, Melle, Belgium

Erik Fransen
StatUa Center for Statistics, University of Antwerp, Antwerp, Belgium

Stefaan De Smet
Laboratory for Animal Nutrition and Animal Product Quality (LANUPRO), Department of Animal Production, Faculty of Bioscience Engineering, Ghent University, Melle, Belgium

Brandy M. Jacobs, John F. Patience and Kenneth J. Stalder
Department of Animal Science, Iowa State University, Ames 50010, USA

Merlin D. Lindemann
Department of Animal and Food Sciences, University of Kentucky, Lexington 40546, KY, USA

Brian J. Kerr
USDA-ARS-National Laboratory for Agricultural and the Environment, Ames 50010, IA, USA

Chanwit Kaewtapee
University of Hohenheim, Institute of Animal Science, Emil-Wolff-Strasse 10, 70599 Stuttgart, Germany
Department of Animal Science, Faculty of Agriculture, Kasetsart University, 50 Ngam Wong Wan Rd, Chatuchak, Bangkok 10900, Thailand

Katharina Burbach, Georgina Tomforde, Amélia Camarinha-Silva, Sonja Heinritz, Jana Seifert, Rainer Mosenthin and Pia Rosenfelder-Kuon
University of Hohenheim, Institute of Animal Science, Emil-Wolff-Strasse 10, 70599 Stuttgart, Germany

Thomas Hartinger
University of Hohenheim, Institute of Animal Science, Emil-Wolff-Strasse 10, 70599 Stuttgart, Germany
University of Bonn, Institute of Animal Science, Endenicher Allee 15, 53115 Bonn, Germany

Markus Wiltafsky
Evonik Nutrition and Care GmbH, Rodenbacher Chaussee 4, 63457 Hanau-Wolfgang, Germany

Zhongchao Li, Yakui Li, Zhiqian Lv, Hu Liu, Jinbiao Zhao, Fenglai Wang, Changhua Lai and Defa Li
State Key Laboratory of Animal Nutrition, Ministry of Agriculture Feed Industry Centre, China Agricultural University, No. 2, Yuanminyuan west road, Haidian district, Beijing 100193, China

Jean Noblet
Inra, Umr Pegase, 35590 Saint-Gilles, France

Damian Knecht, Anna Jankowska-Mąkosa and Kamil Duziński
Institute of Animal Breeding, Wroclaw University of Environmental and Life Sciences, Chelmonskiego 38C, 51– 630 Wroclaw, Poland

Yating Su and Xiaohua Guo
Provincial Key Laboratory for Protection and Application of Special Plants in Wuling Area of China, College of Life Science, South-Central University for Nationalities, No. 182, Minyuan Road, Hongshan District, Wuhan, Hubei Province 430074, China

Xingjie Chen
Guangxi Yang-Xiang Animal Husbandry Co. Ltd., Guigang, Guangxi Province 537100, China

Ming Liu
Beijing China-agri Hong-Ke Biotechnology Co., Ltd., Beijing 102206, China

Changyou Shi, Yu Zhang, Zeqing Lu and Yizhen Wang
Institute of Feed Science, College of Animal Science, Zhejiang University, Yuhangtang Road 866#, Hangzhou, Zhejiang Province 310058, People's Republic of China

L. M. A. Barroso, M. Nascimento, A. C. C. Nascimento and C. F. Azevedo
Department of Statistics, Federal University of Viçosa, Av. P H Rolfs, s/n, University Campus, Viçosa, MG 36570-000, Brazil

F. F. Silva, P. S. Lopes and S. E. F. Guimarães
Department of Animal Science, Federal University of Viçosa, Av. P H Rolfs, s/n, University Campus, Viçosa, MG 36570-000, Brazil

N. V. L. Serão
Department of Animal Science, Iowa State University, Kildee Hall 50011 Ames, Iowa, USA

C. D. Cruz
Department of General Biology, Federal University of Viçosa, Av. P H Rolfs, s/n, University Campus, Viçosa, MG 36570-000, Brazil

M. D. V. Resende
Department of Statistics, Federal University of Viçosa, Av. P H Rolfs, s/n, University Campus, Viçosa, MG 36570-000, Brazil
Embrapa Forestry, Estrada da Ribeira, km 111, Colombo, PR, Brazil

F. L. Silva
Department of Plant Science, Federal University of Viçosa, Av. P H Rolfs, s/n, University Campus, Viçosa, MG 36570-000, Brazil

Johannes Gulmann Madsen and Camilo Pardo
Agroscope Posieux, la tioleyre 4, 1725 Posieux, Switzerland
ETH Zurich, Institute of Agricultural Sciences, Universitätsstrasse 2, 8092 Zurich, Switzerland

Michael Kreuzer
ETH Zurich, Institute of Agricultural Sciences, Universitätsstrasse 2, 8092 Zurich, Switzerland

Giuseppe Bee
Agroscope Posieux, la tioleyre 4, 1725 Posieux, Switzerland

Miao Yu, Chuanjian Zhang, Yuxiang Yang, Chunlong Mu, Yong Su, Kaifan Yu and Weiyun Zhu
Jiangsu Key Laboratory of Gastrointestinal Nutrition and Animal Health, Laboratory of Gastrointestinal Microbiology, College of Animal Science and Technology, Nanjing Agricultural University, Nanjing, Jiangsu 210095, China

Jeffrey L. Vallet, Jeremy R. Miles and Bradley A. Freking
USDA, ARS, U.S. Meat Animal Research Center (USMARC), Clay Center NE, Nebraska 68933, USA

Shane Meyer
Plymouth Ag Group, Diller NE, Nebraska 68342, USA

Sisi Li, Haichao Wang, Xinxia Wang, Yizhen Wang and Jie Feng
Key Laboratory of Animal Nutrition and Feed, Zhejiang Province, College of Animal Science, Zhejiang University, Hangzhou, China

Wenting Li and Kejun Wang
College of Animal Science and Technology, China Agricultural University, Beijing 100193, China
College of Animal Science and Veterinary Medicine, Henan Agricultural University, Zhengzhou 450002, China

Yujun Quan
College of Animal Science and Technology, China Agricultural University, Beijing 100193, China
Institute of Zoology, Chinese Academy of Sciences, Beijing 100101, China

Mengmeng Zhang, Muzhen Zhu and Keliang Wu
College of Animal Science and Technology, China Agricultural University, Beijing 100193, China

Ye Chen
The Department of Animal Husbandry, Rongchang Campus, Southwest University, Rongchang, Chongqing 402460, China

Qiuyan Li
State Key Laboratory for Agrobiotechnology, China Agricultural University, Beijing 100193, China

Xiaojian Yang
Southern Research and Outreach Center, University of Minnesota, Waseca, MN 56093, USA

Carissa Nath
Agricultural Utilization and Research Institute, Marshall, MN 56258, USA

Alan Doering
Agricultural Utilization and Research Institute, Waseca, MN 56093, USA

John Goihl
Agri-Nutrition Services, Inc., Shakopee, MN 55379, USA

Samuel Kofi Baidoo
Southern Research and Outreach Center, University of Minnesota, Waseca, MN 56093, USA
Department of Animal Science, University of Minnesota, St. Paul, MN 55108, USA

Diana L. Turpin and John R. Pluske
School of Veterinary and Life Sciences, Murdoch University, Murdoch, WA 6150, Australia

Pieter Langendijk
Trouw Nutrition, Veerstraat 38, JN, The Netherlands

Kate Plush
South Australian Research and Development Institute, Roseworthy Campus, JS Davis Building, Roseworthy, SA 5371, Australia
SunPork Farms, 563 Coleman Road, Pinkerton Plains, SA 5400, Australia

Haifeng Wan, Jiatao Zhu, Caimei Wu, Pan Zhou, Yong Shen, Yan Lin, Shengyu Xu, Lianqiang Che, Bin Feng, Jian Li, Zhengfeng Fang and De Wu
Institute of Animal Nutrition, Sichuan Agricultural University, No. 211, Huimin Road, Wenjiang District, Chengdu 611130, Sichuan, People's Republic of China

Index

www.ingramcontent.com/pod-product-compliance
Lightning Source LLC
Chambersburg PA
CBHW082022190326
41458CB00010B/3240